Electrons in Finite and Infinite Structures

NATO ADVANCED STUDY INSTITUTES SERIES

A series of edited volumes comprising multifaceted studies of contemporary scientific issues by some of the best scientific minds in the world, assembled in cooperation with NATO Scientific Affairs Division.

Series B: Physics

RECENT VOLUMES IN THIS SERIES

Volume 15 – Nuclear and Particle Physics at Intermediate Energies
edited by J. B. Warren

Volume 16 – Electronic Structure and Reactivity of Metal Surfaces
edited by E. G. Derouane and A. A. Lucas

Volume 17 – Linear and Nonlinear Electron Transport in Solids
edited by J. T. Devreese and V. van Doren

Volume 18 – Photoionization and Other Probes of Many-Electron Interactions
edited by F. J. Wuilleumier

Volume 19 – Defects and Their Structure in Nonmetallic Solids
edited by B. Henderson and A. E. Hughes

Volume 20 – Physics of Structurally Disordered Solids
edited by Shashanka S. Mitra

Volume 21 – Superconductor Applications: SQUIDs and Machines
edited by Brian B. Schwartz and Simon Foner

Volume 22 – Nuclear Magnetic Resonance in Solids
edited by Lieven Van Gerven

Volume 23 – Photon Correlation Spectroscopy and Velocimetry
edited by H. Z. Cummins and E. R. Pike

Volume 24 – Electrons in Finite and Infinite Structures
edited by P. Phariseau and L. Scheire

Volume 25 – Chemistry and Physics of One-Dimensional Metals
edited by Heimo J. Keller

The series is published by an international board of publishers in conjunction with NATO Scientific Affairs Division

A	**Life Sciences**	Plenum Publishing Corporation
B	**Physics**	New York and London
C	**Mathematical and Physical Sciences**	D. Reidel Publishing Company Dordrecht and Boston
D	**Behavioral and Social Sciences**	Sijthoff International Publishing Company Leiden
E	**Applied Sciences**	Noordhoff International Publishing Leiden

Electrons in Finite and Infinite Structures

Edited by

P. Phariseau and L. Scheire
Seminarie voor Theoretische Vaste Stof- en Lage Energie Kernfysica
Rijksuniversiteit-Gent
Ghent, Belgium

PLENUM PRESS • NEW YORK AND LONDON
Published in cooperation with NATO Scientific Affairs Division

Library of Congress Cataloging in Publication

Nato Advanced Study Institute on Electrons in Finite and Infinite Structures, Ghent, 1976.
Electrons in finite and infinite structures.

(Nato advanced study institutes series: Series B, Physics; v. 24)
Includes index.
1. Energy-band theory of solids–Congresses. 2. Electrons–Congresses. 3. Molecular theory–Congresses. I. Phariseau, P. II. Scheire, L. III. Title. IV. Series.
QC176.8.E4N34 1976 530.4 77-5020
ISBN 0-306-35724-0

Lectures presented at the NATO Advanced Study Institute on Electrons in Finite and Infinite Structures held in Ghent, Belgium, August 30–September 11, 1976

A Division of Plenum Publishing Corporation
227 West 17th Street, New York, N.Y. 10011

Printed in the United States of America

Preface

This book contains the transcripts of the lectures presented at the NATO Advanced Study Institute on "Electrons in Finite and Infinite Structures," held at the State University of Ghent, Belgium, August 30-September 11, 1976.

Over the last few years substantial progress has been made in the description and the understanding of the behavior of electrons in extended bodies. This includes the study of the energy spectrum of electrons in large molecules, perfect as well as imperfect crystals, and disordered alloys. Not only local potential techniques but also the many-body aspects are discussed in detail. As atomic, molecular, and solid state physics involve common techniques and insights, we believe that physicists and chemists active in these fields have benefited from these lectures and the interchange of ideas during the course.

The aim of the Institute was to familiarize young scientists in the field with the current state of the art and to indicate in which areas advances may be expected in the near future. The A.S.I. consisted of two parts: detailed instructional and review lectures over the whole period and some evening sessions where the participants were offered the opportunity to present their own work and discuss their ideas with senior scientists.

Since the Institute took place a few weeks after Prof. Dr. John C. Slater was suddenly taken from our scientific community, it was a great honor for us to dedicate this course, on behalf of the organizing committee, to the late John C. Slater.

The Advanced Study Institute was financially sponsored by the NATO Scientific Affairs Division (Brussels, Belgium). Co-sponsors were the National Science Foundation (Washington, D. C., U.S.A.), the Department of Higher Education and Scientific Research of the Ministry of National Education and Culture (Brussels, Belgium), and the Faculty of Sciences of the University of Ghent. In particular we are indebted to Dr. T. Kester of the NATO Scientific Affairs Division for his interest and constant encouragement.

The editors would also like to thank all lecturers for their most valuable contribution and their collaboration in preparing the manuscripts. Thanks are also due to the members of the International Advisory Board: Professors J. Callaway (Baton Rouge, La., U.S.A.), W. Dekeyser (Ghent, Belgium), F. E. Harris (Salt Lake City, Utah, U.S.A.), F. Herman (San Jose, Calif., U.S.A.), K. H. Johnson (Cambridge, Mass., U.S.A.), A. B. Lidiard (Harwell, U.K.), P. O. Löwdin (Uppsala, Sweden and Gainesville, Fla., U.S.A.), and N. H. March (London, U.K.).

The Institute itself could not have been realized without the enormous enthusiasm of all participants and lecturers and without the untiring efforts of our co-workers Dr. F. Dhoore, Dr. R. Nuyts, and Mr. R. Rotthier at the "Seminarie voor Theoretische Vaste Stofe- en Lage Energie Kern Fysica" of the "Rijksuniversiteit - Gent." Also, Mrs. A. Goossens-De Paepe's help in typing the manuscripts is gratefully acknowledged.

P. Phariseau

L. Scheire

Ghent, December 1976

Contents

THE SCF-Xα SCATTERED-WAVE METHOD WITH APPLICATIONS TO MOLECULES AND SURFACES

Notker Rösch

Lehrstuhl für Theoretische Chemie

Technische Universität München, Germany

1. INTRODUCTION

Currently theoreticians in molecular science and solid state physics are confronted with a growing demand for a quantitative description of systems consisting of a large, but finite number of electrons. Experiments on systems of increasing complexity are being performed, and very often such additional information is necessary to fully interpret the measurements. Systems with a finite number of electrons may exist as isolated entities, like polyatomic molecules in the gas phase, but in most cases they are models to study more or less localized phenomena in extended systems. To mention a few examples, we may think of a molecular complex in solution, e. g. a transition metal ion surrounded by a number of ligand atoms or molecules, of impurities or defects in an otherwise perfect crystal or of an atom or molecule chemisorbed on a metal surface. The size of such a model cluster will certainly depend on the interactions between the constituent atoms. But our choice may also be influenced by the kind of question we would like to have answered and, more often than not, our computing facilities will enforce some compromise. These questions will be taken up in more detail later on and in the lectures of Prof. Lidiard and Dr. Herman.

The solution of the Schrödinger equation, even for such a finite cluster, is a formidable task and we have to look for approximate methods taking us as far as possible. The self-consistent field is an important idea in this context which dates back to the earliest days of quantum mechanics /1,2/. Thereby the problem is reduced to the determination of one-electrons states in an averaged potential set up by the other electrons and the nuclei of the system. The most prominent method, at least in molecular science, is the <u>ab initio</u> Hartree-Fock (HF) method, based on representing molecular orbitals (MO) as linear combinations of atomic orbitals /3,4/ (SCF-HF-LCAO method). This method, however, has been limited in practice to systems up to N~80 electrons (to give a rough estimate) because one has to adopt large basis sets and has to compute very many multicenter integrals ($\sim N^4$). Simplified LCAO-type molecular-orbital methods, for instance those based on the "neglect of differential overlap"(NDO methods /5/) have found many useful applications, mainly in organic chemistry. Attempts to apply the same approximations to molecular clusters containing atoms of higher atomic number (e.g. transition metals) have met with considerable difficulties (see sec. 5.2).

In spite of these computational problems arising in the implementation of HF theory, there can be no doubt about its great success in many atomic and molecular problems. However, limitations inherent to the method should not be overlooked. Well known is the failure of a simple HF-MO description for the dissociation of a hydrogen molecule whereby the two electrons are put into the symmetrical orbital /6/. To remedy this difficulty one must add another determinant to the HF wave function leading to a simple case of configuration interaction transcending the self-consistent field level. For more atoms this procedure becomes impractical from a computational point of view even faster than HF theory as it implies wave functions built out of a thousand or more determinants. For large clusters or even crystals a completely different approach to this correlation problem, as it is called, has to be looked for /7/. Some methods starting from HF theory will be discussed in Prof.Harris' lectures. A careful analysis of the physical nature of the self-consistent field led Slater to advocate a different self-consistent field method, namely the Xα method /8/.

The self-consistent potential felt by an electron consists of two contributions, the "coulomb potential" and the "exchange-correlation potential". The electrostatic potential energy of an electron may be calculated by classical electrostatics from the charge distribution of all occupied spin orbitals plus the nuclear charges. This potential describes not only the electrostatic interaction of an electron with all charges, but also with itself. The exchange potential therefore has to correct for this self-interaction and it has further to take into account effects arising from the antisymmetry of the wave function which tends to keep apart those electrons having the same spin. The Xα exchange potential is intended to provide a local potential approximation to these effects. In contrast, the HF potential is non-local - one reason why it is often less convenient to handle.

A second feature of the Xα exchange potential is its statistical nature /8/ in the sense that it should be most appropriate for systems with a large number of electrons. On the other hand, it may be shown /9/ that the overlap of the HF wave function with the true wave function decreases exponentially with the number N of the electrons in the system. The statistical character brings the Xα method in close connection to the theory of the electron gas and the density functional theory of Kohn and Sham /14/ which will be dealt with in Prof. March's lectures.

The local exchange model, as originally proposed by Slater in 1951 /8/, was first applied to atoms /10/. These potentials were then used in calculations to determine the energy band structure of perfect crystalline solids /11/. When these calculations could be carried to selfconsistency /12/, it was found that agreement with experimental data could sometimes only be obtained by reducing the exchange potential by a uniform scaling factor $\alpha < 1$. This observation was supported theoretically by the Gaspar-Kohn-Sham theory /13,14/ of the inhomogeneous electron gas which through a variation of the statistical total energy leads to an exchange potential 2/3 as large as the original form proposed by Slater. A systematic study of all atoms in the periodic table by Herman and Skillman /10/ showed generally correct trends in a comparison of the Xα energy eigenvalues (calculated with the original $\alpha = 1$) and the X-ray spectra. However, the core ionization potentials were almost always underestimated, a discrepancy that should become worse if the (attractive) exchange potential is reduced. On the other hand, much better radial charge distributions were obtained with the value $\alpha = 2/3$. This puzzle was resolved by the observation that Koopmans' theorem was not satisfied in the Xα model /15/ and therefore the eigenvalues could not be compared directly with experimental spectra. We shall discuss this point extensively in sec. 2.4.

The Xα method was finally developed by choosing the scaling parameter α in such a way as to optimize the statistical exchange approximation for systems with only a few electrons /16/, i.e. for atoms. In this way, one hopes to have a method for a unified treatment of systems with widely varying character ranging from atoms to molecules and clusters to crystalline solids.

The application of the model to molecules and clusters of atoms did not come until Johnson, following a suggestion by Slater /18/ in 1965, developed the scattered-wave (SW) method /18/ to effectively solve the one-particle Schrödinger equation for such systems which had neither the translational symmetry of an ideal crystal nor the rotational symmetry of an atom. The scattered-wave formalism is based on the multiple scattering technique like its periodic analogue, the Korringa-Kohn-Rostocker (KKR) /19/ method used for energy bands in solids or like methods applied in the theory of electron bands of alloys (see the lectures by Prof. Gyorffy). Although other methods have been used to solve the Xα one-electron equations (some of them will be discussed in chap. 4) for cluster problems, Johnson's SW formalism may be compared in its importance to Roothaan's LCAO formalism /3/ in HF theory. The combined SCF-Xα-SW method will be the central topic of these lectures, but we will also discuss recent developments of the theory.

There have been a number of recent review articles on the SCF-Xα-SW method itself /20-22/ and its applications /23/ and I will draw freely from these treatments. A definitive exposition of the Xα method has been given by Slater in the fourth volume of his "Quantum Theory of Molecules and Solids." /24/

2. THE Xα METHOD

2.1 The Statistical Exchange Approximation

The nonrelativistic hamiltonian of a many-electron system is of the form

$$H = \sum_i f_i + \sum_{i<j} g_{ij} \tag{1}$$

where f_i is the one-electron operator including the kinetic energy and the potential energy in the field of the nuclei (with charges Z_μ at the position $\underset{\sim}{R}_\mu$):

$$f_i = -\nabla_i^2 - 2\sum_{\mu} Z_\mu / r_{i\mu} \qquad (2)$$

with $r_{i\mu} = |\underset{\sim}{r}_i - \underset{\sim}{R}_\mu|$ giving the distance from the position $\underset{\sim}{r}_i$ of electron i to that of the nucleus μ. The repulsive coulomb interaction between the i-th and the j-th electrons is given by

$$g_{ij} = 2/r_{ij} = 2/|\underset{\sim}{r}_i - \underset{\sim}{r}_j| \,. \qquad (3)$$

As is costumary in the Xα theory, we are using Rydberg energy units throughout: 1 Ry = 0.5 a.u. For a molecule or a solid, we assume the Born-Oppenheimer approximation of fixed nuclei to be valid. There will then be an additional constant term in the hamiltonian representing the repulsive interaction between the nuclei,

$$U_{NN} = \sum_{\mu \neq \nu} Z_\mu Z_\nu / R_{\mu\nu} \qquad (4)$$

This term does not effect the following discussion and may therefore be added to the total energy at the end of the calculation.

If one takes the wavefunction as a determinant formed from orthonormal spin orbitals u_i, one obtains from (1) the familiar Hartree-Fock expression for the energy expectation value /4,24/ :

$$\langle H \rangle = E_{HF} = \sum_i n_i \int d1\, u_i^*(1)\, f_1\, u_i(1)$$
$$+ \sum_{i<j} n_i n_j \int d1 d2\, g_{12}\, u_i^*(1) u_j^*(2) \left[u_i(1) u_j(2) - u_j(1) u_i(2) \right] \qquad (5)$$

The occupation numbers n_i have the values of either 1 or 0, according to whether the spin orbital u_i is occupied in the determinantal function or not. The symbolic integration is meant to include the summation over the spin variable.

It is convenient to break up the total electronic charge density,

$$\rho(1) = \sum_i n_i\, u_i^*(1) u_i(1), \qquad (6)$$

into two parts, the spin density $\rho\uparrow$, coming from the summation over those orbitals in (6) with spin up,

$$\varrho\uparrow(1) = \sum_i (i\uparrow)\, n_i u_i^*(1) u_i(1), \tag{7}$$

and $\varrho\downarrow$, defined in an analogous fashion. The HF energy (5) may then be written as

$$E_{HF} = \sum_i n_i \langle i|f_1|i\rangle + \frac{1}{2}\int \varrho(1)\varrho(2)\, g_{12}\, d1\, d2$$

$$-\frac{1}{2}\int [\varrho\uparrow(1) U_{XHF}\uparrow(1) + \varrho\downarrow(1) U_{XHF}\downarrow(1)]\, d\underset{\sim}{r}_1\, d\underset{\sim}{r}_2 \tag{8}$$

where

$$U_{XHF}\uparrow(1) = -\sum_{ij}(i\uparrow, j\uparrow) n_i n_j \int u_i^*(1) u_j^*(2) u_i(2) u_j(1) g_{12}\, d\underset{\sim}{r}_1 / \varrho\uparrow(1) \tag{9}$$

with a similar formula for spin down. Now we can see that the second term in (8) represents the classical coulomb energy of a charge distribution ϱ, including the self-energy. The third term cancels this self-interaction and furthermore introduces the characteristic exchange effects. This exchange potential $U_{XHF}\uparrow$ (the same thing is true for $U_{XHF}\downarrow$, of course) may formally be interpreted as to arise from an "exchange charge density" $\varrho_X(1,2)\uparrow$: /2,24/

$$\varrho_X\uparrow(1,2) = -\sum_{ij}(i\uparrow, j\uparrow) n_i n_j u_i^*(1) u_j^*(2) u_i(2) u_j(1)/\varrho\uparrow(1) \tag{10}$$

This charge density has the same spin as the spin density it interacts with and its magnitude is such that it exactly neutralizes the spin density at that point:

$$\varrho_X\uparrow(1,1) = -\varrho\uparrow(1). \tag{11}$$

From its integral,

$$\int \varrho_X\uparrow(1,2)\, d\underset{\sim}{r}_2 = -1, \tag{12}$$

we see that the exchange charge represents an electron hole of charge one, the so-called "Fermi hole". The Fermi hole travels around with an electron representing a sphere of influence from which electronic charge with the same spin is excluded. Hereby it is assured that each electron is acted on only by N-1 electrons, as it should be, if there are N electrons in the system.

As the integral of the Fermi hole is independent of the position of electron 1, it is tempting to assume that its shape may be described by a uniform shape function /22/, i.e.

$$\rho_X\uparrow(1,2) = -\rho\uparrow(1)\, f(r_{12}/a) \tag{13}$$

Equation (11) requires that f(0) = 1 and (12) relates the range parameter a to the density:

$$4\pi a^3 \rho\uparrow(1) \int_0^\infty x^2 f(x)\,dx = 1. \tag{14}$$

Using (9), (10) and (13), we obtain for the exchange potential:

$$\begin{aligned} U_X\uparrow(1) &= \int \rho_X\uparrow(1,2)\, g_{12}\, dr_2 \\ &= -8\pi a^2 \rho\uparrow(1) \int_0^\infty x f(x)\,dx \end{aligned} \tag{15}$$

which reduces to

$$U_X\uparrow(1) = -C\left[4\pi\rho\uparrow(1)\right]^{1/3} \tag{16}$$

where

$$C = 2\int_0^\infty x f(x)\,dx \Big/ \left[\int_0^\infty x^2 f(x)\,dx\right]^{2/3} \tag{17}$$

Therefore, we find in this approximation that the exchange potential is proportional to the cube root of the spin density. The proportionality constant (17) depends on the first and second moment of the hole shape function. Let us mention here that it is possible to obtain the exchange potential (16) by applying approximation (13) directly to an exchange charge density defined in terms of the second order density matrix of the system /22/. In this way, one avoids viewing the Xα method as an approximation to HF theory, but rather as a different self-consistent field solution of the general many-electron problem. The derivation of the Xα method given here is intended to underline the common physical content in both self-consistent field methods.

We can get an idea of the extension and the shape of the Fermi hole by considering two limiting cases. If we assume a spherical hole of constant density, i.e. if the shape function were a theta function $f(x) = \Theta(1 - x)$, equation (14) gives a simple relation between the density and the radius of the Fermi sphere:

$$r_s = \left[3/4\pi \rho\uparrow(1)\right]^{1/3} \tag{18}$$

Fig. 1 shows the radius r_s of the Fermi hole for a copper ion Cu^+, as a function of r. The Fermi hole is obviously large enough to cover regions of widely varying electron density, so that this procedure gives only an order-of-magnitude estimate for its size.

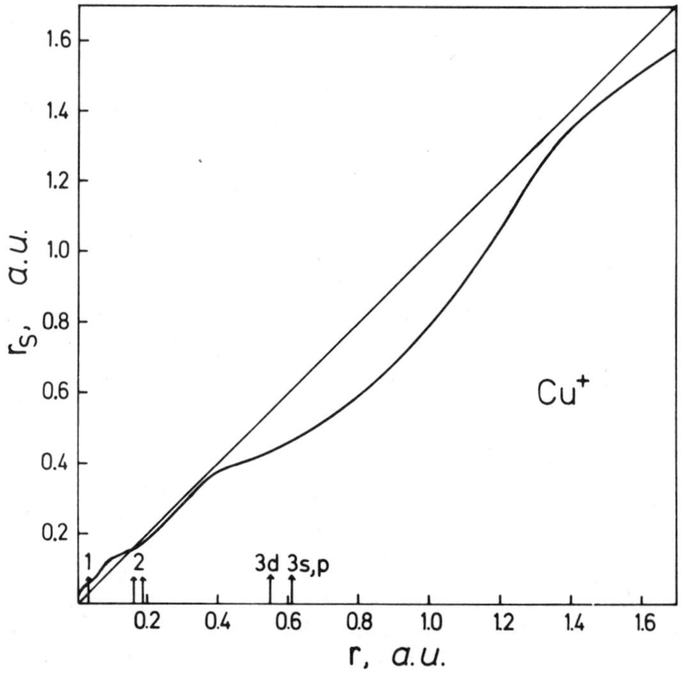

Fig. 1 Radius r_s of the Fermisphere for Cu^+, as a function of r, using Xα orbitals. The radii of maximum radial charge density of the various shells are indicated.

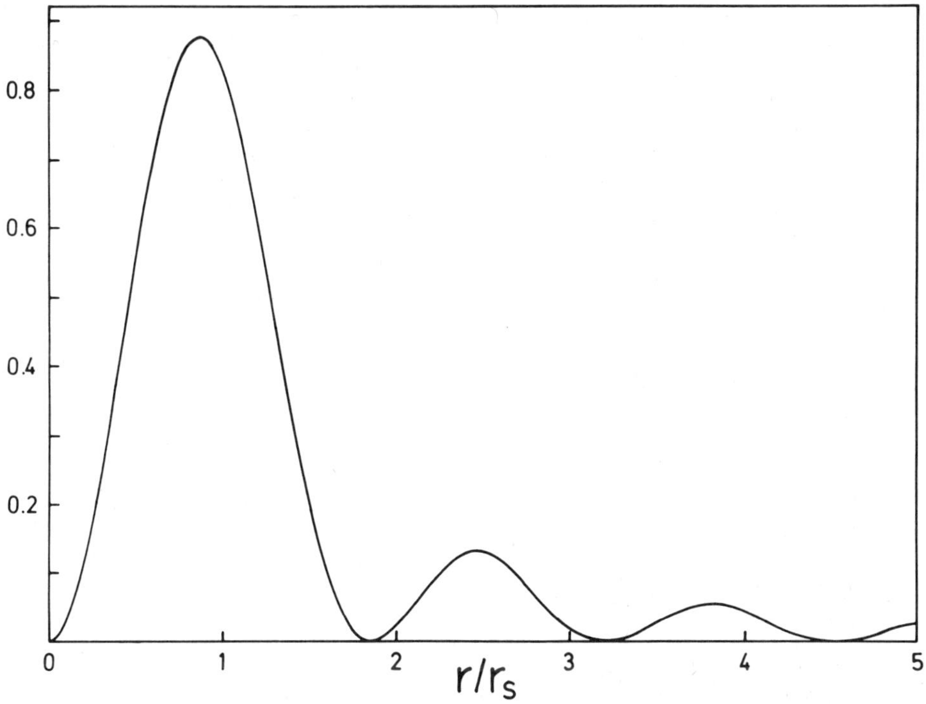

Fig. 2 Radial exchange charge density for the free-electron gas plotted as a function of r/r_s, where r_s is radius of a sphere containing one electron.

The free-electron gas is a system for which the exchange charge can be calculated exactly /25-27/. The shape function is given by

$$f(x) = \frac{9}{x^6}\,(\sin x - x\cos x)^2 \tag{19}$$

with $x = (9\pi/2)^{1/3}\, r/r_s$. The radial exchange charge density $4\pi r^2 \rho_X\!\uparrow(r)$ is plotted in Fig. 2 as a function of the reduced quantity r/r_s.

A somewhat lengthy calculation /27/ yields the following exchange potential from the shape function (19):

$$U_{XFE}\!\uparrow(1) = -6\left[\frac{3}{4\pi}\,\rho\!\uparrow(1)\right]^{1/3} \tag{20}$$

This expression is expected to provide a reasonable statistical approximation for systems with many electrons and a weakly varying charge density. For other systems we introduce a scaling parameter α in this formula:

$$U_{X\alpha}\uparrow(1) = -9\alpha\left[\frac{3}{4\pi}\,\rho\uparrow(1)\right]^{1/3} \tag{21}$$

For historical reasons, the scaling is chosen in such a way that $\alpha = 2/3$ corresponds to the case of the free-electron gas.

The $X\alpha$ approximation is then fully characterized by an energy functional of the form

$$E_{X\alpha} = \sum_i n_i \int u_i^*(1)\, f_1\, u_i(1)\, d1 + \frac{1}{2}\int \rho(1)\,\rho(2)\, g_{12}\, d1d2$$
$$-\frac{9}{2}\alpha\left[\frac{3}{4\pi}\right]^{1/3}\int\left\{[\rho\uparrow(1)]^{4/3} + [\rho\downarrow(1)]^{4/3}\right\} d\underset{\sim}{r}_1 \tag{22}$$

The next step is to apply the variational principle to this energy functional. If we vary $E_{X\alpha}$ with respect to one spin orbital u_i, keeping the occupation numbers fixed, we obtain

$$\delta E_{X\alpha} = n_i \int \delta u_i^*(1)\, f_1\, u_i(1)\, d1 + n_i \int \delta u_i^*(1)\, u_i(1)\,\rho(2)\, g_{12}\, d1d2$$
$$-\frac{9}{2}\alpha\left[\frac{3}{4\pi}\right]^{1/3}\frac{4}{3}\int \delta u_i^*(1)\, u_i(1)\left[\rho\uparrow(1)\right]^{1/3} d\underset{\sim}{r}_1 \tag{23}$$

Introducing the normalization condition through a Lagrangian multiplier ε_i, we arrive at the $X\alpha$ one-electron equations:

$$\left[-\nabla_1^2 + V_C(1) + V_{X\alpha}\uparrow(1)\right] u_i(1) = \varepsilon_i\, u_i(1) \tag{24}$$

with the corresponding equations for spin down orbitals. The spin orbitals u_i turn out to be orthogonal if they belong to different eigenvalues as they are solutions of the same eigenvalue problem. Degenerate spin orbitals can be chosen orthogonal to each other /24/. V_C is the total coulomb potential felt by one electron:

$$V_C(1) = -2\sum_\mu Z_\mu / r_{1\mu} + 2\int \rho(2)/r_{12}\, d2 \tag{25}$$

and $V_{X\alpha}\uparrow = \frac{2}{3} U_{X\alpha}\uparrow$ is a local exchange potential simplifying the solution of (24) as compared to the corresponding non-local HF equations.

It is important to note that the eigenvalues ε_i differ fundamentally from those which arise in the HF method. If no orbital relaxation is allowed, i.e. if we describe the ion with the same set of orbitals as the ground state except for a change of occupation of one orbital u_i (from $n_i = 1$ to $u_i = 0$), Koopmans' theorem states for the ionization potential I_i /4,28/:

$$-I_i = E_{HF}(n_i = 1) - E_{HF}(n_i = 0) = \varepsilon_{i\,HF} \tag{26}$$

The HF eigenvalue ε_{iHF} is related to a finite difference of energies E_{HF} for two states whereas the eigenvalue $\varepsilon_{iX\alpha}$ method turn out to be given by a partial derivative of the Xα total energy:

$$\varepsilon_{i\,X\alpha} = \int d\underset{\sim}{r}_1\, u_i^*(1)\left[f_1 + V_C(1) + \tfrac{2}{3} U_{X\alpha}\uparrow(1)\right] u_i(1) = \frac{\partial E_{X\alpha}}{\partial n_i} \tag{27}$$

as may be seen by differentiating (22), keeping the spin orbitals u_k fixed, and by comparing the result to (24). However, the same expression holds for relaxed orbitals determined self-consistently for varying occupation numbers (see sec. 2.4).

In this context, we meet for the first time the concept of continuously varying occupation numbers in the Xα method. Some comments on their rôle seem to be appropriate here. In the HF theory the spin orbital occupation numbers are either 0 or 1 according to the occurrance of the orbital in the determinantal wave function. In contrast, we find no conceptual restriction of the occupation numbers to integer values in the Xα method. Inspection of (22) and (24) shows that the theory works equally well for fractional values which, of course, have to be restricted to the interval $0 \leq n_i \leq 1$ by the exclusion principle. The state of lowest energy may then be determindes by requiring that

$$\delta\left[E_{X\alpha} - \mu \sum_i n_i\right] = 0 \tag{28}$$

where Lagrangian parameter μ is introduced to keep the total number of electrons constant. We then find the state of minimum energy as the one where

$$\varepsilon_i = \frac{\partial E_{X\alpha}}{\partial n_i} = \mu, \tag{29}$$

except for those levels with $n_i = 1$ or $n_i = 0$. It would therefore be impossible to lower $E_{X\alpha}$ by an infinitesimal transfer of charge δn from one spin orbital i to any other spin orbital k:

$$\delta E_{X\alpha} = -\frac{\partial E_{X\alpha}}{\partial n_i}\delta n + \frac{\partial E_{X\alpha}}{\partial n_k}\delta n = (\varepsilon_k - \varepsilon_i)\,\delta n \geq 0,$$

i.e. charge may not be transferred to lower lying levels. From this we conclude that levels for which $n_i = 1$ correspond to $\varepsilon_i < \mu$ and those for which $n_i = 0$ have $\varepsilon_i > \mu$. Only spin orbitals with $\varepsilon_i = \mu$ can have fractional occupation numbers. In other words, the Xα method exactly fulfills the Fermi statistics /16,24/ and the parameter μ corresponds to the Fermi level. The important consequences of this fact for band structure calculations have been discussed in ref. 29.

Fractional occupation numbers are not all that unusual as it might seem on first sight. The third row transition elements from Ti to Cu have their highest two levels, 3d and 4s, almost degenerate and the question how to populate these levels arises. Let us discuss Cu $(3d^{10}4s)$, to take a specific example. In the configuration $(3d\uparrow)^5\ (3d\downarrow)^4\ 4s\uparrow\ 4s\downarrow$, one finds both 3d levels below the 4s levels and therefore this cannot be the ground state according to the discussion above. We will therefore start transferring charge from $4s\downarrow$ into $3d\downarrow$. Fig. 3 shows the four eigenvalues as a function of the occupation variable x in a configuration $(3d\uparrow)^5\ (3d\downarrow)^{4+x}\ 4s\uparrow\ (4s\downarrow)^{1-x}$. Actually, these values E_i' have been obtained as relaxed derivatives from the total energy of a Hyper-Hartree-Fock (HHF) calculation. This method has been shown to give results in close agreement with the Xα method, especially for $\varepsilon_i(X\alpha)$ and $E_i'(\mathrm{HHF})$ (see the discussion in 2.3). As can be seen from Fig. 3, the configuration $(3d\uparrow)^5\ (3d\downarrow)^5\ 4s\uparrow$ (x = 1) is not selfconsistent either, because the empty level 4s lies below the filled 3d levels. The minimum total energy is obtained for x = 0.59 where the levels $3d\downarrow$ and $4s\downarrow$ cross and where the Fermi statistics is obeyed. The configuration $(3d\uparrow)^5\ (3d\downarrow)^{4.59}\ 4s\uparrow\ (3s\downarrow)^{0.41}$ is closer to $(3d\uparrow)^5\ (3d\downarrow)^5\ 4s\uparrow$ than to $(3d\uparrow)^5\ (3d\downarrow)^4\ 4s\uparrow\ 4s\downarrow$ from which one may conclude /24/ that the former is the ground state of Cu, which is known to be true from experiment.

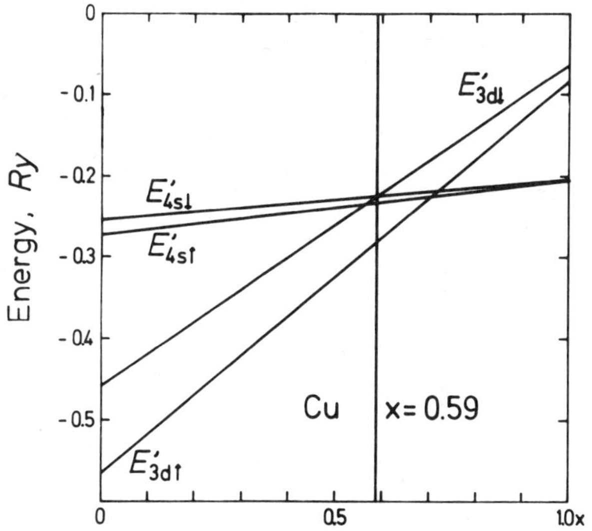

Fig. 3 Eigenvalues of 3d↑, 4s↑, 4s↓, 3d↓ of Cu as functions of the occupation variable x, in the configuration

$$(3d\uparrow)^5\ 4s\uparrow\ (4s\downarrow)^{1-x}\ (3d\downarrow)^{4+x}$$

Eigenvalues E_i' from HHF calculation of ref. 29.

2.2 The Hellmann-Feynman and the Virial Theorem

The Hellmann-Feynman theorem and the virial theorem are rarely used with the exact hamiltonian despite of their conceptual importance in the theory of electronic structure because very accurate wave functions are needed. The Xα energy functional (22) on the other hand, has been shown to satisfy both theorems /30/. From the Hellmann-Feynman theorem we can calculate the change in the total energy with respect to a displacement of some nucleus in terms of three-dimensional integrals instead of the six-dimensional integrals in the total energy. The theorem states that the force on the μ-th nucleus is simply given from the explicit dependence of the functional (22) on R_μ :

$$-\nabla_\mu E_{X\alpha} = \sum_i n_i \int d\underset{\sim}{r}_1 u_i^*(1) \nabla_\mu \frac{2Z_\mu}{|\underset{\sim}{r}_1 - \underset{\sim}{R}_\mu|} u_i(1) - \sum_{\mu \neq \nu} \nabla_\mu \frac{Z_\mu Z_\nu}{|\underset{\sim}{R}_\mu - \underset{\sim}{R}_\nu|} \qquad (30)$$

The terms arising from the implicit dependence of the spin orbitals u_i on the position $\underset{\sim}{R}_\mu$ may be obtained from (23) by formally replacing δu_i^* by ∇u_i^* and by using (24):

$$-\nabla_\mu E_{X\alpha} = -\sum_i n_i \int d\underset{\sim}{r}_1 \left[\varepsilon_i u_i(\underset{\sim}{r}_1) \nabla_\mu u_i^*(\underset{\sim}{r}_1) + u_i^*(r_1) \varepsilon_i \nabla_\mu u_i(r_1) \right]$$

$$+ \text{ terms of eq. (30)} \qquad (31)$$

$$= -\sum_i n_i \varepsilon_i \nabla_\mu \int d\underset{\sim}{r}_1 \, u_i^*(r_1) u_i(r_1)$$

$$+ \text{ terms of eq. (30)}$$

The additional terms vanish on account of the normalization of the u_i's and (30) has been proven as a consequence of the variational character of the spin orbitals.

The prove of the virial theorem follows from the method used by Fock, based on a scaling argument of the wave function /31/. If we scale all coordinates (including those of the nuclei) by a factor η , we get for the normalized spin orbitals:

$$u_i' = \eta^{3/2} u_i(\eta \underset{\sim}{r}_1, \ldots, \eta \underset{\sim}{R}_\mu, \ldots) \qquad (32)$$

For the kinetic energy we have

$$T' = \eta^2 T(\eta \underset{\sim}{R}_\mu) \qquad (33)$$

whereas all the potential energy terms scale as

$$V' = \eta V(\eta \underset{\sim}{R}_\mu). \qquad (34)$$

For the Xα exchange energy this follows from quantities like

$$\int \left[u_i^{*\prime}(\underset{\sim}{r}_1) u_i'(r_1) \right]^{4/3} d\underset{\sim}{r}_1 = \eta \int \left[u_i^*(\eta \underset{\sim}{r}_1) u_i(\eta \underset{\sim}{r}_1) \right]^{4/3} d(\eta \underset{\sim}{r}_1). \qquad (35)$$

Now we invoke the variational principle:

$$0 = \frac{\partial}{\partial \eta} E'_{X\alpha} = \frac{\partial}{\partial \eta}(T' + V')$$

$$= 2\eta T + V + \frac{1}{\eta} \sum_{\mu} \underset{\sim}{R}_{\mu} \cdot \nabla_{\mu} E_{X\alpha}(\eta \underset{\sim}{R}_{\mu}) \tag{36}$$

which leads to the general virial theorem:

$$\sum_{\mu} \underset{\sim}{R}_{\mu} \cdot \nabla_{\mu} E_{X\alpha} = -2T - V \tag{37}$$

The virial theorem has been used to optimize the parameter α and to discuss the quality of the Xα spin orbitals (cf. the next section). Two remarks may be appropriate here concerning the Xα-SW method as applied to molecules and clusters /30/. (1) It is common practice to use different α-values in different regions of space. This would lead to small correction terms in both theorems. (2) The sphere radii occuring in the muffin-tin potential should be scaled in the same way as the interatomic distances.

2.3 The Choice of the Exchange Parameter

The exchange parameter α has been left undetermined and we now take up the subject of choosing its value to obtain optimal results. Unfortunately, we cannot determine α by minimizing the total energy because $E_{X\alpha}$ (22) is to a very good approximation a linear function of α. The larger the value of α, the lower the total energy will be, as may be seen from the negative sign of the exchange-correlation energy in (22). The next obvious choice would be to link up the Xα method to HF theory, especially for atomic systems because these are convenient to handle in both methods and additional insight can be expected from a systematic study of the periodic table.

Digression: the Hyper-Hartree-Fock method. Despite of the simplicity of an atomic system, a characteristic difficulty arises, as most isolated atoms have only a partially filled outer shell of electrons. As a consequence of the interaction between the electrons of this partially filled shell, we will have to solve the familiar problem of multiplets associated with a certain configuration /4/. To describe such a multiplet we will need wavefunctions built up from several determinantal functions. The energy separation between the various multiplet levels is therefore beyond a one-electron picture where the electrons are supposed to have degenerate one-electron energies. (Actually, this statement holds strictly true only if we neglect a spin polarization that might exist, i.e. if we enforce $\rho\uparrow = \rho\downarrow = \rho/2$.) In the Xα method we find degenerate levels in a spherically symmetric potential and are led to fractional occupation numbers. It would be very convenient if we had an extension of the HF method to open-shell atoms where the multiplet separations were neglected and an average energy were assigned to a configuration. Such a theory has been developed by Slater who called it Hyper-Hartree-Fock (HHF) theory /4/. The total energy E_{HHF},

$$E_{HHF} = \sum_i \left[q_i I(i) + \tfrac{1}{2} q_i (q_i - 1)(i,i) \right] + \sum_{i<j} q_i q_j (i,j) \quad (38)$$

is written as a sum over shells i each holding q_i electrons. The various integrals represent the contributions from the one-electron operators,

$$I(i) = \int u_{ik}^*(\underset{\sim}{r}_1) f_1 u_{ik}(\underset{\sim}{r}_1) d\underset{\sim}{r}_1 \quad (39)$$

(the value of the integral is independent of the particular spin orbital u_{ik} of shell i) and from the averaged interaction within one shell, (i,i) and between two shells, (i,j). For a more detailed discussion and a reduction of this interaction terms to radial atomic integrals see ref. 4. In the present context, it is important to note that a relation analogous to (26) holds for the eigenvalues ε_{iHHF} and that the HHF theory reduces to the HF case for closed-shell systems. Self-consistent solutions are available for all atoms of the periodic table /32/. The results of these calculations have been used to determine the value of the exchange parameter α and to judge the quality of Xα spin orbitals.

With this information at hand, the most obvious procedure in the case of an atom would be to choose α such that the energy functional $E_{X\alpha}$ is exactly equal to the energy determined from the HHF method (38),

$$E_{X\alpha}[\alpha_{HF}] = E_{HHF} \quad (40)$$

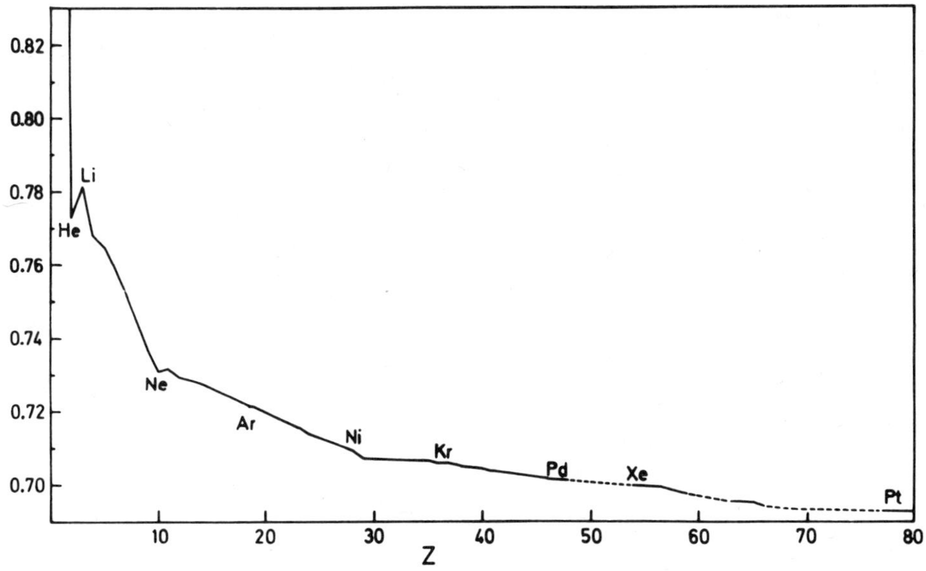

Fig. 4 Exchange parameter α as a function of the atomic number.

Table I Optimized exchange parameters for the atoms of the first row

Z	Atom	α^a	$\alpha'^{\,b}$
1	H	.9780	.7763
2	He	.7730	.7730
3	Li	.7815	.7716
4	Be	.7682	.7682
5	B	.7653	.7621
6	C	.7593	.7533
7	N	.7520	.7452
8	O	.7445	.7419
9	F	.7373	.7359
10	Ne	.7308	.7308

[a] non-spin polarized (ref. 33)
[b] spin-polarized (ref. 22)

This program was carried out by Schwarz /33/ who determined the α_{HF}-values for most atoms of the periodic table. The results are plotted in Fig. 4 as a function of the atomic number Z. The values fall monotonically from about 0.78 for light atoms to 0.71 for 3d transition elements and decrease slowly thereafter to approximately 0.69 for heavy atoms ($Z > 70$). We also note that for heavy atoms α comes rather close to the free-electron gas value 2/3. In his study, Schwarz used the non-spin-polarized version of the Xα model. The spin-polarized version gives a more accurate description of an open-shell system and leads to values α'_{HF} differing somewhat from α_{HF} /22/. In Table I these two parameters are compared for some lighter atoms. The hydrogen atom furnishes an extreme case because the non-spin-polarized version with $\rho\uparrow = \rho\downarrow = \rho/2$ gives a very bad description of the system where $\rho\uparrow = \rho$ and $\rho\downarrow = 0$. For this one-electron system we do not have any exchange effects so that $U_{X\alpha}$ serves only to cancel the self-interaction of the one electron. Thus we must have /22/

$$-9\alpha'\left(\frac{3}{4\pi}\rho\right)^{1/3} = -9\alpha\left(\frac{3}{8\pi}\rho\right)^{1/3} \tag{41}$$

from which follows $\alpha' = 2^{-1/3}\,\alpha \approx 0.794\,\alpha$. For atoms heavier than those shown in Table I, the difference becomes quite negligible.

The resulting spin orbitals u_i from a self-consistent calculation with such an optimized value of α are found to be very good approximations to HF spin orbitals. One way to check this statement is to calculate E_{HF}(5) using Xα spin orbitals determined from (24). Because of the variation principle, this value $E_{HF}(X\alpha)$ has to be higher than E_{HF} computed from self-consistent HF orbitals, but is found to lie only slightly above this value. The quality of the Xα orbitals tested in this way has been compared to the so-called double-zeta basis sets proposed by Clementi, which are used in quite accurate molecular-orbital calculations /33a/.

Following Slater, there is yet another way to compare Xα and HF orbitals /24/. We have shown that the solution to the Xα model satisfies the virial theorem for any value of α. This means for an isolated atom that the kinetic energy is exactly equal to the negative of the total energy. But the HF solution also satisfies the virial theorem, hence the kinetic energies as determined by either method agree precisely. This is a sensitive test for the orbitals. We further conclude that the potential energies have to be equal, and that the coulomb energies for similar charge distributions must give values close to each other, as they are calculated from the same expression (cf. eqs. (5) and (22)). Therefore, by subtraction we expect close agreement for the exchange energy, too. Slater has compared the exchange potentials $U_{XHF}\uparrow$ and $U_{X\alpha}\uparrow$ for Cu^+ (see Fig. 5) and found very good agreement indeed.

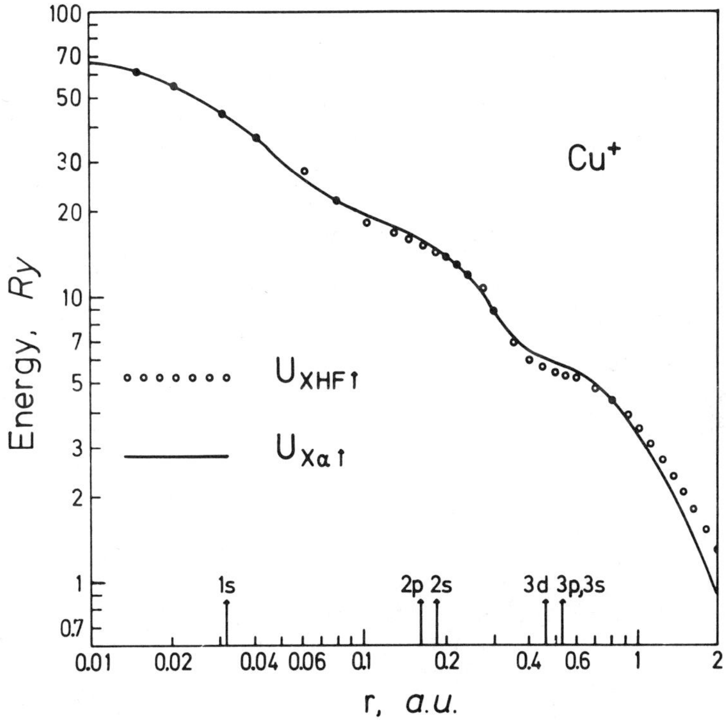

Fig. 5 Exchange Potentials $U_{XHF}\uparrow$ and $U_{X\alpha}\uparrow$ for Cu^+. Radii of the maximum radial charge density of the various shells are shown (from ref. 24)

Two other procedures for the optimization of α have been suggested for atoms. (1) Lindgren /34/ proposed to use the value α_{min} found by minimizing the above mentioned HF energy expression $E_{HF}(X\alpha)$, evaluated with self-consistent $X\alpha$ spin orbitals. These values, determined by Kmetko /35/ and Wood /36/, fall in the same range as α_{HF} but do not show a smooth variation as a function of the atomic number. (2) Following Berrondo and Goscinski /37/, one evaluates the kinetic energy $T_{HF}(X\alpha)$ and the HF potential energy $V_{HF}(X\alpha)$ separately with $X\alpha$ orbitals and chooses α_{VT} such that the virial theorem is satisfied:

$$V_{HF}(X\alpha) = -2\, T_{HF}(X\alpha) \tag{42}$$

The parameters α_{VT} so obtained agree closely with α_{HF} /32/.

2.4 The Transition State

After we have convinced ourselves that the $X\alpha$ method is capable of giving quite an accurate approximation to the spin orbitals and to the total energy of HF theory, at least for atoms, we return to a discussion of the remarkable fact that the $X\alpha$ eigenvalue is given as a partial derivative (22) of the total energy with respect to the occupation number. To study the functional dependence of the energy on the occupation numbers and to fully exploit its consequences we follow Slater and Wood /38/ in setting up a power-series expansion. We use as our variables the shell occupation numbers q_i and the number of net spins per shell μ_i:

$$q_i = q_i\uparrow + q_i\downarrow$$

$$\mu_i = q_i\uparrow - q_i\downarrow \tag{43}$$

Let us further take a non-spin-polarized reference state with energy E_o. Then all odd powers of the μ_i's vanish, as the energy must be independent of the spin direction, and we obtain:

$$E = E_o + \sum_i (q_i - q_{i_o}) \frac{\partial E}{\partial q_i} + \frac{1}{2!} \sum_{ij} (q_i - q_{i_o})(q_j - q_{j_o}) \frac{\partial^2 E}{\partial q_i \partial q_j}$$

$$+ \frac{1}{3!} \sum_{ijk} (q_i - q_{i_o})(q_j - q_{j_o})(q_k - q_{k_o}) \frac{\partial^3 E}{\partial q_i \partial q_j \partial q_k}$$

$$+ \frac{1}{2!} \sum_{ij} \mu_i \mu_j \frac{\partial^2 E}{\partial \mu_i \partial \mu_j}$$

$$+ \frac{1}{3!} \sum_{ijk} (q_i - q_{i_o}) \mu_j \mu_k \frac{\partial^3 E}{\partial q_i \partial \mu_j \partial \mu_k} \quad (44)$$

The energy is to a good approximation a quadratic function of the q_i's because of the dominant coulomb term and therefore we do not have to carry the series beyond the third power terms for most purposes. The derivatives are to be understood as "relaxed" ones, including the dependence of the spin orbitals on the occupation numbers. The first derivatives stay unaffected by this relaxation because the energy is stationary to first order with respect to any changes of the spin orbitals. Higher derivatives, however, cannot be found by straightforward ("unrelaxed") differentiation of (22) or (38), but only by fitting the total energy values of several self-consistent calculations with varying occupation numbers to the power-series (44).

One of the major objectives in studying the electronic structure of a system is the determination of the ionization spectrum. We take the standard state with energy E_o to be the ground state and remove one electron from shell i. Then only $q_i - q_{io} = -1$ is different from zero in (44) and we obtain:

$$E(q_i = q_{i_o}) - E(q_i = q_{i_o} - 1) = \frac{\partial E}{\partial q_i} - \frac{1}{2} \frac{\partial^2 E}{\partial q_i^2} + \frac{1}{6} \frac{\partial^3 E}{\partial q_i^3} \quad (45)$$

The first term is the Xα eigenvalue and we expect the second derivative to give the leading correction. Higher terms are in fact found to be rather small, as is illustrated for chromium in the configuration $3d^5 4s$ /24/. Numerical values for the individual terms are given in Table II. Also given are the HHF eigenvalues and experimental values which have been determined from x-ray data and have been corrected for relativistic effects by using a method of Herman and Skillman /10/.

Table II Binding energies for the orbitals of Cr in the $3d^5 4s$ state according to eq. (45), Koopmans' theorem and the Xα transition state procedure (from ref. 24)

Orbital	$\frac{\partial E}{\partial q_i}$	$-\frac{1}{2}\frac{\partial^2 E}{\partial q_i^2}$	$+\frac{1}{6}\frac{\partial^3 E}{\partial q_i^3}$		Experiment[a]	Koopmans[b]	Xα transition state
1s	-428.546	- 10.103	+ .038	= -438.611	-437.4	-440.81	-438.619
2s	- 48.347	- 1.383	- .077	= -49.807	-51.0	-52.54	-49.787
2p	- 41.187	- 1.692	- .086	= -42.965	-42.9	-44.40	-42.944
3s	- 5.269	- 0.402	- .062	= -5.733	-5.9	-6.66	-5.718
3p	- 3.276	- 0.392	- .059	= -3.727	-3.6	-4.19	-3.712
3d	- 0.190	- 0.323	- .055	= -0.568	-0.75	-0.645	-0.554
4s	- 0.253	- 0.204	- .008	= -0.465	-0.57	-0.415	-0.464

[a] From J. C. Slater, Phys. Rev., 98, 1039 (1955) corrected to remove relativistic effects by ref. 10.

[b] HHF eigenvalues from ref. 32.

We notice considerable differences between the Xα and the HHF eigenvalues and the different order of the 3d and 4s levels. In a spin-polarized Xα calculation one finds the level ordering as $3d\uparrow < 4s\uparrow < 4s\downarrow < 3d$ which gives the correct ground state for Cr. The Xα eigenvalues undergo a remarkable relaxation, but the overall agreement between the experimental values and those calculated from (45) is quite satisfacotry. According to Koopmans' theorem /28/, the total energy difference should be equal to the HHF eigenvalue. Especially the core levels come out too high in absolute value. This is a consequence of the "frozen" orbital approximation of the ion. Let us apply a power-series expansion to E_{HHF}(38), analogous to (44), but with unrelaxed derivatives in the spirit of Koopmans' theorem. Only terms up to second order are obtained and equation (45) reads upon rearrangement /4/:

$$\frac{\partial E_{HHF}}{\partial q_i} = \varepsilon_{i\,HHF} + \frac{1}{2}(i,i) \tag{46}$$

This quantity is actually identical to E_i' used above in the discussion of spin-polarization of Cu (see sec. 2.3) and plotted in Fig. 3. The close similarity of $E_{X\alpha}$ and E_{HHF} as a function of the occupation number may serve as an argument for the former substitution.

Relaxation effects have been included in the determination of an ionization potential according to (45) at the prize of several self-consistent total energy calculations. Therefore using the HF eigenvalue on the basis of Koopmans' theorem is much more attractive computationally. Slater /16,38/ succeeded in designing a procedure which combines both features, computational simplicity and high accuracy: the transition state procedure. For this we take the standard state halfway between the ground state ($q_i = q_{io} + 1/2$) and the ion ($q_i = q_{io} - 1/2$) and apply (44) twice to obtain:

$$E(q_i = q_{io} \pm \tfrac{1}{2}) = E_o \pm \frac{1}{2}\frac{\partial E}{\partial q_i} + \frac{1}{8}\frac{\partial^2 E}{\partial q_i^2} \pm \frac{1}{48}\frac{\partial^3 E}{\partial q_i^3} \tag{47}$$

Thus we deduce

$$E_{X\alpha}(\text{initial}) - E_{X\alpha}(\text{final}) = \varepsilon_i^t + \frac{1}{24}\frac{\partial^3 E_t}{\partial q_i^3} \tag{48}$$

The i-th ionization potential is therefore approximately equal to $-\varepsilon_i^t$ in the transition state, where half an electron has been removed. Comparing the first and the last column in Table II, the correction term is seen to be quite negligible. Experience has shown, in cases of large relaxation (such as deep core levels) that the Xα transition state gives ionization potentials superior to the HF eigenvalues. For large systems, like molecules and clusters, the transition state offers a direct route to ionization energies. One avoids the numerically very inaccurate subtraction of total energies which may easily be several orders of magnitude larger. The transition state concept is one of the biggest assets of the Xα method, especially in the application to molecular ionization potentials. Many examples will be given later on and in the lectures of Dr. Herman.

We are now in a position to solve the puzzle mentioned in the Introduction: the success of Xα ground state eigenvalues in a direct comparison with atomic X-ray spectra. The missing downward shift of the eigenvalues when going over into the transition state had been (partly) overcompensated by the use of a more attractive potential due to an exchange parameter of $\alpha = 1$.

Recently, Schrieffer /39/ made an interesting comment on the difference between the eigenvalues in HF theory and the Xα method which also sheds some light on the original choice $\alpha = 1$. Although the self-interaction of an electron is cancelled approximately in the total energy $E_{X\alpha}$ (22) due to the exchange potential $U_{X\alpha}$ (21), this is no longer true for the exchange potential $V_{X\alpha}$ in the one-particle Schrödinger equation (24). However, within the HF method, the direct and the exchange self-interaction of an electron cancel exactly in both cases:

$$\langle ij|g_{12}|ij\rangle - \langle ij|g_{12}|ji\rangle = 0 \qquad \text{for } i = j \tag{49}$$

In the Xα model, the self-interaction (direct + exchange) $E_{SI}(n_i)$ varies for a localized orbital u_i roughly as

$$E_{SI}(n_i) \simeq \frac{1}{2} U_i (n_i^2 - c\, n_i^{4/3}), \tag{50}$$

where U_i is the self-Coulomb interaction. E_{SI} contributes little to the total energy ($n_i = 1$) because of the choice $C \simeq 1$. But the potential determining the eigenvalue ε_i involves

$$V_{SI}(n_i) = \frac{\partial E_{SI}}{\partial n_i} \simeq U_i (n_i - \tfrac{2}{3} c\, n_i^{1/3}) \tag{51}$$

which for $n_i = 1$ is $U_i/3$. In the transition state, however, this contribution becomes again quite negligible, since $V_{SI}(1/2) \simeq -0.03\ U_i$. The original choice for the exchange parameter $\alpha = 1$ (or $C \simeq 3/2$) was made to cancel the self-interaction in the one-particle Schrödinger equation. The Xα method shares this problem with other local density schemes /14/. However, by the transition state concept of the Xα method, one is provided with a well-defined procedure to obtain a physical interpretation for the eigenvalues.

This concept may also be applied to estimate energies of one-electron excitations. If we are interested in the transition from an initial state with occupation numbers (q_i, q_j) to a final state with occupancy (q_i-1, q_j+1), we define the transition state having the occupation numbers $(q_i-1/2,\ q_j+1/2)$ and derive from (44):

$$\Delta E(i\rightarrow j) = (\varepsilon_j - \varepsilon_i)_t + \frac{1}{24}\left[\frac{\partial^3 E}{\partial q_i^3} - 3\frac{\partial^3 E}{\partial q_i^2 \partial q_j} + 3\frac{\partial^3 E}{\partial q_i \partial q_j^2} - \frac{\partial^3 E}{\partial q_j^3}\right] \quad (52)$$

The third-order terms have again been found to be small.

According to Slater /24/, one may partially uncover the multiplet structure by applying the transition state concept in its spin-polarized version. Let us illustrate the argument by the simple case of a closed shell molecule where we look at the transition from the highest occupied level i to the lowest unoccupied level j, both of which are assumed to be non-degenerate. The two lowest lying excitations correspond to transitions from the singlet ground state (S) to the first excited singlet (S′) and triplet (T) states. The eigenvalue difference (52), on the other hand, approximates the excitation energy to the multiplet average A which according to the multiplicities may symbolically be expressed as

$$A = \frac{1}{4}\,(3T + S') \quad (53)$$

Besides this non-spin-polarized final states, two spin-polarized excited states O and T may be formed. The various situations are best characterized by explicitly listing their occupation numbers:

	i↑	i↓	j↑	j↓
S	1	1	0	0
T	1	0	1	0
O	1	0	0	1
A	1/2	1/2	1/2	1/2

From this table we see that T has two parallel spins and therefore belongs to the triplet, whereas O has no net spin and thus is an average over two states, one from the triplet and the singlet:

$$O = \frac{1}{2}(T + S') \tag{54}$$

The singlet-triplet separation may thus be obtained by performing two spin-polarized transition state calculations ST and SO (the meaning of the symbols should be obvious) and by using (54):

$$\Delta E\ (\text{singlet-triplet}) = S' - T = 2(O - T)$$

$$\simeq 2\left[(\varepsilon_j - \varepsilon_i)_{SO} - (\varepsilon_j - \varepsilon_i)_{ST}\right] \tag{55}$$

This approach has been criticized by Bagus and Bennett /40/ who suggest that (55) is at most an upper bound to the singlet-triplet splitting. There are, of course, third order terms in both cases which may affect the value of such a small energy difference. By inspection of the occupancy of the various transition states as well as by the results of the cited analysis, one is lead to recommend eq. (55) for estimations of the singlet-triplet splitting and to reject other combinations involving transition states to the multiplet average A.

The success of the Xα transition state concept stimulated the construction of similar procedures in other one-electron theories, like HF theory. With the use of a transition operator /41,42/ it was possible to get ionization potentials that show significant improvement over results from Koopmans' theorem.

2.5 Comparison to Related Local Potential Models

The Xα method is only one out of a large number of local potential models for the self-consistent field problem. Some of them will be discussed in the lectures by Prof. March. Before we make some remarks on their relation to the Xα method, a collection of its main advantages and disadvantages may be helpful /22/. This will be done mainly in comparison to HF theory, so some points will similarly apply to other local potential models.

(1) Simplicity and computational speed may be considered as the main advantages of the Xα method. The HF method requires a greater computational effort because it uses a different local potential for every orbital (which is just another phrase for a non-local potential) instead of one local potential as in the Xα method.

(2) HF theory treats occupied and unoccupied (virtual) orbitals differently. Virtual orbitals correspond to an electron which sees a potential set up by N occupied orbitals. They are therefore not suited to describe excited states, rather they belong to a N+1 electron system. In the Xα method, both occupied and virtual orbitals are under the influence of N-1 other electrons, which leads to a better description of electronic excitations. Xα ground state eigenvalue differences are often a fair description of excitation energies, even without going to the transition state. Xα virtual orbitals are less diffuse than those from HF theory. One may hope that they provide a better starting point for theories beyond the self-consistent field, like configuration interaction etc.

(3) Both the Hellmann-Feynman and the virial theorem hold for the Xα model. Furthermore, Fermi statistics is strictly fulfilled. The levels in the ground state are filled from the bottom upwards.

(4) The transition state concept provides a simple and quite accurate approximation to excitation energies which may be either calculated directly (ionization potentials) or as a difference of two eigenvalues (optical excitations).

(5) These advantages are contrasted by several disadvantages, the most serious of which is that no total wave function can be rigorously defined. Except for the total energy, only quantities represented by one-particle operators may be calculated.

(6) A further disadvantage of the method is the introduction of a parameter α. There are several reasonable ways to choose this parameter for atomic systems (as we have discussed in sec. 2.3), but it is not immediately clear how to procede in the case of molecules or solids. As an example consider the hydrogen molecule. If the two atoms are very far apart they both carry one unpaired electron and the spin-polarized exchange parameter $\alpha = .7763$ should be used. However, as the two atoms

approach each other, a molecular orbital is formed which is able to hold both electrons, the spins paired. No spin density is found in the molecule and one would expect a different α value. A possible answer to this problem may be discussed within the density functional formalism to which we turn in a moment.

(7) There seem to be several more indications that the Xα method has trouble in handling spin-dependent effects in an accurate way. For instance, the polarization of core orbitals through an unpaired spin in a valence orbitals is underestimated /43/. Consequently, hyperfine coupling constants turn out too small. Furthermore, remember the problems in calculating the singlet-triplet splitting.

(8) A numerical disadvantage, although not a serious one, is encountered when one tries to use LCAO techniques in molecular calculations because the exchange potential (21) is not a linear functional of the density. Recent efforts have shown how to overcome this problem. The basic ideas of these procedures will be mentioned in Chapter 4.

Following Slater, we have introduced the Xα method on plausible physical arguments and have presented empirical evidence for its validity. But the problem of interacting electrons has also been formulated in several other ways leading to similar one-electron equations with an effective local potential. A typical example with the appeal of a rigorous approach free of parameters is the density functional formalism by Kohn and Sham /14/. It is based on the Hohenberg-Kohn theorem /44/ which asserts that the total energy can be expressed as a unique functional of the charge density. However, the exchange-correlation portion of this functional is not known. For a slowly varying charge density its form is known to be /14/

$$E_{xc} = \int \varrho(\underset{\sim}{r})\, \varepsilon_{xc}[\varrho]\, d\underset{\sim}{r} \qquad (56)$$

where $\varepsilon_{xc}[\varrho]$ is the exchange-correlation function of the free-electron gas. Only the exchange part of the function is known exactly, it is identical to the Xα form (21) with $\alpha = 2/3$. Various approximations for the correlation part have been used /45/. The resulting one-electron equations are more complicated and have only been solved for diatomic molecules with up to four electrons /46/ (besides many applications to surface problems /47/).

However, from our previous discussion it would seem that the Xα model has a wider range of applicability than models on the basis of a slowly varying density. Certainly, this assumption does not apply to atoms of higher atomic number and molecules in general. A further quite severe disadvantage of models like those of Kohn-Sham /14/ or Hedin-Lundqvist /45/ is the lack of simple correspondence betwen excitation energies and the eigenvalues of the one-particle equation.

Several extensions of the Xα model have been suggested to correct some of its weak points, like variations of the exchange parameter in space /48/ (different values for the core and valence region) or with the orbital /49/. A model including inhomogeneity corrections is the so-called Xαβ model /50/ (see (20)):

$$V_{X\alpha\beta}(1) = -\left[\alpha + \beta\, G(\varrho)\right] 6 \left[\frac{3}{8\pi}\, \varrho(1)\right]^{1/3} \tag{57}$$

where

$$G(\varrho) = \left[\frac{4}{3}\left(\frac{\nabla\varrho}{\varrho}\right)^2 - 2\,\frac{\nabla^2\varrho}{\varrho}\right] \varrho^{-2/3} \tag{58}$$

It was possible /51/ to fit the total energies of all atoms with fixed parameters, independent of Z : $\alpha = 2/3$, $\beta = 0.003$. This is of some practical advantage for molecules. However, none of these extensions have lead to any significant improvement over the original formulation /21/.

We may summarize the discussion in this chapter by the statement that the Xα model, although in some sense a crude approximation, provides a simple, yet reasonably accurate way to the self-consistent field of interacting electrons.

3. THE SCATTERED-WAVE FORMALISM

In order to apply the Xα method to molecules and clusters we have to look for an efficient and accurate way of solving a one-particle Schrödinger equation,

$$\left[-\nabla^2 + V(\underset{\sim}{r}) - E\right] \Psi(\underset{\sim}{r}) = 0 \tag{59}$$

The scattered-wave (SW) formalism developed by Johnson /18,20/ was the first successful method, and it is still the most popular one in this field of applications /22, 23/. However, finding improved or new methods is still an area of active research. Although we shall present the formalism in its simplest form, we shall do so in a manner that underlines the features it shares with other methods using the powerful multiple-scattering (MS) techniques /52/. This formulation facilitates the discussion of possible generalizations which have been suggested. In the following chapter we will also comment on LCAO-based solution methods. Recently, Williams has given an excellent discussion on the multiple-scattering technique and its advantages over LCAO methods /53/.

3.1 The Multiple Scattering Method

The MS method starts from an integral equation equivalent to (59) /52,54,55/:

$$\Psi(\underset{\sim}{r}) = \int G_E(\underset{\sim}{r},\underset{\sim}{r}')\, V(\underset{\sim}{r}')\, \Psi(r')\, d\underset{\sim}{r}' \qquad (60)$$

where $G_E(\underset{\sim}{r},\underset{\sim}{r}')$ is the free-space Green's function defined by

$$(\nabla^2 + E)\, G_E(\underset{\sim}{r},\underset{\sim}{r}') = \delta(\underset{\sim}{r}-\underset{\sim}{r}') \qquad (61)$$

Next, one partitions space into subvolumes (or cells) Ω_p where the wave function is expanded as

$$\Psi(\underset{\sim}{r}) = \sum_L C_L^p(E)\, \varphi_L(\underset{\sim}{r}_p, E) \qquad (62)$$

Here, $\underset{\sim}{r}_p = \underset{\sim}{r} - \underset{\sim}{R}_p$ is the local coordinate with respect to the reference origin $\underset{\sim}{R}_p$ in the subvolume Ω_p and $L = (l,m)$ is a shorthand for the angular momentum quantum numbers. The members $\varphi_L(\underset{\sim}{r},E)$ of the cellular basis are assumed to be solutions of the Schrödinger equation at energy E in the subvolume Ω_p. This implies a predetermination of the radial part of the wave function. The expansion coefficients $C_L^p(E)$ are determined by relating the solutions in the various subvolumes to each other. This can be done by inserting (62) into (60) and by exploiting the resulting identities in $\underset{\sim}{r}$ for the various regions of space /56/. Alternatively, a variational principle equivalent to (58) may be employed /58/:

$$\delta \Lambda = 0 \tag{63}$$

where

$$\Lambda = \int \Psi^*(\underset{\sim}{r}) V(\underset{\sim}{r}) \left[\Psi(\underset{\sim}{r}) - \int G_E(\underset{\sim}{r},\underset{\sim}{r}') V(\underset{\sim}{r}') \Psi(\underset{\sim}{r}') d\underset{\sim}{r}' \right] d\underset{\sim}{r} \tag{64}$$

With the cellular expansions (62), eq. (64) is reduced to an algebraic variational problem where the coefficients depend on the geometrical arrangement of the scattering centers and on "multipole" integrals of the cellular solutions $\varphi_L(\underset{\sim}{r}_p, E)$ /53/. In the case of simply shaped subvolumes, e.g. spheres, one may also solve (57) directly by matching the cellular solutions (62) and their first derivatives across the surface of two adjacent cells /20,21/. However, this approach becomes rather awkward even for simple cell shapes like a spheroid or a cylinder /57/.

At this point, some important differences between the MS approach and the LCAO method may be discussed /53/. The cellular basis (62) is very flexible because radial and angular dependence are decoupled. Together with an optimal energy dependence, a MS basis set renders higher numerical accuracy and better convergence than an LCAO basis, the radial dependence of which is given analytically, but fixed through a specific choice of the orbital exponent. The resulting secular problem has a much smaller dimension. A further important advantage of this procedure is that heavier atoms are almost as easily treated as lighter atoms. These desirable features have to be paid for by the energy dependence of the basis and by the difficulties associated with the partitioning of space. The energy dependence of the basis does not present serious problems. It merely expresses the fact that in an MS calculation the energy levels are found one after the other, rather than simultaneously (by a diagonalization) as in the LCAO method. The efforts necessary to solve the MS equation depend critically on the partitioning of space and the realism of the employed potential. We now turn to a model which may be characterized as fair in this respect, but which nevertheless provides a reasonable first approximation.

3.2 MS Formalism for a Muffin-Tin Potential

The MS method for a muffin-tin potential is essentially an adaption of the Korringa-Kohn-Rostocker (KKR) method of energy band calculations for solids to finite clusters/19/. The method uses the so-called muffin-tin approximation of the potential which greatly simplifies the evaluation of the functional (64).

Only spheres will be allowed as cellular shapes. The space is then partitioned into three regions as shown in Fig. 6 /20/:

I. Atomic: N non-overlapping spheres of radius b_p centered on each constituent atom p at position $\underset{\sim}{R}_p$;

II. Interatomic: the region between the "inner" atomic spheres and an "outer" sphere surrounding the entire cluster;

III. Extramolecular: the region outside the "outer sphere" of radius b_o, centered at $\underset{\sim}{R}_o$.

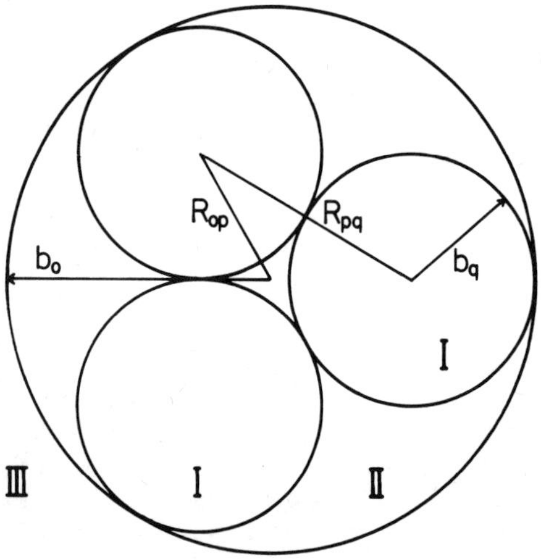

Fig. 6 Partitioning of space for a hypothetical cluster A_3 into I, atomic; II, interatomic; and III, extramolecular regions. Also shown are b_i, sphere radii, $\underset{\sim}{R}_p$, atomic positions; and $\underset{\sim}{R}_{pq}$, the interatomic separations.

The potential inside each atomic region may always be expanded in a series of spherical harmonics:

$$V(\underset{\sim}{r}_p) = \sum_L V_L^p(r_p)\, Y_L(\hat{r}_p) \tag{65}$$

(We use real spherical harmonics throughout.) The first, spherically symmetric term L = (0,0) is the only one retained in the muffin-tin approximation. It is important to emphasize that it does not only include the contribution of the atom located at $\underset{\sim}{R}_p$, but also the spherical averages of all neighboring atoms. In the extramolecular region III a similar spherical average of the potential is performed. In region II, the potential is taken to be a constant $\bar{V}$, equal to the volume average

$$\bar{V} = \int_{\Omega_{II}} V(\underset{\sim}{r})\, d\underset{\sim}{r} \tag{66}$$

Thus, within the muffin-tin approximation the potential is replaced by the model potential of the following form:

$$V(\underset{\sim}{r}) = \begin{cases} V^p(r_p) & r_p \le b_p \qquad p = 1, \dots N \\ \bar{V} & \underset{\sim}{r} \in \Omega_{II} \\ V^0(r_0) & r_0 \ge b_0 \end{cases} \tag{67}$$

For the various regions a cellular basis may be generated quite easily. We have in regions I and III:

$$\Psi(\underset{\sim}{r}) = \sum_L C_L^p(E)\, R_\ell^p(r_p; E)\, Y_L(\hat{r}_p) \tag{68}$$

with $0 \le r_p \le b_p$, p = 1, ..., N and $b_0 \le r_0 < \infty$, p = 0. The functions ${}^pR_\ell^p(r,E)$ are solutions of the radial Schrödinger equation

$$\left[-\frac{1}{r}\frac{d^2}{dr^2} r + \frac{\ell(\ell+1)}{r^2} + V^p(r) - E\right] R_\ell^p(r; E) = 0 \tag{69}$$

The radial functions in an atomic sphere must be regular at the origin r = 0. They are generated by outward numerical integration for each trial energy E and angular momentum ℓ, subject to the boundary condition

$$R_\ell^p(r;E) \sim r^\ell \quad \text{for } r \to 0;\ p>0.$$

Sometimes, interstitial spheres with spherically averaged potentials, so-called empty spheres, are introduced to reduce the volume of region II. They may be treated in the same manner.

The radial functions R_ℓ^0 of region III must decay exponentially at large distances from the molecule in order to represent localized molecular orbitals. These solutions are obtained by inward numerical integration of (69).

The calculations may be simplified by measuring the energy with respect to the constant potential $\overline{V}$. Formally, this corresponds to putting $\overline{V} = 0$. The Schrödinger equation for region II then reduces to the Helmholtz equation /54/:

$$\left[\nabla^2 + E\right] \Psi(\underset{\sim}{r}) = 0 \tag{70}$$

The solutions of this equation are known and we will discuss the form of the wave function Ψ in region II later on. However, we do not need a cellular basis for this region.

The integral (64) is restricted to regions I and III because the potential vanishes in the interatomic region. We exploit the special form (67) of the potential to transform the volume integral (64) into a surface integral. This will be done with the help of Green's theorem applied to the wave function $\Psi(r)$ and the Green's function $G(\underset{\sim}{r},\underset{\sim}{r}')$ /54/. From (59) and (61) we obtain

$$G(\underset{\sim}{r},\underset{\sim}{r}')(\nabla'^2+E)\Psi(\underset{\sim}{r}') - \Psi(r')(\nabla'^2+E)G(\underset{\sim}{r},\underset{\sim}{r}')$$

$$= G(\underset{\sim}{r},\underset{\sim}{r}')V(\underset{\sim}{r}')\Psi(r') - \delta(\underset{\sim}{r}-\underset{\sim}{r}')\Psi(\underset{\sim}{r}). \tag{71}$$

We integrate this identity over regions I and III and apply Gauss' theorem:

$$\int_{S_{II}} dS' \left[\Psi(\underset{\sim}{r}') \frac{\partial}{\partial n'} G(\underset{\sim}{r},\underset{\sim}{r}') - G(\underset{\sim}{r},\underset{\sim}{r}') \frac{\partial}{\partial n'} \Psi(r') \right] + \int_{I+III} d\underset{\sim}{r}' G(\underset{\sim}{r},\underset{\sim}{r}') V(\underset{\sim}{r}') \Psi(\underset{\sim}{r}')$$

$$= \begin{cases} \Psi(\underset{\sim}{r}), & \text{if } \underset{\sim}{r} \text{ is inside region I+III.} \quad (72.a) \\ 0, & \text{if } \underset{\sim}{r} \text{ is inside region II.} \quad (72.b) \end{cases}$$

The symbol $\partial / \partial n'$ denotes the normal derivative along the inside normal of the surface S_{II} enclosing region II. On the other hand, we may choose region II as the volume of integration ($V(\underset{\sim}{r}) \equiv 0$!):

$$\int_{S_{II}} dS' \left[G(\underset{\sim}{r},\underset{\sim}{r}') \frac{\partial}{\partial n'} \Psi(\underset{\sim}{r}') - \Psi(r') \frac{\partial}{\partial n'} G(\underset{\sim}{r},\underset{\sim}{r}') \right]$$

$$= \begin{cases} \Psi(\underset{\sim}{r}), & \text{if } \underset{\sim}{r} \text{ is inside region II.} \quad (73.a) \\ 0, & \text{if } \underset{\sim}{r} \text{ is inside region I+III.} \quad (73.b) \end{cases}$$

The meaning of the integrals (72) and (73) is not immediately clear if $\underset{\sim}{r}$ is on the surface S_{II} and we have to use a limiting procedure to deal with the singularities of G properly. Let us define the discontinuous function (73) as $\Psi(r)$ inside and on the surface S_{II} /54/. The wave function in region II is then given by

$$\Psi_{II}(\underset{\sim}{r}) = \sum_{p=0}^{N} \int_{b_p-\varepsilon_p} dS_p' \left[G_E(\underset{\sim}{r},\underset{\sim}{r}') \frac{\partial}{\partial n'} \Psi(r') - \Psi(\underset{\sim}{r}') \frac{\partial}{\partial n'} G_E(\underset{\sim}{r},\underset{\sim}{r}') \right] \quad (74)$$

The quantity ε_p is defined as

$$\varepsilon_p = \begin{cases} -\varepsilon, & p = 0. \\ +\varepsilon, & p = 1, \dots N. \end{cases}$$

and the variable $\underset{\sim}{r}'$ in (73) is to run over the surfaces of spheres with radii $b_p - \varepsilon_p$. This implies, for instance, a sphere around the center of the cluster, just outside the outer sphere. The free-space Green's function is given by /20,53,55/

$$G_E(\underset{\sim}{r},\underset{\sim}{r}') = \begin{cases} -\dfrac{\cos(k|\underset{\sim}{r}-\underset{\sim}{r}'|)}{4\pi|\underset{\sim}{r}-\underset{\sim}{r}'|}, & E > 0 \\ -\dfrac{\exp(-k|\underset{\sim}{r}-\underset{\sim}{r}'|)}{4\pi|\underset{\sim}{r}-\underset{\sim}{r}'|}, & E < 0. \end{cases} \quad (75)$$

where

$$k = \begin{cases} \sqrt{E} \quad , & E > 0 \\ \sqrt{-E} \quad , & E < 0 \;. \end{cases} \qquad (76)$$

With the choice of the so-called standing-wave representation of G for E > 0, one avoids complex quantities in the final formulas. In the following, we will only carry through the case E > 0. The general solution, also for the out-going-wave representation of G, may be found in the literature /18,20,58/.

The well-known bipolar spherical-wave expansion of the Green's function (actually the real part of the formula which is usually derived) reads /20,59/:

$$-\frac{\cos(k|\underline{r}-\underline{r}'|)}{4\pi|\underline{r}-\underline{r}'|} = -k\sum_L n_\ell(kr_>)\, j_\ell(kr_<)\, Y_L(\hat{r})\, Y_L(\hat{r}') \qquad (77)$$

where $r_> = r$ and $r_< = r'$ if $r > r'$ and vice versa. The functions $j_\ell(x)$ and $n_\ell(x)$ are spherical Bessel functions and spherical Neumann functions, respectively /59/. They form a pair of linear independent solutions to the radial form of the Helmholtz equation (70), which may be obtained from (69) by putting $V(r) \equiv 0$. We recall the asymptotic forms of these functions, for $x \ll \ell$:

$$j_\ell(x) \longrightarrow \frac{x^\ell}{(2\ell+1)!!}$$

$$n_\ell(x) \longrightarrow -\frac{(2\ell-1)!!}{x^{\ell+1}} \;, \qquad (78)$$

and for $x \gg \ell$:

$$j_\ell(x) \longrightarrow \frac{1}{x}\sin(x-\ell\pi/2)$$

$$n_\ell(x) \longrightarrow -\frac{1}{x}\cos(x-\ell\pi/2). \qquad (79)$$

If we use relation (77) and expansion (68), valid on the surfaces (note: $r'_p = b_p - \varepsilon_p$!), and the orthonormality properties of the spherical harmonics, the evaluation of (74) becomes straightforward:

$$\Psi_{II}(\underset{\sim}{r}) = -\sum_L k b_o^2 C_L^o [n_\ell(kb_o), R_\ell^o(b_o;E)] j_\ell(kr_o) Y_L(\hat{r}_o)$$

$$+\sum_{p,L} k b_p^2 C_L^p [j_\ell(kb_p), R_\ell^p(b_p;E)] n_\ell(kr_p) Y_L(\hat{r}_p), \tag{80}$$

with the Wronskian notation

$$[f(\alpha r), g(\beta r)] = f(\alpha r)\frac{d}{dr}g(\beta r) - g(\beta r)\frac{d}{dr}f(\alpha r). \tag{81}$$

Introducing the coefficients

$$A_L^p = k b_p^2 [j_\ell(kb_p), R_\ell^p(b_p;E)] C_L^p, \tag{82}$$

$$A_L^o = -k b_o^2 [n_\ell(kb_o), R_\ell^o(b_o;E)] C_L^o, \tag{83}$$

we obtain for the wave function in region II /20/:

$$\Psi_{II}(\underset{\sim}{r}) = \sum_L A_L^o j_\ell(kr_o) Y_L(\hat{r}_o) + \sum_{p,L} A_L^p n_\ell(kr_p) Y_L(\hat{r}_p) \tag{84}$$

We may interpret the first term as an "incoming" spherical wave directed towards the center of the molecule, which has been scattered by the extramolecular region. The second term may be interpreted as "outgoing" spherical waves which have been scattered by the atomic spheres centered at positions $\underset{\sim}{R}_p$. This interpretation of the wave function $\Psi_{II}(\underset{\sim}{r})$ (84) follows naturally only in the energy range above the constant potential $\bar{V}$ (66). For energies below this value the wave function in the intersphere region is a linear combination of spherical waves, characterized by the modified spherical Bessel functions $i_\ell(x)$ and $k_\ell^{(1)}(x)$ /20/ which decay exponentially towards the interior of region II.

Let us evaluate (73.b) to give an example of the general techniques to be used in the evaluation of the functional Λ, eq. (64). We choose a position $\underset{\sim}{r}$ outside the outer sphere and obtain by the same token as above:

$$0 \equiv - \sum_{L} k b_o^2 C_L^o \left[j_\ell(kb_o), R_\ell^o(b_o; E) \right] n_\ell(kr_o) Y_L(\hat{r}_o)$$

$$+ \sum_{p,L} k b_p^2 C_L^p \left[j_\ell(kb_p), R_\ell^p(b_p; E) \right] n_\ell(kr_p) Y_L(\hat{r}_p) \tag{85}$$

In order to exploit this identity, valid for all r in region III, we have to reduce this multicenter expansion to an expansion around one center. We choose the center of the molecule $\underset{\sim}{R}_o$. The argument of the spherical waves going out from center p may then be written as

$$\underset{\sim}{r}_p = \underset{\sim}{r}_o - \underset{\sim}{R}_{op} , \tag{86.a}$$

where the vector

$$\underset{\sim}{R}_{op} = \underset{\sim}{R}_p - \underset{\sim}{R}_o \tag{86.b}$$

locates the center of the atomic region p with respect to the center of the molecule (see Fig. 6). We need a bipolar expansion of a Helmholtz solid harmonics $n_\ell(r)$ $Y_L(r)$ with respect to the two "centers" $\underset{\sim}{r}_o$ and $\underset{\sim}{R}_{op}$. A theorem accomplishing just that may be found in the extensive literature dealing with bipolar expansions /20,60/:

$$n_\ell(kr_p) Y_L(\hat{r}_p) = \sum_{L'} n_{\ell'}(kr_o) Y_{L'}(\hat{r}_o)\, 4\pi\, i^{\ell'-\ell} \times$$

$$\times \sum_{L''} i^{-\ell''} I_{L''}(L,L')\, j_{\ell''}(kR_{op}) Y_{L''}(\hat{R}_{op}) . \tag{87}$$

The integrals

$$I_{L''}(L,L') = \int Y_{L''}(\hat{r}) Y_L(\hat{r}) Y_{L'}(\hat{r})\, d\hat{r} \tag{88}$$

over triple products of real spherical harmonics are related to the Gaunt coefficients and may easily be computed in terms of Clebsch-Gordan coefficients /61/. They are nonzero only for the following conditions on the angular momentum indices:

$$|\ell - \ell'| \leq \ell'' \leq \ell + \ell'$$

$$\ell + \ell' + \ell'' \quad = \text{even integer} \tag{89}$$

If we use (83) and (87), then the identity (85) implies the set of equations among the modified expansion coefficients A_L^p:

$$A_L^o = -\frac{[n_\ell, R_\ell^o]}{[j_\ell, R_\ell^o]} \sum_{pL'} S_{LL'}^{op}(E)\, A_{L'}^p \tag{90}$$

in which we define the quantities /20/

$$S_{LL'}^{op}(E) = 4\pi\, i^{\ell-\ell'} \sum_{L''} i^{-\ell''} I_{L''}(L,L')\, j_{\ell''}(kR_{op})\, Y_{L''}(\hat{R}_{op}). \tag{91}$$

After these preparatory developments we turn to our main subject, the evaluation of Λ. We do so by applying eq. (72) twice, first (72.a) to transform the square bracket in (64) into a surface integral, and then the complex conjugate of (72.b) to do the same with the remaining volume integral. And again, we use a limiting process to deal with the surfaces properly. We put /19a, 58/

$$\Lambda = \lim_{\varepsilon \to 0} \Lambda_\varepsilon, \tag{92}$$

where

$$\Lambda_\varepsilon = \sum_{p,q=0}^{N} \int_{b_p - 2\varepsilon_p} dS_p \left[\frac{\partial \Psi^*(\underset{\sim}{r})}{\partial n} - \Psi^*(\underset{\sim}{r}) \frac{\partial}{\partial n} \right] \times$$

$$\times \int_{b_q - \varepsilon_q} dS_q' \left[\Psi(r') \frac{\partial}{\partial n'} G(\underset{\sim}{r},\underset{\sim}{r}') - G(\underset{\sim}{r},\underset{\sim}{r}') \frac{\partial}{\partial n'} \Psi(r') \right] \tag{93}$$

For convenience, we collect various bipolar expansions of the Green's function needed in the evaluation of these surface integrals /20/:

$$\begin{aligned} G_E(\underset{\sim}{r},\underset{\sim}{r}') &= -k \sum_{LL'} j_\ell(kr_p) Y_L(\hat{r}_p)\, G_{LL'}^{pq}(E)\, j_{\ell'}(kr_q') Y_{L'}(\hat{r}_q') \\ &= -k \sum_{LL'} j_\ell(kr_p) Y_L(\hat{r}_p)\, S_{LL'}^{po}(E)\, n_{\ell'}(kr_o') Y_{L'}(\hat{r}_o') \\ &= -k \sum_{LL'} n_\ell(kr_o) Y_L(\hat{r}_o)\, S_{LL'}^{op}(E)\, j_{\ell'}(kr_p') Y_{L'}(\hat{r}_p') \end{aligned} \tag{94}$$

where $p \neq q$ and $p \neq 0$. The quantities $S_{LL'}^{po}(E)$ are defined in analogy to (83) and

$$G_{LL'}^{pq}(E) = (1-\delta_{pq})\,4\pi\, i^{\ell-\ell'} \sum_{L''} i^{-\ell''} I_{L''}(L,L')\, n_{\ell''}(kR_{pq})\, Y_{L''}(\hat{R}_{pq}) \tag{95}$$

with the relative position vector $\underset{\sim}{R}_{pq} = \underset{\sim}{R}_q - \underset{\sim}{R}_p$. Substitution of (94) into (93) yields after some algebraic manipulations using (91):

$$\Lambda = \frac{1}{k} \sum_{pL} \sum_{qL'} A_L^{p*} \left\{ \frac{[n_\ell, R_\ell^p]}{[j_\ell, R_\ell^p]} \delta_{pq}\, \delta_{LL'} + G_{LL'}^{pq}(E) \right\} A_{L'}^{q}$$

$$+ \frac{1}{k} \sum \sum_{L'} \left\{ A_L^{p*}\, S_{LL'}^{po}(E)\, A_{L'}^{o} + A_{L'}^{o*}\, S_{L'L}^{op}(E)\, A_L^{p} \right\}$$

$$+ \frac{1}{k} \sum_{L} A_L^{o*} \frac{[j_\ell, R_\ell^o]}{[n_\ell, R_\ell^o]} A_L^{o}\,. \tag{96}$$

Eq. (96) may be simplified by defining the scattering factors

$$t_\ell^p(E) = -\frac{[j_\ell(kb_p), R_\ell^p(b_p;E)]}{[n_\ell(kb_p), R_\ell^p(b_p;E)]} \tag{97}$$

$$t_\ell^o(E) = -\frac{[n_\ell(kb_o), R_\ell^o(b_o;E)]}{[j_\ell(kb_o), R_\ell^o(b_o;E)]} \tag{98}$$

and by collecting the various quantities into vectors and matrices, e.q.

$$\mathbb{A}_p = (\ldots.\; A_L^p \;\ldots.)^t \tag{99}$$

With these definitions, eq. (90) reads

$$\mathbb{A}_o = \mathbb{t}_o(E) \sum_p \mathbb{S}_{op}(E)\, \mathbb{A}_p \tag{100}$$

and it may be used to write Λ in its final form:

$$\Lambda(E) = \frac{1}{k} \sum_{pq} \mathbb{A}_p^{+}\, \mathcal{L}_{pq}(E)\, \mathbb{A}_q, \tag{101}$$

where

$$\mathcal{L}_{pq}(E) = \mathbb{G}_{pq}(E) - \mathbb{t}_p^{-1}(E)\,\delta_{pq} + \mathbb{S}_{po}(E)\,\mathbb{t}_o(E)\,\mathbb{S}_{oq}(E) \tag{102}$$

Variation of the expression (101) with respect to C_L^p (or equivalently with respect to A_L^p) yields a homogeneous system of linear equations to determine the expansion coefficients A_L^p (or C_L^p, see (82)):

$$\sum_q \mathcal{L}_{pq}(E)\, \mathbb{A}_q = 0. \tag{103}$$

Non-trivial solutions of these equations are determined from the secular equation

$$\det\left| \mathbb{G}_{pq}(E) - \mathbb{t}_p^{-1}(E)\delta_{pq} + \mathbb{S}_{po}(E)\mathbb{t}_o(E)\mathbb{S}_{oq}(E)\right| = 0. \tag{104}$$

The molecular orbital energies correspond to the zeros of this determinant. The coefficients for the expansion in the outer sphere, $\mathbb{A}_o$, can be obtained from (100), once (103) has been solved. Because real spherical harmonics are used throughout and because of the conditions (81), all the matrix elements are real and symmetric:

$$\mathbb{G}_{pq}(E) = \mathbb{G}_{pq}(E)^* = \mathbb{G}_{qp}(E)^\dagger \tag{105}$$

$$\mathbb{S}_{op}(E) = \mathbb{S}_{op}(E)^* = \mathbb{S}_{po}(E)^\dagger. \tag{106}$$

From this we conclude that the system of linear equations (103) may be solved with real coefficients A_L^p leading to a real wave function $\Psi(\underset{\sim}{r})$.

All quantities in the secular determinant (96) have a direct physical interpretation in terms of an electron being scattered in the potential $V(\underset{\sim}{r})$ /20,52,58/. The "scattering factor" $t_\ell^p(E)$ gives the tangent of the phase-shift experienced by a partial wave of angular momentum ℓ and energy E from the scattering at the atomic potential p. The "propagation" of the partial waves between any two atoms p and q is described by the matrix elements $G_{LL'}^{pq}(E)$, the "free-space" single-particle Green's function. However, it describes only direct scattering. Waves may also scatter from center p to center q indirectly via the outer surface, where they experience an additional phase shift, given by $t_\ell^o(E)$. The "propagators" $S_{LL'}^{op}(E)$ describe these scattering processes. The various quantities in the secular determinant fall obviously into either one of two categories. The "propagators" $G_{LL'}^{pq}(E)$ and $S_{LL'}^{op}(E)$ are solely dependent on the structure of the cluster whereas the potential only affects the "scattering factors" $t_\ell^p(E)$ and $t_\ell^o(E)$.

The muffin-tin potential (67) is only a first approximation to the full potential $V(\underset{\sim}{r})$ of the system. Later on, we will discuss examples for which an improved potential is desirable. In sec. 3.2 we have already mentioned that general potentials and cell shapes other than spheres can be handled by the multiple-scattering formalism. It is quite remarkable that the form of the secular equation (104) is retained in these cases /53,56/. In these extensions, the "structure factors" G and S remain unchanged, but the "scattering factors" t become non-diagonal in L. Also, the radial functions of the cellular basis (62) become dependent on m and are obtained from a system of coupled differential equations. The coupling is caused,(1) by higher L-components of the potential (65) and (2) by deviations of the cell boundaries from a spherical shape. For certain molecules, like linear or large flat clusters, a suitable choice of the outer surface might lead to a more realistic description /62/. If the muffin-tin form of the potential is retained for the atomic spheres, only moderately higher efforts are needed to solve this improved model as only the third term in the secular determinant (96) is affected.

The potential (67) is not of the most general muffin-tin form one could think of. There we have only a two-step hierarchy of spheres, atomic spheres inside one enclosing sphere and a constant potential in between. One could also devise a general muffin-tin partitioning /63/ where assemblies of atoms are localized inside an enclosing sphere and several such assemblies, possibly together with other atomic spheres, are surrounded by an outer sphere and so on. The potential would be constant outside the lowest-level atomic spheres, but have different constant values in different assemblies etc. Such a generalized molecular partitioning would possibly provide a better description for systems like an ion surrounded by several solvent molecules. The SW formalism has been set up for such a generalized partition scheme. The structure of the coupling conditions between the various assemblies allows an iterative approach to the exact solution, focusing one part of the system at a time and alternating between interacting parts until self-consistency is attained. Such a generalized SW procedure has been applied to a water molecule trimer /64/.

The multiple-scattering formalism has also been generalized to solve the one-electron equations resulting from a relativistic Xα model. The method starts with an energy functional of the form (22) the spin orbitals u_i now being four-component spinors. The one-electron operators are replaced by the equivalent Dirac matrices

$$f_i = c\,\underset{\sim}{\alpha}_i \cdot \underset{\sim}{p}_i + \beta_i c^2 \qquad (107)$$

with

$$\underset{\sim}{\alpha} = \begin{pmatrix} 0 & \underset{\sim}{\sigma} \\ \underset{\sim}{\sigma} & 0 \end{pmatrix} \quad \text{and} \quad \beta = \begin{pmatrix} I & 0 \\ 0 & -I \end{pmatrix} \qquad (108)$$

σ_k are the Pauli spin matrices. The application of the variational principle leads to a one-electron Dirac equation with a self-consistent potential, defined in the same way as in the non-relativistic case. The SW equations are of the same structure as those derived previously. The various elements of the secular determinant bear a close relation to the corresponding non-relativistic quantities. Applications of the relativistic SW formalism are in progress and an improved description is to be expected for systems containing heavier atoms ($Z \gtrsim 55$).

Let us conclude this section with a comment on the use of the two terms "multiple-scattering" vs. "scattered-wave" formalism /20,53/. In Johnson's scattered-wave formalism, one starts from solutions of the one-particle Schrödinger equation in the three regions of space and matches the wave function and its derivative across the sphere boundaries. In the multiple-scattering method, heavy use is made of the Schrödinger equation in its integral equation equivalent (60). For a simple muffin-tin potential, they both lead to identical formulas. One can easily show that the wave function determined from the equations (100), (103) and (104) fulfills the continuity conditions across the sphere boundaries. Let us demonstrate this for the wave function (84) in region II in the vicinity of a specific atomic sphere q. All waves going out from other spheres $p \neq q$ and reflected from the outer sphere add up to an incoming wave of amplitude B_L^q with respect to center q:

$$\Psi_{II}(\underset{\sim}{r}) = \sum_L \left[A_L^q n_\ell(kr_q) + B_L^q j_\ell(kr_q) \right] Y_L(\hat{r}_q). \qquad (109)$$

From expansions similar to (87) and to those underlying (94) one obtains

$$\mathbb{B}_q = \sum_p \mathbb{G}_{qp}(E)\mathbb{A}_p + \mathbb{S}_{qo}(E)\mathbb{A}_o \tag{110}$$

which by virtue of (100) and (103) may be reduced to

$$\mathbb{B}_q = \mathbb{t}_q^{-1}(E)\mathbb{A}_q . \tag{111}$$

Using (82) and (97), the wave function just outside the q-th atomic sphere is given by

$$\Psi_{II}(\underset{\sim}{r}) = \sum_L k b_q^2 C_L^q Y_L(\hat{r}_q) \times$$

$$\times\left\{ n_\ell(kr_q)\left[j_\ell(kb_q), R_\ell^q(b_q)\right] - j_\ell(kr_q)\left[n_\ell(kb_q), R_\ell^q(b_q)\right]\right\} \tag{112}$$

With the help of /59/

$$\left[j_\ell(kb), n_\ell(kb)\right] = 1/kb^2 \tag{113}$$

it is quite easy to show that the wave function (112) and its radial derivative match with the corresponding quantities of the cellular basis (68).

3.3 Practical Aspects of the SCF-Xα-SW Method

We have now completed the presentation of the Xα model and of the SW formalism to solve the resulting one-electron Schrödinger equation. To implement an SCF-Xα-SW calculation for a molecule or cluster, the question how selfconsistency is obtained remains to be discussed. Furthermore, several critical points in such a calculation warrant a general discussion. We shall take up these topics in this section.

To start the SCF-Xα-SW calculation for a cluster, one first performs SCF-Xα calculations for the constituent atoms or ions with a program of the type developed by Herman and Skillman /10/. For each atom p one obtains a spherically symmetric charge distribution (cf. eq.(68)):

$$\sigma_p(r) = \frac{1}{4\pi} \sum_{i\ell} q_{i\ell}\left[C_{i\ell}^p R_{i\ell}^p(r)\right]^2, \tag{114}$$

where $q_{i\ell}$ is the occupation of the i-th sub-shell characterized by the angular momentum ℓ. The atomic coulomb potential is then given by

$$V_C^p(r) = -\frac{2Z_p}{r} + \frac{1}{r}\int_0^r 4\pi r'^2 \sigma_p(r')dr' + \int_r^\infty 4\pi r' \sigma_p(r')dr' \qquad (115)$$

This is simply the solution of Poisson's equation for atom p:

$$\nabla^2 V_C^p(r) = 8\pi \left(Z_p \delta(r) - \sigma_p(r) \right) \qquad (116)$$

To obtain the starting potential for a cluster calculation, one has to superimpose contributions from all constituent atoms and to perform the muffin-tin average (67). Mattheis suggested /66/ treating the coulomb and the exchange part of the total potential separately. As Poisson's equation is linear, the atomic potentials may be added in a straightforward way:

$$V_C(r_p) = V_C^p(r_p) + \sum_{q \neq p} V_C^q(r_p | R_{pq}). \qquad (117)$$

The potentials from all other centers q have to be expanded about a center at distance R_{pq}. For a general spherical-harmonic expansion this can be achieved using Löwdin's α-expansion technique /67/. In our case, one only needs the L = (0,0) component (see eq. (65)) of a spherically symmetric function (116) which may be calculated from simple geometrical considerations /68/:

$$V(r|R) = \frac{1}{2Rr}\int_{|R-r|}^{R+r} r' V(r')dr' \qquad (118)$$

The total charge distribution may be expanded around the center p in complete analogy to (117) by summing the various contributions (114). The calculation of the exchange potential

$$V_{X\alpha}^p(r_p) = -6\alpha_p \left[\frac{3}{8\pi} \varrho(r_p) \right]^{1/3} \qquad (119)$$

is then an easy task. The constant potential of region II is obtained in a similar manner by adding the spatially integrated potentials. In general, atomic exchange parameters α are used in regions I and a (weighted) average over these values in regions II and III.

Having assumed a certain geometry for the cluster we still must decide on the various sphere radii to fully determine the muffin-tin potential (67). The realism of the muffin-tin model depends critically on a reasonable choice of these parameters which therefore requires some care. The constant potential in region II is obviously the most severe approximation involved in the muffin-tin model. Consequently, the volume of this region is to be minimized by requiring the muffin-tin spheres to touch, if possible. Then all sphere radii are fixed in certain highly symmetric cases, like homonuclear diatomics. In all other systems one has to choose the radii according to one of several criteria.

One might consider the sphere radii as variational parameters and minimize the total energy (22). This approach has two severe disadvantages. First, one has to carry out several calculations for one cluster. However, more important, the total energy of a cluster is commonly evaluated only for a charge density which has been brought to muffin-tin form. The resulting total energy is very unreliable in most cases, as will be discussed in chap. 4, and therefore is not suitable as the basis for such an important choice. The "muffin-tin" total energy has been studied extensively as a function of the sphere radii for some simple systems, such as the water molecule H_2O /69,70/ or germane, GeH_4 /71/. In a favourable situation like the tetrahedral GeH_4 (an homologue of methane, CH_4), this method yields "optimized" sphere radii not too far from those obtained by the following criteria.

It is more reasonable to divide a heteronuclear bond according to the relative size of the involved atoms. A partitioning through points of minimal charge density through which paths of largest $|\nabla\rho|$ pass, separating 'virial fragments', would be satisfactory from a theoretical point of view /72/. But this choice to be feasible the charge density has to be known. With an appeal to physical or chemical intuition, one might use Slater's covalent atomic radii /73/ to measure the relative size of atoms and to divide the bond length according to the ratio of Slater's radii. This criterion has been used successfully in many cluster calculations. For some systems, ratios of ionic radii have been used /74/. If the relative size of the atoms measured in this way differs greatly, better results are obtained by dividing the bond length such that the potential is continuous along the bond /75/.

Having discussed how to obtain the initial muffin-tin potential we now turn to the determination of the orbital energies from the secular equation (104). As the energy E is a parameter occurring in all elements of the determinant (104), it is necessary to evaluate the latter in the vicinity of a molecular orbital energy and to interpolate for the zero of this function. It should be noted that the value of the determinant, and consequently the zero location, depend on the number of partial waves (68) included for each atomic sphere. Experience has shown that this expansion is rapidly convergent /20/. Partial waves up to $\ell = 2$ per atom (even for transition metals) are sufficient to reach convergence of the orbital energies to $\pm$ 0.001 Ry. The ℓ cut-off may be compared to the necessity of using a finite basis set in an LCAO calculation. However, the resulting secular problems are of much smaller dimension in a SW calculation due to the greater flexibility and better convergence of a SW 'basis' as has already been explained in sec. 3.1.

The dimension of the secular problem (104) may, in practice, be reduced by exploiting any symmetry the cluster might possess. This can be done with the help of group theoretical projection operators /76/. For an octahedrally coordinated molecule like SF_6, which will be discussed in sec. 4.1, molecular orbitals transforming as the irreducible representations a_{1g}, e_g, t_{1g}, t_{1u}, t_{2g} and t_{2u} of the point group O_h are found /77/. If we include partial wave up to $\ell = 2$ on the S atom, up to $\ell = 1$ on F atoms and up to $\ell = 4$ in the extramolecular region, the largest symmetrized secular matrix is only of dimension 6 x 6 (t_{1u}). Hardly any savings through symmetry, especially for large systems, are possible in an LCAO-MO solution of the HF equations /78/.

Additional numerical benefits are obtained when searching for the zeros of a symmetrized determinant. They are easier to locate because they are, in general, further apart, and no multiple roots will occur when we project on one column of a irreducible representation /77/.

The search for deeper bound levels, say below -5 Ry, is facilitated if we take into account the fact that these orbitals hardly extend beyond the corresponding muffin-tin sphere. They may then be treated like atomic levels, completely by-passing the SW problem. Very often, we are only interested in the valence molecular orbitals which exclusively determine the chemical properties of a molecule. To a very good approximation, we may then take core charge distributions from free atoms obtaining a simple Xα-SW analogue of a pseudo-potential method /79/.

No orthogonality problems arise in this 'frozen core approximation' as we are always solving the differential equation (50) directly in space.

Even after we have located all orbital energies and solved (103) for the coefficients of the multicenter expansions, the molecular orbital wave functions are determined only up to a normalization constant. In order to generate the resulting charge distribution, and consequently the new potential as input to the next iteration in a self-consistency cycle, we still have to normalize the orbitals.

Integrals over the charge density inside the atomic spheres and in the extramolecular region are easily obtained:

$$Q_p = \int_0^{b_p} 4\pi r^2 \sigma_p(r)\, dr \tag{120.a}$$

$$Q_o = \int_{b_o}^{\infty} 4\pi r^2 \sigma_o(r)\, dr \tag{120.b}$$

with the spherically averaged charge distribution (see eq. (68)):

$$\sigma_p(r_p) = \frac{1}{4\pi} \int \Psi^*(\underset{\sim}{r})\, \Psi(\underset{\sim}{r})\, d\hat{r}_p$$

$$= \frac{1}{4\pi} \sum_L \left[C_L^p R_L^p(r_p)\right]^2, \qquad p = 0, 1, \ldots N. \tag{121}$$

It is somewhat harder to perform the integral of $|\Psi_{II}(\underset{\sim}{r})|^2$ (84) over the awkwardly shaped interatomic region. Two procedures are available.

If we shift the value of the constant potential in region II by a small amount $\Delta\bar{V}$ /80/, we expect the change ΔE in the orbital energy to be given by perturbation theory (see eq. (59)):

$$\langle \Psi | \Psi \rangle \Delta E = \langle \Psi | \Delta V(\underset{\sim}{r}) | \Psi \rangle = \Delta\bar{V} \langle \Psi_{II} | \Psi_{II} \rangle, \tag{122}$$

where $\langle \Psi_{II} | \Psi_{II} \rangle$ is the integral we are looking for:

$$\langle \Psi_{II} | \Psi_{II} \rangle = \int_{\Omega_{II}} \Psi^*(\underset{\sim}{r})\, \Psi(\underset{\sim}{r})\, d\underset{\sim}{r} \; . \tag{123}$$

The normalization constant may then be calculated from

$$\langle \Psi | \Psi \rangle = \sum_{p=0}^{N} Q_p / (1 - \Delta E / \Delta \bar{V}). \tag{124}$$

Another way /81/, which is more accurate and more elegant, exploits the fact that $\Psi_{II}(\underset{\sim}{r})$ (84) fulfills the Helmholtz equation (70):

$$(\nabla^2 + E)\, \Psi_{II} = 0. \tag{125}$$

We differentiate with respect to the energy:

$$(\nabla^2 + E) \frac{\partial}{\partial E} \Psi_{II} = - \Psi_{II} \,. \tag{126}$$

If we multiply the complex conjugate of (125) by $\partial \Psi_{II} / \partial E$ and (126) by Ψ^*_{II}, subtract the two equations and integrate over region II, we obtain after applying Green's theorem:

$$\langle \Psi_{II} | \Psi_{II} \rangle = \int_{S_{II}} dS \left[\Psi^*_{II} \frac{\partial}{\partial n} - \frac{\partial}{\partial n} \Psi^* \right] \frac{\partial}{\partial E} \Psi_{II} . \tag{127}$$

Again, we have used the derivative along the inside normal of the surface S_{II}. We use the eqs. (74) and (93) and the symmetry of the Green's function (75) to obtain

$$\langle \Psi_{II} | \Psi_{II} \rangle = \frac{\partial}{\partial E} \Lambda(E) \tag{128}$$

The energy differentiation is only with respect to the energy dependence introduced explicitly by the Green's function $G_E(\underset{\sim}{r}, \underset{\sim}{r}')$ (75). It may therefore be performed analytically after all the surface integrals in $\Lambda(E)$ have been calculated. With (101) and (103) we obtain as our final result:

$$\langle \Psi_{II} | \Psi_{II} \rangle = \frac{1}{k} \sum_{pq} A^{+}_{p} \frac{\partial}{\partial E} \mathcal{L}_{pq}(E) A_q . \tag{129}$$

The normalization may then be performed as above.

The new coulomb potential is generated from the charge distribution by applying a muffin-tin average to

$$V_c(\underset{\sim}{r}) = 2 \int \frac{\rho(\underset{\sim}{r}')}{|\underset{\sim}{r} - \underset{\sim}{r}'|} d\underset{\sim}{r}'. \tag{130}$$

As a simplification, the charge distribution ϱ is usually replaced by its muffin-tin average. Thus the charge density is constant in region II and a sum over the occupied orbitals of contributions (121) in regions I and III. The calculations necessary to obtain the muffin-tin part of (130) are straightforward, but lengthy. The exchange potential generated from a muffin-tin charge density has muffin-tin form as well. The resulting formulas may be found in the literature /82,21,69/.

The two averaging procedures (charge and potential) involved in the calculation are connected /83/. A variation of $E_{X\alpha}$ (22) evaluated with a muffin-tin charge density leads to a one-particle Schrödinger equation with a muffin-tin potential. In this sense, there is only one averaging procedure.

Let us mention here briefly that any step beyond the muffin-tin potential (67) in the MS formalism also asks for extra efforts in the self-consistency cycle. However, similar expansion techniques may be employed, as Poisson's equation may be regarded as Schrödinger's equation in the limit of zero energy /53/. Taking the limit $E \rightarrow 0$ in the Green's function (75) verifies this statement:

$$\lim_{E\to 0} G_E(\underset{\sim}{r},\underset{\sim}{r}') = -\left[4\pi|\underset{\sim}{r}-\underset{\sim}{r}'|\right]^{-1}. \qquad (131)$$

The complete SCF-Xα-SW procedure is illustrated schematically in Fig. 7, summarizing the discussion of this section.

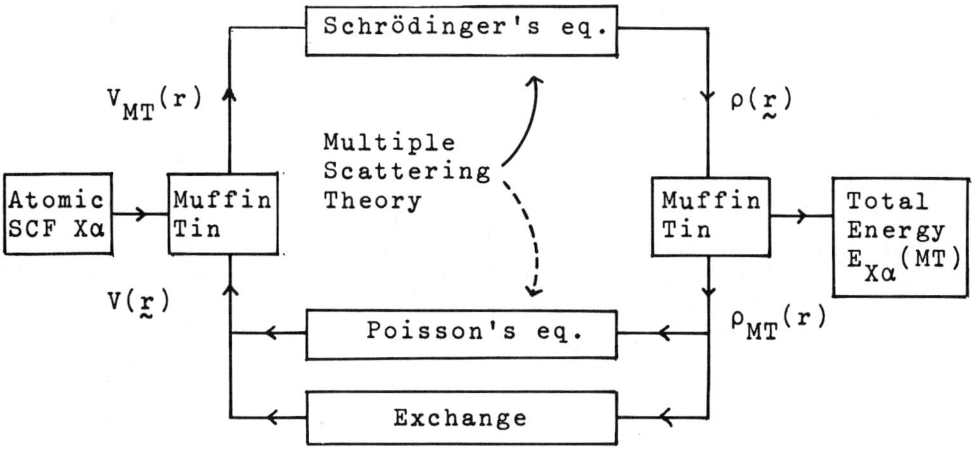

Fig. 7 Schematic self-consistent cycle (adapted from ref. 53).

4. APPLICATIONS TO MOLECULES

After we have outlined the theoretical framework of the SCF-Xα-SW method, we shall now discuss applications to problems in chemistry and molecular physics. The purpose of this chapter is twofold. First, we want to test the reliability of the method by comparing the results to those from conventional ab initio calculations and to experimental data. Furthermore, we would like to show how an Xα-SW molecular orbital calculation may be used to enlarge our understanding of the electronic structure of specific molecules or classes of molecules. Of course, this can only be done for illustrative examples. The rate of newly appearing work using the Xα-SW method is still increasing. Listings of all applications up to 1975 have been compiled in ref. 21, 22 and 23.

4.1 The Hypervalent Molecules SF_4 and SF_6

The interpretation of the photoelectron spectrum of sulfur hexafluoride, SF_6, in 1971, was one of the first applications of the Xα-SW method /75/, and is by now already a classical one. A recent large-scale ab initio calculation /84/ confirmed the success of this work. However, theoretical chemists have been interested in SF_6 since long, because it belongs to a class of molecules with interesting bonding properties. Among these molecules are sulfuranes, SF_4 being the prototype, interhalogen compounds like ClF_3, or noble gas compounds like XeF_4, to mention a few examples.

The common characteristic feature of these systems is a central atom of the main groups V to VIII which forms additional bonds beyond the number allowed by the Lewis-Langmuir theory of paired valence electrons forming a "stable octett". For sulfur, with the configuration $3s^2 3p^4$, only a molecule like SH_2 observes the "octett rule". More than two bonds are formed with electronegative atoms, primarily fluorine. But these bonds are weaker and longer than normal covalent bonds, leading to the remarkable phenomenon of two different bonds between the same kind of atoms within one molecule. The resulting geometry is illustrated for SF_4 in Fig. 8. The structure is best described as idealized trigonal bipyramid with a sulfur lone electron pair in one of the equatorial positions. The equatorial S-F bonds are 0.1 Å shorter than the axial ones /85/. From SF_4, we formally arrive at SF_6 by attaching two more fluorine ligands at the side of

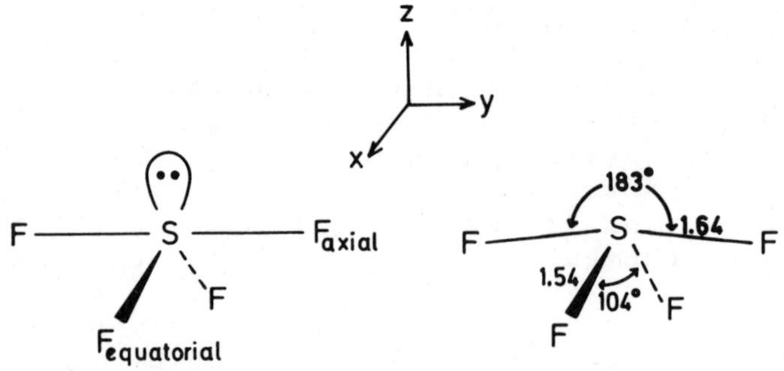

Fig. 8 Structure of sulfur tetrafluoride, SF_4 (bond lengths in Å).

the lone pair. The coordination of S is that of an ideal octahedron with a bond length d_{S-F} = 1.564 a.u., intermediate between the two bonds of SF_4.

For molecules which contain an atom with its valency expanded in this manner, Musher coined the descriptive name hypervalent molecules /86/. There have been quite a number of suggestions to qualitatively understand their electronic structure and their geometry /86-90/. In valence bond theory one has to invoke d-orbital participation on the central atom in making the appropriate orthogonal hybrid orbitals /87/ or one has to consider ionic valence bond structures /88/. However, a qualitative description of the bonding may also be given by simply using non-orthogonal hybrid orbitals removing the necessity of d orbital participation /86a/. In the language of molecular orbital (MO) theory, a hypervalent bond, like F_{ax}-S-F_{ax} in SF_4, may be characterized as electron-rich three-center bond /90/. The hypervalent bonds have been further classified according to whether only p orbitals on the central atom are necessary to explain the bonding, as in SF_4 (type I), or whether sp mixing is supposed to play a significant role, as in SF_6 (type II) /86a/.

Quite a few calculations have been carried out recently, with semiempirical and ab initio methods, in order to assess the relative importance of the various electronic factors for the structure of these compounds /91/. Out of the number of systems studied with the Xα-SW method we have chosen SF_4, SF_6 and SCl_6 to discuss how and to what extent one is able to correlate the information from Xα-SW MO's with the chemical features of these systems /92,93/.

Let us first compare the electronic structure of SF_6 and of the hypothetical system SCl_6 which has not yet been synthesized. For SF_6, the experimental octahedral structure was assumed. The muffin-tin sphere radii were determined by requiring the potential to be continuous along the S-F bonds. The ratio of the resulting sphere radii is rather close to that of Slater's covalent radii (S: 1.00 Å; F: 0.50 Å). For SCl_6, the bond length was determined by scaling that found experimentally for SCl_2 in the ratio of the corresponding bond lengths in SF_2 and SF_6: d_{S-Cl} = 1.968 Å. The sulfur muffin-tin sphere had to be enlarged in comparison to SF_6 to account for the differing relative size of S and Cl (covalent radii 1:1). The parameters used in the calculation are collected in Table III. In Fig. 9, the resulting muffin-tin structure is illustrated.

The convergence of the calculation is such that a basis set up to l = 2 (including d orbitals) on sulfur, up to l = 1 on the ligands F and Cl and up to l = 4 in

Table III Structure parameters used in the Xα-SW calculation on SF_6, SCl_6 and SF_4 (all values in Å)

	SF_6	SCl_6	SF_4
bond length	1.564	1.968	1.542[a], 1.643[b]
sulfur sphere radius	0.942	0.984	0.929
ligand sphere radius	0.622	0.984	0.613[a], 0.714[b]
outer sphere radius	2.186	2.952	2.357

[a] equatorial fluorine ligands.
[b] axial fluorine ligands.

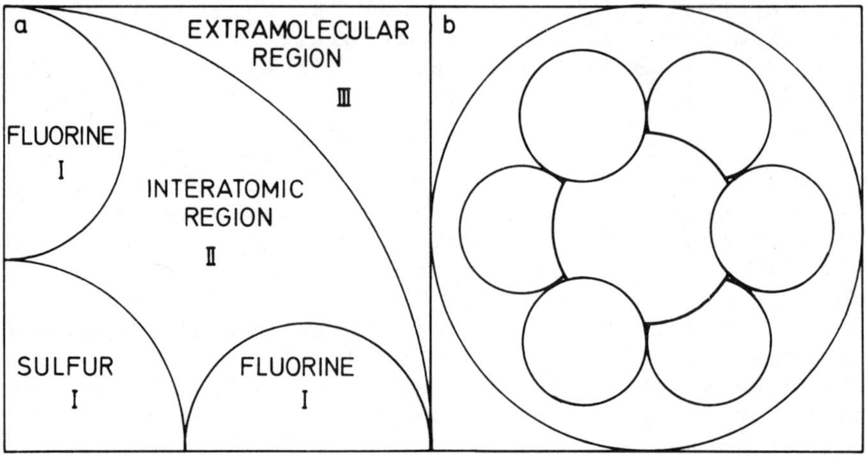

Fig. 9 a) A quadrant of projection of the muffin-tin spheres on the xy plane, for SF_6.
b) A projected view of the various spheres for the partitioning of the SF_6 molecular space (O_h symmetry).

the outer sphere is sufficient. All calculations described in this section were carried out in the "frozen core" approximation (1s, 2s and 2p on S and Cl, 1s on F). For SF_6, about 12 iterations were required to achieve convergence to ≈ 0.001 a.u. in the one-electron energies. Each iteration requires approximately 2.5 seconds of CPU time on a 360/91 IBM computer. The whole calculation takes about 45 seconds, including the preparatory steps.

For each valence orbital of both compounds, the calculated energy and the charge distribution over the various regions of space normalized to unity are shown in Table IV. The orbitals are labeled according to the irreducible representations of the O_h symmetry group and numbered with respect to the valence region only to facilitate the comparison. In Fig. 10, the orbital energies are shown together with the SCF-Xα orbital energies for the various free atoms.

Table IV and Fig. 10 reveal a fair similarity between the energy level spectra and spatial charge distributions of SF_6 and SCl_6. The energy levels of SCl_6 are shifted upwards compared with those of SF_6.This is due to the higher valence levels of Cl and bears important consequences on

Table IV SCF-Xα-SW orbital energies and normalized charge distributions in the various molecular regions for the valence orbitals of SF_6 and SCl_6

Orbital	SF_6					SCl_6				
	$-\varepsilon_i$ [a]	S	F	II	III	$-\varepsilon_i$ [a]	S	Cl	II	III
$1a_{1g}$	2.633	0.213	0.083	0.271	0.015	2.111	0.455	0.050	0.244	0.002
$1t_{1u}$	2.413	0.062	0.112	0.242	0.025	1.749	0.083	0.113	0.225	0.015
$1e_g$	2.337	0.026	0.122	0.209	0.032	1.652	0.014	0.131	0.172	0.025
$2a_{1g}$	1.665	0.361	0.072	0.161	0.057	1.484	0.231	0.093	0.172	0.040
$2t_{1u}$	1.310	0.232	0.079	0.244	0.048	1.123	0.299	0.070	0.239	0.041
$1t_{2g}$	1.123	0.031	0.094	0.365	0.041	0.913	0.018	0.088	0.415	0.036
$2e_g$	1.006	0.091	0.112	0.166	0.070	0.747	0.081	0.110	0.184	0.075
$1t_{2u}$	0.962	0.0	0.113	0.289	0.032	0.760	0.0	0.106	0.331	0.035
$3t_{1u}$	0.961	0.018	0.110	0.281	0.042	0.772	0.032	0.097	0.343	0.042
$1t_{1g}$	0.889	0.0	0.122	0.236	0.033	0.669	0.0	0.117	0.257	0.039
$3a_{1g}$ [b]	0.324	0.279	0.095	0.088	0.066	0.478	0.161	0.092	0.202	0.087
$4t_{1u}$	0.200	...	...	...	...	0.218	0.352	0.066	0.187	0.063

[a] Energies in Rydberg units (1 Ry = 13.6 eV)

[b] Orbitals listed below the dashed line are unoccupied in the ground state of the molecule.

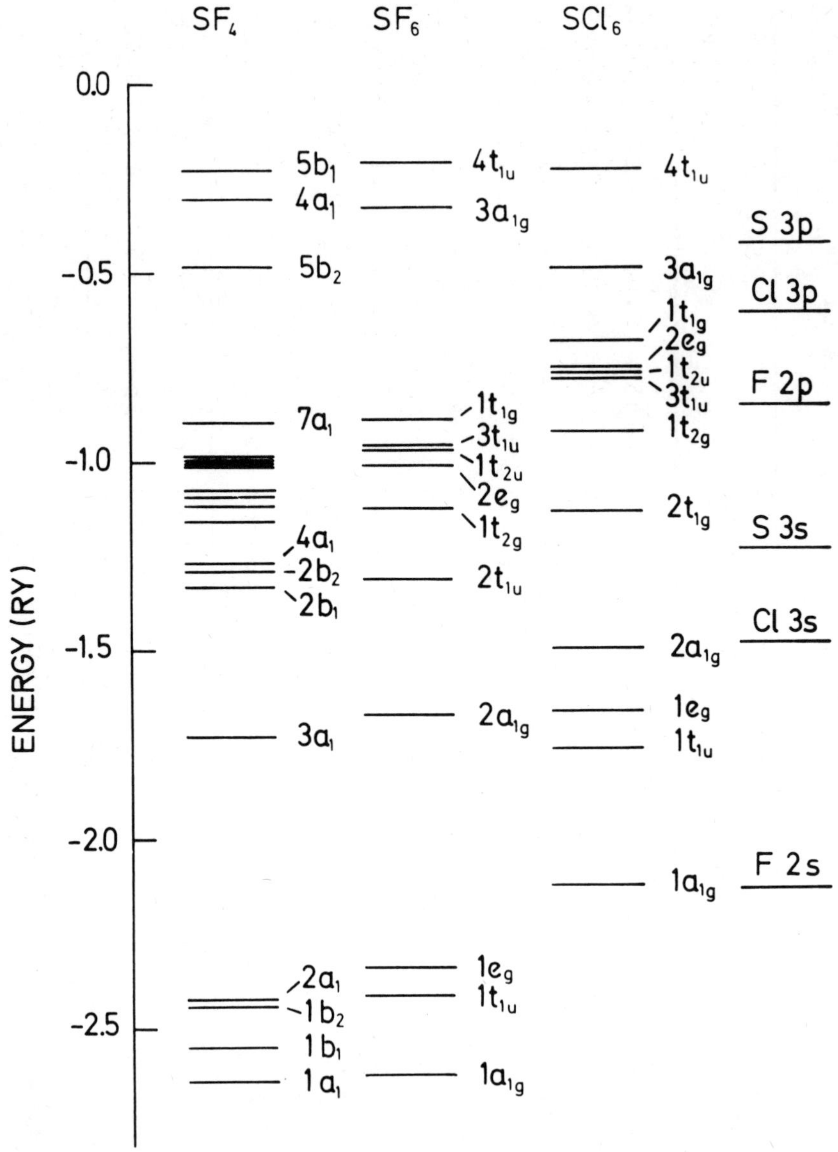

Fig. 10 Comparison of SCF-Xα-SW orbital energies for SF_4, SF_6 and SCl_6. The highest occupied orbital are the levels $7a_1$ for SF_4 and $1t_{1g}$ for SF_6 and SCl_6. The SCF-Xα orbital energies for the various free atoms are also shown.

the electronic structure of SCl_6. The very low lying $1a_{1g}$, $1t_{1u}$ and $1e_g$ valence orbitals, mainly attributed to ligand s orbitals, are normally not included in the qualitative discussions of the bonding of hypervalent molecules, yet do participate in sulfur-ligand bonding. The $1a_{1g}$ level plays the prominent role due to its large localization within the sulfur sphere. The sulfur 3s atomic orbital contributes significantly to the bond formation in SF_6 through the interaction with fluorine 2p orbitals as postulated for a hypervalent molecule of type II /86a/. The orbital contour map of the $2a_{1g}$ level of SF_6, shown in Fig. 11a, clearly confirms this view. This is, however, in strong constrast to the character of the corresponding $2a_{1g}$ level of SCl_6 (Fig. 11b) which may be described as S 3s - Cl 3s antibonding orbital. It is obviously the antibonding partner to the $1a_{1g}$ level. In SF_6, the antibonding character is overcome by the p bonding contribution of the F 2p levels lying energetically much closer to the S 3s orbital than the Cl 3p orbitals. Therefore, in contrast to the situation in SF_6, we find that the contributions of the $1a_{1g}$ and $2a_{1g}$ levels cancel giving no net contribution to the sulfur ligand bonding. This effect of the ligand s orbitals is not included in the usual discussion of the electronic structure of hypervalent molecules. The present results show that it cannot always be neglected.

The main bonding interaction between S 3p and ligand p orbitals is represented in the $2t_{1u}$ level, of course. The levels $1t_{2g}$ and $2e_g$ contain small admixtures of sulfur d orbitals. But even this small influence decreases when going from SF_6 to SCl_6. The $2e_g$ level is always found above the $1t_{2g}$ level, despite the larger bonding contribution from S 3d orbitals. As this ordering is found even in an F_6 cluster of octahedral symmetry /91/, it might be attributed to ligand-ligand interaction. The remaining occupied levels $1t_{2u}$, $3t_{1u}$ and $1t_{1g}$ are so-called non-bonding orbitals being localized mostly on the ligand spheres.

Orbital contour maps for all occupied orbitals of SF_6 and SCl_6 except the $2a_{1g}$ shown in Fig. 11) are rather similar after proper scaling to account for the differing bond lengths. If one measures the bond strength by overlap populations, as is done in the LCAO formalism, this observation would imply a reduction by a factor of approximately $(d_{S-F}/d_{S-Cl})^3 \sim 0.5$ from SF_6 to SCl_6, again pointing to weaker bonds in SCl_6. Unfortunately, we are not able to use total energies in our argument on the existence of SCl_6 (see sec. 4.2). Since the bond strength is

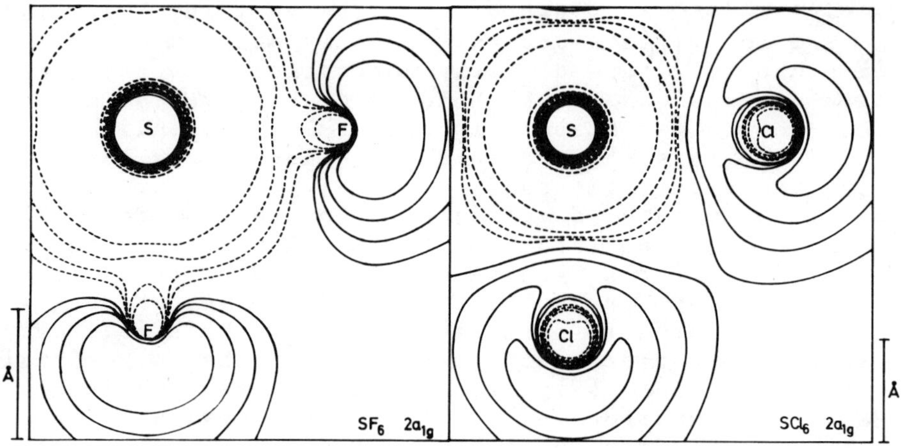

Fig. 11 Contour plots of the $2a_{1g}$ orbital wave function of SF_6 and SCl_6. The continuous lines represent values of the orbital of one sign, the dashed lines values of the opposite sign. The contours near the atomic centeres have been omitted. Note the different scale!

58 kcal/mole in Cl_2 (vs. 37 kcal/mole in F_2), it appears improbable that the total energy of SCl_6 lies below that of SCl_4 + Cl_2 /94/. In turn, SCl_4, dissociates already at ~-30°C into SCl_2 + Cl_2 /95/.

Let us now compare the electronic structure of SF_6 and SF_4 /93/. A number of MO calculations for SF_4 or related sulfuranes (SH_4 etc.) have appeared /91/. Here we are interested in possible comments on the distinction between a type-I and a type-II hypervalent electronic structure on the basis of an Xα-SW calculation. There is some ambiguity how to choose the sphere radii in this case. The equatorial S-F bonds have approximately the same length as the bonds in SF_6. We therefore required continuity of the potential along these bonds. For touching spheres, all other radii are then determined by geometry. The values are also listed in Table III. The resulting energy levels are shown in Fig. 10. They are labeled according to the irreducible representations a_1, a_2, b_1 and b_2 of C_{2v} symmetry group, appropriate for SF_4. A detailed description of the Xα-SW MO's may be found elsewhere /93/. Here, we only describe the main features.

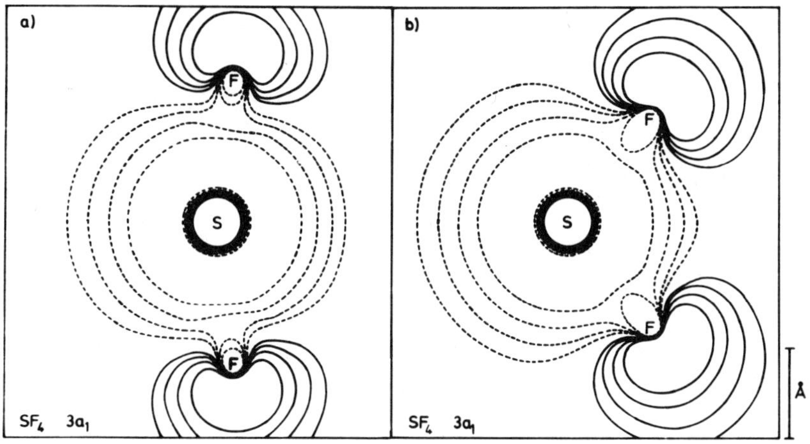

Fig. 12 Contour plots of the $3a_1$ orbital in SF_4 showing the interaction of the S 3s orbital with the F 2p orbitals. The set of contour values is the same as that in Fig. 11.
a) plane defined by the axial ligands: F_{ax} SF_{ax},
b) plane defined by the equatorial ligands: F_{eq} SF_{eq}.

Fig. 10 reveals a close similarity between the energy level schemes of SF_4 and SF_6. The levels $1a_1$, $1b_1$ and $1b_2$ are mainly derived from F 2s orbitals and contribute to S-F bonding, the $1a_1$ level most strongly. The $2a_1$ level is mainly a non-bonding in-phase linear combination of the two axial F 2s orbitals. The $3a_1$ level shows a strong localization in the sulfur sphere (0.455) in agreement with the expectation for type-I hypervalent bonding. Nevertheless, it participates in S-F bonding much in the same way as does the corresponding $2a_{1g}$ level in SF_6. This may be seen by comparing the contour map of this orbital (see Fig. 11a) to the two contour maps in Fig. 12 which show the $3a_1$ orbital in the planes defined by F_{ax} SF_{ax} and F_{eq} SF_{eq}. From these findings, two different bonding models /85a/ (hypervalent type I and II) do not seem to be justified. The next three levels, $2b_1$, $2b_2$ and $4a_1$ may formally be derived from the $2t_{1u}$ level of SF_6 when going from O_h symmetry to C_{2v} in SF_4. The highest occupied MO is the $7a_1$ level which lies somewhat separated in energy. There has been some controversy whether it should contain mainly the sulfur lone pair /96/ or be delocalized over the axial ligands containing the extra electrons of the electron-rich hypervalent bonds /86c, 97/. In the Xα-SW calculation, it turns out to be of mixed character /93/. Of course, the sulfur lone pair will be smeared out over several levels in an MO approach. One finds an appreciable localization of sulfur p_z character in the $4a_1$ (0.192) and in the $7a_1$ orbitals (0.116). With this discussion, we hope to have demonstrated that Xα-SW MO's may be used with profit to elucidate the electronic structure of a molecule much in the same way as MO's from an LCAO calculation.

The Mulliken population analysis may be considered as an information reduction scheme for an LCAO wave function. The resulting populations (or atomic charges) may often be correlated with the chemical behaviour of the atoms in a given compound. The population analysis is based on the arbitrary division of the overlap density between the two contributing atoms. Furthermore, it is sensitive to the basis set chosen and frequently one has to face negative populations. These disadvantages do not impair the population analysis as a tool for tracing qualitative features of a wave function. In an Xα-SW calculation, on the other hand, one obtains a distribution of the electronic charge over different regions of space (see Table IV) in a way inherent to the method /92, 93/. Such a SW population analysis, too, is hampered by several disadvantages. Atomic sphere radii are to a certain extent arbitrary parameters of the model and may only be loosely tied to the size of the atoms. A comparison between different compounds is often rendered difficult by different sphere sizes, even in related compounds (cf. the sulfur sphere radius in Table III). This factor may only be partially eliminated by considering the changes between superposed atomic charges and the self-consistent molecular charge distribution /93/. Furthermore, quite a substantial fraction of the valence electronic charge is found in the intersphere region and cannot uniquely be assigned to the various atoms (see Table IV). This is, of course, a difficulty intrinsic in any description of the electronic structure using delocalized MO's. As a crude assumption, one might devide the charge in region II equally among all atoms /20, 92/. With this procedure one obtains atomic charges correlating with the trends of other chemical properties, such as electronegativities, but only within a series of structurally related molecules. No meaning can be attributed to the absolute values of these total charges, in general. This limitation finds its counterpart in the current practice of the LCAO-type population analysis. For a detailed discussion of the

SW population analysis as well as applications to hypervalent molecules see refs. 92 and 93.

Further information on the electronic structure of SF_6 may be obtained from an interpretation of the photoelectron spectrum excited by UV light (He I at 21.21 eV and He II at 40.8 eV) /99/. The assignment of the five non-bonding F 2p levels ($1t_{2g}$ to $1t_{1g}$ in Table IV) remained an unresolved puzzle for quite a while /100/. The attempts to clarify the photoelectron spectrum of SF_6 have been based on the vibrational structure of the spectrum, on considerations of Jahn-Teller splittings, intensities and on a CNDO calculation /99/. Also the X-ray absorption and emission spectrum has been used /99c/. Practically no agreement could be obtained, even with the help of several LCAO-HF calculations /101, 102/. A specific problem is posed by the very intense line at 17 eV which one attributed to two orbitals. Therefore, in a calculation, not only the ordering but also the relative distance of the orbital energies has to be reproduced correctly. A final solution to the problem was presented recently by calculating the ionization potential directly using an elaborate many-body technique based on an LCAO-HF solution /84/. Beside an interpretation based on measured and computed intensities /99d/, the assignment derived from Xα-SW transition state energies /75, 92/ turned out to be the only correct one (for a complete listing of other assignments see ref. 84). In Table V, the HF, many-body and Xα-SW results are compared to experiment. One notices that Koopman's theorem gives too large ionization potentials as a consequence of the neglected orbital relaxation and a wrong level ordering. On the other hand, the SCF-Xα-SW ground state orbitals are seen to relax by about 3.8 eV when going to the transition state. However, the ordering and the relative distances between the Xα-SW orbitals remain unchanged. The agreement of the seven lowest Xα-SW transition energies with experiment is as good as the values from the many-body treatment and much better than those from Koopmans' theorem. The agreement with ionization potentials from more strongly bound orbitals is as good as that for a HF calculation. This striking success of the Xα-SW method becomes even more impressive when one compares the necessary computer times. A total time of 2 minutes had to be spent on an IBM 360/91 for the complete Xα-SW calculation including the transition state calculations. This has to be compared to about 2 hours on the same machine for the HF calculation and 1.5 hours additional time for the many-body calculation /84/. The molecule SF_6

Table V Comparison of theoretical and experimental ionization energies (in eV) for SF_6

Orbital	ab initio[a]		SCF-Xα-SW		Experiment[d]
	Koopmans	Many-body	ground state[b]	transition state[c]	
$1t_{1g}$	19.2	16.7	12.1	15.9	15.7
$1t_{1u}$	20.0	17.7	13.1	16.8	17.0
$3t_{1u}$	20.4	17.9	13.1	16.8	17.0
$2e_g$	20.4	18.5	13.7	17.5	18.6
$1t_{2g}$	23.4	20.8	15.3	18.8	19.8
$2t_{1u}$	25.7	23.4	17.8	21.8	22.9
$2a_{1g}$	30.3	28.0	22.7	26.7	27.0
$1e_g$	45.2	...	31.8	35.6	39.3
$1t_{1u}$	46.9	...	32.8	36.5	41.5
$1a_{1g}$	51.0	...	35.8	39.3	44.2

[a] Ref. 84.

[b] Ref. 75.

[c] Ref. 92.

[d] Levels $1t_{1g}$-$1t_{2g}$: ref. 99d, lebels $2t_{1u}$-$1a_{1g}$: ref. 99c.

is certainly a favorable system for the application of the Xα-SW method due to its high symmetry and the globular shape (see Fig. 10). The errors introduced by the muffin-tin approximation seem to be quite negligible. This will not be true in general, but the present results underline once more the power of the Xα method, especially for the interpretation of photoelectron spectra.

Let us conclude the discussion of the hypervalent molecules by commenting on the importance of d orbital participation in molecules containing second-row atoms /87, 89, 90, 101/. Since d, f, g ... functions are all members of a complete set, their inclusion in the basis set will certainly improve the wave function. However, if their contributions are small, no chemical significance can be attached to them. In SF_6 and SF_4, the contribution of d waves to the total valence electron charge in the sulfur is rather small: 0.653 (18%) and 0.298 (9%), respectively /93/. However, in agreement with other calculations /101, 102/, the $2e_g$ level of SF_6 is found to increase in energy from the fourth to the topmost position, if the d orbitals are omitted in the sulfur sphere. The changes in all other levels are not significant. Thus, one has to conclude that d orbital participation is important as far as the assignment of the SF_6 photoelectron spectrum is concerned. The charge distribution within the molecule, and consequently the forces between the atoms are far less sensitive to d orbital components.

4.2 The Total Energy

The total energy is a quantity of central interest in electronic structure theory. The binding energy of a molecule can be derived from it and it may be used as a criterion for the quality of a variationally determined wave function. From the variation of the total energy with the positions of the nuclei information is obtained on the forces within the molecule, i.e. on the relative stability of various geometrical configurations and on the vibrations about the equilibrium configuration.

In the preceding section we tried to explore the relative stability of the compounds SF_6 and SCl_6 using only the character of the valence electron orbitals. Why not compare the binding energies of the two molecules? First, it is our aim not only to calculate certain numbers, but to understand why they come out the way they do. Second, binding energies of large polyatomic molecules are hard to calculate to chemical accuracy, as they are obtained as a small difference of two large numbers, molecular total energy vs. limit of separated atoms. In SF_6, the binding energy amounts only to 0.02 percent of the total energy. Even the best HF calculation with a very large basis set /84/ gives a negative binding energy of about the same amount, i.e. an error of 200%.

As has been mentioned already (see sec. 3.3), the usual way to calculate the total energy in the Xα-SW method is by assuming a muffin-tin form both for the potential $V(\underset{\sim}{r})$ and the total charge density $\varrho(\underset{\sim}{r})$ when evaluating the functional (22) (cf. Fig. 7). Although this is a consistent procedure /83/, it is expected to introduce major errors in the value of the total energy. As the bonding of two atoms is attributed to the "overlap" charge which is mostly situated in region II and is treated worst in the muffin-tin model, the total energy calculated in such a way does not seem to be very reliable. Let us consider several examples to find out about these errors.

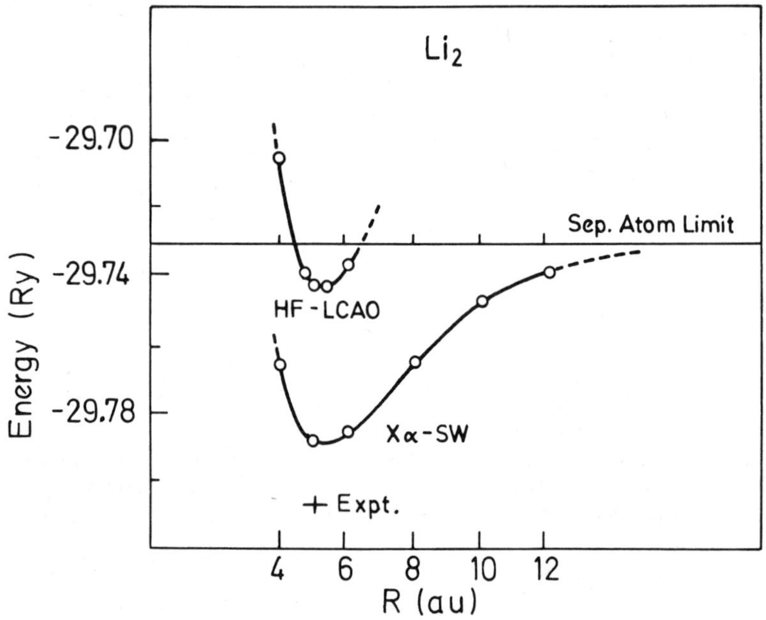

Fig. 13 Total energy of the lithium diatomic molecule as a function of internuclear distance, calculated by the SCF-Xα-SW and the restricted HF-LCAO methods (from ref. 103).

In Fig. 13 the total energy of the lithium diatomic molecule, Li_2, is shown as a function of internuclear distance. Both, the SCF-Xα-SW and the SCF-HF-LCAO method give a reasonably good bond length. The equilibrium internuclear distance derived from the Xα-SW potential curve is 2.69 Å, instead of the experimental bond length 2.67 Å. On the other hand, the HF binding energy is rather poor, amounting to only 15% of the experimental value 1.12 eV (= 0.082 Ry). The Xα-SW method gives a binding energy of 0.79 eV, about 70% of the experimental dissociation energy. Furthermore, it can be seen that the Xα-SW total energy curve dissociates properly to the limit of separated atoms whereas the HF curve does not. Of course, there are very accurate configuration interaction calculations for this simple system of only six electrons showing excellent agreement with experiment /105/.

For homonuclear diatomics, the Xα model always dissociates properly due to the local nature of the exchange /4/. For heteronuclear molecules, such as NaCl, the problem is more difficult, since the Xα model obeys Fermi statistics rigorously /106/ (see sec. 2.1). In the neighborhood of the equilibrium the NaCl molecule is predominantly ionic. The sum of the total energies of isolated ions, however, is higher than that of the neutral atoms. A crossing of the covalent and the ionic potential curves occurs at an internuclear sparation of 21.6 a.u. /106/. This happens only for the case of integer occupation numbers. Just as in the case of transition metal atoms (see sec. 2.1), one has to vary the occupation numbers in order to get the most favorable state fulfilling Fermi statistics. One then finds the ionicity reduced by a partial electron transfer from Cl^- to Na^+, for internuclear distances R larger than 8 a.u., leading to a potential curve which smoothly approaches the state $Na^{+0.405}\ Cl^{-0.405}$ at infinity. This is the state, where the Na 3s orbital and the Cl 3p orbital have equal energies /106/.

Many more potential curves have been calculated using the muffin-tin total energy expression /23/, but in most cases the results are rather disappointing /69, 70, 107/. Bond lengths turn out too large, binding energies too small. Consider, for instance, the nitrogen molecule /107/, where a bond length of 4.1 a.u. and a dissociation energy of 1.79 eV are obtained. The experimental values are 2.07 a.u. and 9.91 eV, respectively. In a critical analysis of the muffin-tin errors Mitzdorf /70/ has shown that only for the simultaneous enlargement of all bonds reasonable results may be obtained. Bonding angles have to be viewed with caution. For the water molecule the most stable configuration turns out to be linear /69/, when one chooses the outer sphere to be centered on oxygen; a bent configuration results if the outer sphere touches all atomic spheres /70/. On the other hand, in cases where the muffin-tin errors cancel approximately, good results may be obtained. The rotational barrier in ethylene, experimental value 2.82 eV, is predicted to be 3.52 eV, whereas the HF value is 5.58 eV /108/.

Danese /83, 109/ has achieved dramatic improvements of the total energy. He estimates the non-muffin-tin corrections with a procedure comparable to first-order perturbation theory. The muffin-tin potential defines the zeroth-order problem. The full charge density generated from the zeroth-order orbitals is then used in the evaluation of the Xα total energy expression (22). The muffin-tin corrections to the total energy and to the potential depend on the difference between the charge density and its muffin-tin average. The resulting integrals have to be evaluated by a Monte-Carlo sampling technique which unfortunately does not lend itself to a routine evaluation irrespective of the cluster treated. Notice that the value of the kinetic energy is already correct to first-order in this procedure. Using (24), we have

$$\begin{aligned} T &= \sum_i n_i \int u_i^*(\underset{\sim}{r}) \, (-\nabla^2) \, u_i(\underset{\sim}{r}) \, d\underset{\sim}{r} \\ &= \sum_i n_i \varepsilon_i - \int \varrho(\underset{\sim}{r}) \, V(\underset{\sim}{r}) \, d\underset{\sim}{r} \end{aligned} \tag{132}$$

As the potential $V(\underset{\sim}{r})$ is of muffin-tin form, the value of the integral is independent of whether the charge density ϱ or its muffin-tin average is used /83/. Only small molecules, like C_2 and Ne_2 have been treated with this procedure. For example, the ground state potential curve of C_2 (see Fig. 14) shows a repulsive behaviour in the vicinity of the experimental equilibrium internuclear distance. But the non-muffin-tin corrected Xα total energy is bound with reasonable values for the bond length, the dissociation energy and the frequency of the stretching vibration, the quality being comparable to that from a CI calculation (see Table VI). Once again, the Xα model gives better results than HF theory. The main reason for the success of the non-muffin-tin corrections is that the total Xα energy near its absolute minimum differs

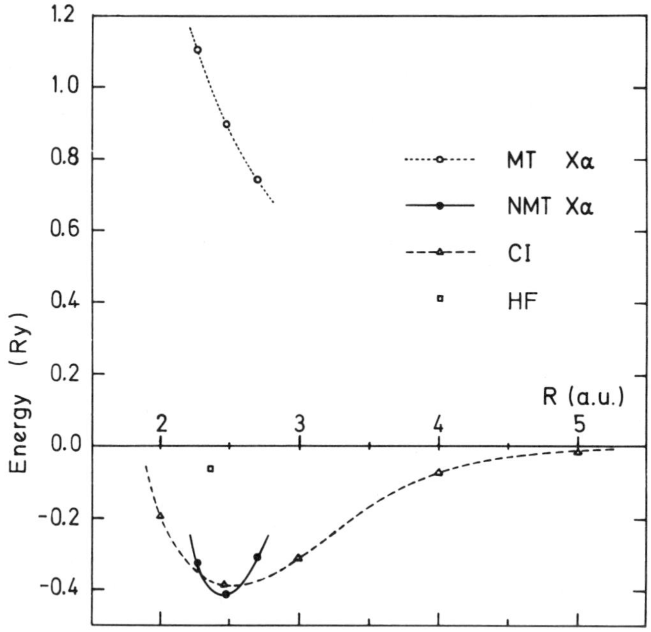

Fig. 14 The muffin-tin (MT), non-muffin-tin (NMT) and configuration interaction (CI) potential curve for the $^1\Sigma_g^+$ ground state of C_2 (adapted from ref. 109).

Table VI Comparison of spectroscopic constants for C_2

Source	Dissociation energy (eV)	Equilibrium separation (Å)	Frequency (cm^{-1})
Xα-SW [a] non-muffin-tin	5.58	1.31	2038
Configuration interaction[b]	5.39	1.36	1503
Experiment[c]	6.36	1.24	1856

[a] Ref. 109.
[b] P.F. Fougere and R.K. Nesbet, J.Chem.Phys., 44, 285 (1966).
[c] E.A. Balik and D.A. Ramsay, Astrophys.J., 137, 84 (1963).

from the exact Xα energy only by second-order effects in the orbitals because of the variational principle. The work of Danese has shown definitely that the discrepancies of the Xα-SW model to experiment are caused mainly by the muffin-tin approximation and not by the Xα method.

4.3 Solution Methods Beyond the Muffin-Tin

In the last section we have encountered a severe limitation of the Xα-SW method. Even though transition state energies are much less affected by the muffin-tin approximation than the total energy /70/, one anticipates, and indeed finds molecules, where the agreement with experimental ionization potentials is far from satisfactory. Large planar molecules, like benzene or other aromatic compounds, are extreme examples; they will be discussed specifically in Dr. Herman's lectures.

From our discussion in chap.3 the discussion to go is clear when one wishes to improve the solution methods for the Xα method beyond the muffin-tin approximation. However, no applications of the general multiple-scattering cellular technique to molecules have appeared up to now. But the necessary effort to treat clusters deviating strongly from a globular shape will be large. An approximate version of the cellular technique has been implemented for SF_6 /110/ with very promising results showing almost perfect agreement for the ionization potentials.

For small molecules like diatomics there is, of course, the possibility to use a brute force method when solving the one-particle Schrödinger equation /46/. One transforms the differential equation into a difference equation and solves the resulting algebraic eigenvalue problem with conventional methods. In order to avoid the difficulties associated with these numerical solutions of the problem - both the multiple scattering technique and the explicit approach - it seems worthwhile trying the traditional strategy of expanding the one-electron wave functions as linear combinations of atomic basis functions. The great merit of the Xα model is its reduction of the numerous two-electron integrals to a simple local potential. The trade-off lies in the extra effort required to calculate the rather awkward integrals arising from the $\rho^{1/3}$ dependence of the exchange potential. Two methods have been suggested to tackle this problem.

One such scheme is the discrete variational (DV)

method suggested by Ellis and Painter /111/ and implemented for chemical problems by Baerends and Ros /112/. In this method, one selects a set of sample points $\underset{\sim}{r}_k$ in configuration space associated with weights $w(\underset{\sim}{r}_k)$ and defines an error matrix

$$\Delta_{ij} = \sum_{k} w(\underset{\sim}{r}_k)\, u_i^*(\underset{\sim}{r}_k)(h - E)\, u_i(\underset{\sim}{r}_k), \tag{133}$$

where h is the one-electron hamiltonian occuring in eq. (24). Minimization of the error functional over the sample points through variation of the parameters C_i in the trial function (with atomic Slater-type orbitals χ_j).

$$u_i = \sum_j \chi_j C_{ji} \tag{134}$$

leads to the familiar secular equation of the Rayleigh-Ritz procedure:

$$\mathbb{H}\,\mathbb{C} = E\,\mathbb{S}\,\mathbb{C}. \tag{135}$$

The elements of the matrices $\mathbb{H}$ and $\mathbb{S}$ are defined with respect to the atomic basis functions:

$$H_{ij} = \sum_{k} w(\underset{\sim}{r}_k)\, \chi_i^*(\underset{\sim}{r}_k)\, h\, \chi_j(\underset{\sim}{r}_k) \tag{136.a}$$

$$S_{ij} = \sum_{k} w(\underset{\sim}{r}_k)\, \chi_i^*(\underset{\sim}{r}_k)\, \chi_j(\underset{\sim}{r}_k). \tag{136.b}$$

Although the DV method does not require the point summations to be approximations to integrals, its usefulness is greatly enhanced if expectation values of observables can be calculated by using the same set of sample points. The Haselgrove technique /113/ for performing multidimensional integrals is therefore used to generate the set of sample points and weights. A further approximation involved is the expansion of the density $\rho(\underset{\sim}{r})$ in one-center functions

$$\rho(\underset{\sim}{r}) \approx \sum_i a_i f_i(\underset{\sim}{r}) \tag{137}$$

to facilitate the evaluation of the Coulomb potential (130). One-electron properties, such as ionization potentials, may be calculated with high accuracy (see Table VII). The estimation of total energies to chemical accuracy is much more difficult, but has been shown to be feasible for small systems /114/.

Recently, Sambe and Felton have proposed another basis set technique /115/ using gaussian expansion functions in (134). In addition to the one-center expansion (137) for the density they introduce a second expansion for the exchange potential:

$$-6\alpha\left[\frac{3}{8\pi}\,\rho(\underset{\sim}{r})\right]^{1/3} \simeq \sum_i b_i\, g_i(\underset{\sim}{r}). \tag{138}$$

Only s-type expansion functions f_i and g_i have been employed. The matrix elements of the one-electron hamiltonian h may then be evaluated in terms of integrals over the functions χ_i, f_i and g_i. The expansion coefficients a_i and b_i are determined by a least square procedure over a small set of sample points for each self-consistent cycle. Typically about 100 sample points are sufficient for a small molecule like H_2O, whereas at least 1000 sample points are necessary in the DV method.

This so-called LCAO-Xα method has been quite successful in calculating both eigenvalues and total energies (see Table VIII). For instance, it predicts the bond angle and length in the H_2O molecule to within 2% of experiment /115/.

Both these methods share with LCAO *ab initio* methods the critical dependence on the choice of the basis set which in order to keep computational times short, must be limited in size and flexibility (see sec. 3.2). However, they both provide a way of finding general solutions to the Xα equations (24) and are still much faster than *ab initio* calculations using the same basis set.

As a further procedure to improve the muffin-tin version of the Xα-SW model let us finally mention the method of overlapping spheres /116-119/. It has been found empirically /116, 117/ that allowing the atomic spheres to overlap without generalizing the formalism results in better agreement with experiment for certain types of molecular clusters. Based on an intuitive physical argument, one aims at a better description of the electronic structure by reducing the volume of the intersphere region and by picking-up major parts of the intersphere charge situated just outside the non-overlapping spheres. This approach has been analysed for its implicit assumptions /118/: (1) an approximate projection of the

Table VII Comparison of experimental and theoretical ionization potentials for ethylene (all energies in eV)

Orbital	MT[a]	OS1[a]	OS2	DVM[b]	Experiment[c]
$1b_{3u}$	13.3	12.1	10.1	10.7	10.5
$1b_{3g}$	13.7	11.7	12.6	12.3	12.4
$3a_g$	16.4	14.7	15.4	14.3	14.5
$1b_{2u}$	17.4	15.5	15.9	15.4	15.7
$2b_{1u}$	19.9	18.1	18.6	18.1	18.9
$2a_g$	25.4	24.0	23.9	23.4	~23

[a] Ref. 116.
[b] Ref. 112.
[c] A.J. Merer and R.S. Mulliken, Chem. Rev.,69,639 (1969).

overlapping spherical component out of a general function, applied to the charge density and part of the potential, (2) the use of trial orbitals generated from a single spherical potential within the overlapping cells, and (3) the neglect of the overlap charge density in constructing the exchange potential. In this sense, one may consider the overlapping sphere model as a first approximation to the general multiple scattering cellular technique /53, 110/. However, it introduces the flavour of a semi-empirical approach into the Xα-SW model as a consequence of the arbitrary degree of overlap.

Norman /119/ has proposed a "non-empirical" criterion for the choice of overlapping spheres. He measures the size of an atom in a molecular cluster by the "atomic number radius" defined as the radius of that sphere which encloses the atomic number of electrons out of the initial superposed-atom charge distribution. Ratios of atomic number radii are essentially independent of the atomic charges assumed in constructing the initial Xα-SW potential. Norman therefore recommends to choose the radii according to atomic number ratios with absolute values determined to satisfy the virial theorem at self-consistency /116, 119/. Even simpler, yet quite successful is the approach to divide a bond length according to

Table VIII Equilibrium bond distances (in a.u.) for N_2, F_2 and CO

Molecule	HF LCAO	Xα LCAO[d]	Xα DV[e]	SW-Xα MT[f]	SW-Xα OS[g]	Expt.[h]
N_2	2.01[a]	2.07	2.18	4.1	2.4	2.07
F_2	2.51[b]	6.0	2.42	4.3	2.8	2.68
CO	2.08[c]	2.10	2.23	...	2.4	2.13

a-h: References 122, 123, 124, 115, 114, 107, 121, 125.

the atomic number ratio and to enlarge all atomic radii uniformly by 25 to 30% /121/. With this procedure, ionization potentials for CO, N_2, F_2 and H_2O have been calculated using the transition state procedure that are in as good an agreement with experiment as those obtained from the DV or LCAO version of the Xα method /121/.

Let us finally give several examples to show how successful the various methods mentioned in this section overcome the deficiencies of the muffin-tin Xα model. The first molecule to be studied by overlapping spheres was ethylene, C_2H_4 /116/. The π electrons in the highest occupied orbital extend beyond the caron sphere which in this case is determined by symmetry (radii: r_C = 0.672 Å, r_H = 0.418 Å, r_{OUT} = 1.961 Å). The interatomic region is filled by the atomic spheres to a fraction of only 12 percent, but holds 5.74 valence electrons out of 12. As a consequence, the constant potential and all valence orbital energies turn out too low by 2~3 eV. In Table VII, the transition state energies of the simple muffin-tin (MT) calculation and two different overlapping sphere (OS) calculations (OS1, OS2: r_C = 0.820 (0.889) Å, r_H = 0.418 (0.609) Å, r_{OUT} = 1.961 (2.152) Å) are compared to the results of the DV method and to experiment. The parameter set OS2 is seen to correct most of the discrepancies to experiment which are caused by the muffin-tin potential of the simple Xα-SW model.

In Table VIII, a comparison is given for the equilibrium bond length of several diatomic molecules as determined from several versions of the Xα method, from

LCAO-HF theory and from experiment. The results show a definite improvement over the muffin-tin model with one exception (F_2: Xα-LCAO) caused probably by an inappropriate basis set.

From these and many other applications one may conclude that the overlapping sphere model at least partly corrects for the muffin-tin approximation, yet retains the computational attractiveness of the original Xα-SW method. This is especially true for one-electron properties. The total energy, however, will meet the requirements of chemical accuracy only in a very limited sense. Besides in cases of large planar organic molecules to be discussed by Dr. Herman, one could think of organometallic complexes as typical applications of the overlapping sphere model. There one first aims at a reasonable description of the organic ligand molecules before using the same parameters in studying the whole complex.

4.4 A Case Study: Ferrocene

Since its discovery in 1951, ferrocene, $(h^5\text{-}C_5H_5)_2Fe$, has played an important role, both theoretically and experimentally in developping an understanding of the electronic structure of organometallic compounds. Its remarkable structure, an iron atom sandwiched symmetrically between two cyclopentadienyl (Cp) rings,

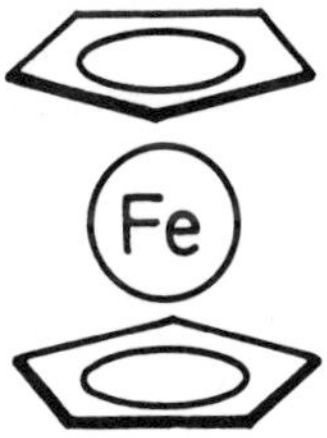

as well as its extraordinary thermal stability provided the initial impetus for a still continuing research effort on transition metal cyclopentadienyl compounds /126/. The basic features of the metal-ring bonding are now considered to be well understood, as can be seen from reviews of this work /126-128/. However, none of the qualitative discussions of the electronic structure and of the numerous semi-empirical MO calculations /129/ has been successful in explaining all the experimental facts as provided by optical absorption measurements /130/, ESR /131/

and photoelectron spectroscopy /132, 133/. Several recent calculations have tried to answer these controversial questions utilizing the HF-LCAO method /134-136/, the Xα-DV /137/ and the Xα-SW method /138/.

Before proceeding to the details of these calculations it seems to be useful to recall the qualitative MO bonding scheme which has been employed hitherto in rationalizing the chemistry of metallocenes /95, 126/. This MO scheme does not depend critically on whether the preferred rotational orientation of the rings is staggered (D_{5d}, see the diagram above) or eclipsed (D_{5h}). The latter conformation is found in the gas phase /139/, the former in the solid phase /125/. Also, no significant differences between the two structures have been found in all calculational methods /134-138/. To conform with the early work /129/, the notation for D_{5d} symmetry group is commonly used while the eclipsed gas phase structure is assumed in the calculations. In any case, the energy difference in the gas phase between the two structures is estimated to be very small ($\sim$1.1 kcal/mole) /139/.

Ferrocene is usually regarded as an iron (II) compound, made up of Fe^{2+} and two cyclopentadienide ions $C_5H_5^-$. However, ferrocene is not an ionic compound; the size of the charge should not be taken literally. Iron (II) has a $3d^6$ electronic configuration. Only the

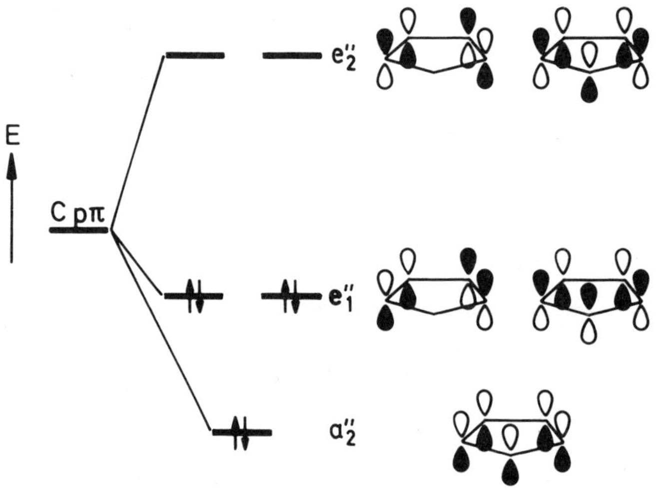

Fig. 15 The molecular orbitals for planar $C_5H_5^-$ (schematic).

electronic structure of the effectively planar rings is relevant for our purpose. The five carbon 2pπ atomic orbitals from a basis for the representation $a_2'' + e_1'' + e_2''$, and according to Hückel theory the relative energies of these orbitals are $e_2'' > e_1'' > a_2''$, as shown in Fig. 15. The a_2'' and e_1'' orbitals are filled with six valence electrons in the anion $C_5H_5^-$. A HF-LCAO calculation on this species /140/ shows indeed that two σ orbitals of the Cp ring are almost degenerate in energy with with the lowest π orbital a_2''. However, this is of no direct consequence for the interaction between the ring and the metal orbitals. The ring π orbitals may be combined to form symmetric and antisymmetric linear combinations suitable for the description of the sandwich system $(C_5H_5^-)_2$: $a_{1g} + a_{2u}$, $e_{1g} + e_{1u}$, $e_{2g} + e_{2u}$ (see Fig. 16). Out of these ring fragment orbitals, only the symmetric ones will interact with the metal d orbitals which span the representations $a_{1g} + e_{1g} + e_{2g}$ under D_{5d}. Invoking perturbation theory, one will expect the main bonding interaction through the e_{1g} level, and some back bonding <u>via</u> the e_{2g} metal orbital into the empty π^* ring orbital. This picture of bonding may be characterized as a two-way donor-acceptor mechanism similar to that proposed by Dewar and Chatt /141/ to describe the transition-metal olefin interaction. The resulting approximate MO diagram is shown in Fig. 16. For simplicity, only d orbitals have been included on the metal. The levels are filled with 18 electrons. All semi-empirical calculations /129/ yield the same general pattern for the one-electron energies with some differences in the ordering, especially for the antibonding MO's (most of which are not shown in Fig. 16). A controversial point is the ordering of the two highest occupied orbitals of predominant metal 3d character. In most semi-empirical calculations, the level ordering $a_{1g} > e_{2g}$ has been found /129/. Also the lowest band in the optical spectrum has been assigned to the ligand field transition $a_{1g} \rightarrow e_{1g}$ /130c/. On the other hand, from ESR measurements it has been concluded that the ground state of the ferricenium ion $FeCp_2^+$ is $^2E_{2g}$ $[a_{1g}^2\ e_{2g}^3]$ /131. The intensity ratio 2 : 1 of the two highest peaks in the photoelectron spectrum /132, 133/ are consistent with this observation.

Let us now describe the results of an Xα-SW calculation of ferrocene. The bond lengths were taken to be Fe-C = 2.106 Å, C-C = 1.43 Å, and C-H = 1.12 Å /139/. The Cp ring was assumed to be planar. In order to achieve a better description of the Cp rings and their interaction with the iron atom, the carbon spheres were enlarged

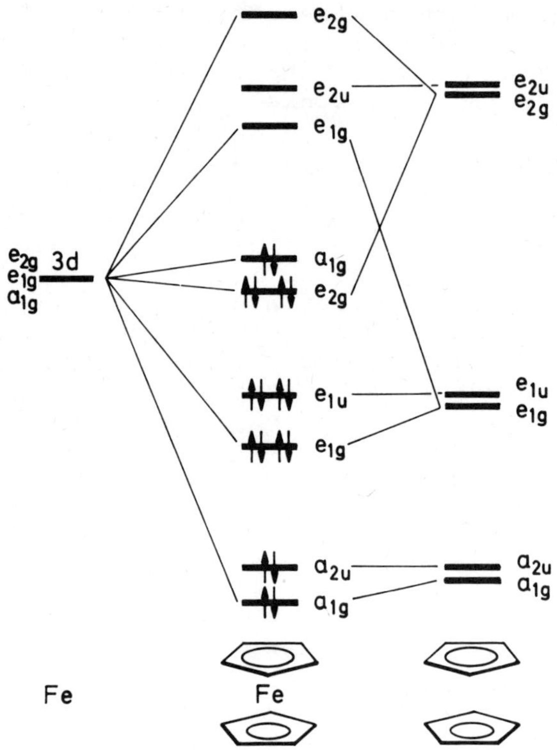

Fig. 16 Approximate MO diagram for ferrocene inferred from simple overlap arguments.

to overlap with each other and with the iron and hydrogen spheres, respectively. The various sphere radii used were: r_{Fe} = 1.391 Å, r_C = 0.820 Å, r_H = 0.405 Å and r_{OUT} = 3.306 Å. The carbon sphere radius is identical to that used in earlier calculations on ethylene and benzene /116/.

The calculation of the ground state orbitals of ferrocene converged to ± 0.001 Ry in 25 iterations. The Fe 1s, 2s, and 2p and the C 1s orbitals, initially frozen, were released after self-consistency in the valence orbitals had been attained, but the resulting shifts in the valence levels were only 0.001 Ry. The calculation required about 20 minutes CPU time on an IBM 370/165 computer. The valence orbital energies are shown in Fig. 17, which includes also the first few unoccupied levels.

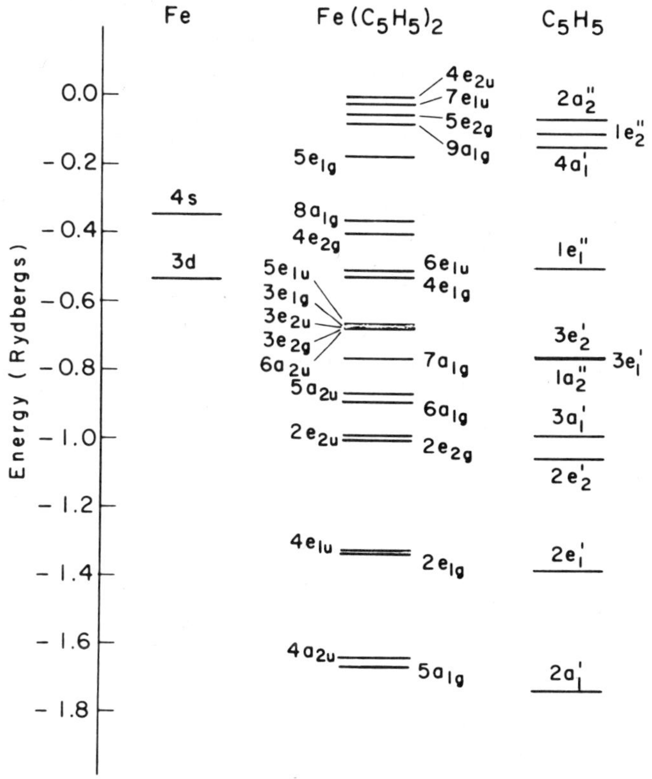

Fig. 17 The SCF Xα-SW orbital energies of ferrocene. The levels are labeled according to the irreducible representations of the point group D_{5d}. Also shown for comparison are the SCF-Xα energies of the free iron atom ($3d^6 4s^2$) and the C_5H_5 moiety.

These orbital energies are compared with the SCF-Xα atomic orbital energies of iron and the MO energies as obtained from a Xα-SW calculation of the Cp moiety. The latter orbitals are labeled by the corresponding representations of the D_{5h} point group. They show the same pattern on the energy scale as found in a HF-LCAO calculation on $C_5H_5^-$ /140/.

The charge distributions and the energies of the Xα-SW orbitals are consistent with the usual description of the bonding in ferrocene outlined above /126-128/. The fact that the two Cp rings exist very much as separate entities in ferrocene is reflected by the energy pairing of those orbitals which are largely non-bonding with respect to the Fe atom (up to the pair $6a_{1g}$ and $5a_{2u}$). The orbitals associated with the bonding of Cp-π electrons to the Fe atom are $4e_{1g}$, $7a_{1g}$, $6e_{1u}$ and $6a_{2u}$, in order ot decreasing charge fraction within the Fe sphere. The two highest occupied orbitals $8a_{1g}$ and $4e_{2g}$ are essentially localized Fe 3d orbitals. The corresponding orbital charge fractions in the Fe sphere are 0.87 and 0.82, respectively, indicating a slight contribution of the $4e_{2g}$ level to back bonding.

The unoccupied orbitals may be briefly characterized as follows. The level $5e_{1g}^*$ corresponds mainly to an Fe 3d-like orbital with significant antibonding character with respect to the Cp rings. It is the antibonding partner to the $4e_{1g}$ orbital which contributes most to the Fe-Cp bonding. The levels $5e_{2g}$ and $4e_{2u}$ are associated with Cp-π^* orbitals, the levels $9a_{1g}$ and $7e_{1u}$ with rather diffuse Fe 4s- and 4p-like orbitals.

We next consider the energies of electronic excitations in ferrocene. The ground state is the singlet $^1A_{1g}$ $(4e_{2g})^4\ (8a_{1g})^2$. The one-electron transition $8a_{1g} \rightarrow 5e_{1g}$ yields excited singlet and triplet states of symmetries $^1E_{1g}$ and $^3E_{1g}$, whereas $^1E_{1g}$, $^1E_{2g}$, $^3E_{1g}$ and $^3E_{2g}$ multiplets are associated with the $4e_{2g} \rightarrow 5e_{1g}$ transition. These are all d-d transitions since initial-state and final-state orbitals are principally Fe 3d-like. The singlet-singlet transitions, which are spin-allowed, are responsible for the most intense optical absorption observed in the visible part of the spectrum /130/. Very weak absorptions in the same spectral range are associated with the spin-forbidden, singlet-triplet excitations /130c/. The Xα transition state procedure has been applied to the transitions $8a_{1g} \rightarrow 5e_{1g}$ and $4e_{2g} \rightarrow 5e_{1g}$, but only in the spin-restricted version (see sec. 2.4). A comparison can therefore be made only with averages of the corresponding measured optical absorption energies. This has been done in Table IX for the data by Sohn et al. /130c/. The bands have been assigned on the basis of a ligand field calculation which also yielded the excitation energies for the unobserved triplets $b\,^3E_{1g}$ and $^3E_{2g}$ /142/. The agreement between Xα-SW transition energies and experiment provides further support for the assignment proposed by Sohn et al. /130c/.

The strongest optical absorption in ferrocene occurs in the UV part of the spectrum /130/ and has been interpreted largely in terms of charge transfer excitations.

Table IX Comparison of d-d transition energies ($\bar{\nu} \times 10^{-3}$ cm^{-1}) calculated by the Xα-SW method with experimental optical data for ferrocene in the visible part of the spectrum

Multiplet	Experiment[a]	Average experimental transition energy[c]	Xα-SW[d]	Orbital transition
a^3E_{1g}	18.9	19.6	20.5	$8a_{1g} \rightarrow 5e_{1g}$
a^1E_{1g}	21.8			
b^3E_{2g}	20.9[b]	23.1	25.2	$4e_{2g} \rightarrow 5e_{1g}$
b^3E_{1g}	22.4[b]			
$^1E_{2g}$	24.0			
b^1E_{1g}	30.8			

[a] Ref. 130c.
[b] Values calculated from a ligand field analysis /130c, 142/.
[c] Weighted according to multiplicity.
[d] Ref. 138.

For example, we associate a shoulder at 37.700 cm^{-1}/130a/ with the first dipole-allowed $6e_{1u}$ (Cp-π)→$5e_{1g}$ (3d) "ligand-to-metal" (L→M) charge-transfer transition. The most intense absorption band has its maximum at 51.600 cm^{-1} and is probably due to charge-transfer transitions from the band of closely spaced ligand levels $5e_{1u}$ (Cp-σ), $3e_{2u}$ (Cp-σ) and $6a_{2u}$ (Cp-π) to the Fe-like level $5e_{1g}$ (3d). A summary of the tentative assignment of the UV optical spectrum of ferrocene is given in Table X. This is only a correlation of optical absorption peaks /130a/ with individual (or groups of) orbital transitions. No attempt has been made to calculate multiplet averages.

A completely different assignment of the optical absorption spectrum has been proposed using the results of an ab initio singly excited CI calculation with a near minimal basis set (split 3d orbitals) /135/. To facilitate a comparison, we reproduce these results in

Table X Comparison of orbital transition energies ($\bar{\nu} \times 10^{-3}$ cm^{-1}) calculated by the Xα-SW method with experimental optical charge-transfer energies in the UV part of the spectrum

Orbital transition	Type of transition	Xα-SW transition energy	Experiment[a]
$6e_{1u} \rightarrow 5e_{1g}$	L → M	36.5	37.7
$8a_{1g} \rightarrow 4e_{2u}$	M → L	39.2	41.0
$4e_{2g} \rightarrow 4e_{2u}$	M → L	43.9	42.8
$6e_{1u} \rightarrow 9a_{1g}$	L → M	47.1	47.5
$6e_{1u} \rightarrow 5e_{2g}$	L → LM[b]	50.1	49.6
$5e_{1u} \rightarrow 5e_{1g}$	L → M	53.9	51.2
$3e_{2u} \rightarrow 5e_{1g}$	L → M	55.1	53.1
$6a_{2u} \rightarrow 5e_{1g}$	L → M	55.5	

[a] Energies of optical absorption peaks /130a/.
[b] Some charge transfer because of the antibonding Fe3d character.

Table XI. The low-energy absorption bands are associated with the orbital transition $4e_{2g} \rightarrow 5e_{1g}$, a major difference to the ligand field calculation /141/ and the Xα-SW assignment. The good agreement for the first three CI transition energies with experiment does not extend to higher absorption energies. Already the transition energy corresponding to the d-d transition $8a_{1g} \rightarrow 5e_{1g}$ ($^1E_{1g}$) would be in error by 50 percent. For the assignment of charge-transfer excitations no new argument may be derived from the calculated transition energies. The fact that they should occur at higher energies than d-d transitions and that they should be associated with the symmetry allowed transition $6e_{1u} \rightarrow 5e_{1g}$, may already be deduced from the qualitative MO scheme in Fig. 16 /130c/.

Although a correlation of Xα-SW orbital transition energies with absorption peaks in the UV spectrum may be fortuitous, it yields energies in the correct range, which should not change drastically when proper

Table XI Assignment of the visible and UV absorption spectrum of ferrocene based on a limited CI calculation[a] (energies in 10^{+3} cm^{-1})

Transition $^1A_{1g}$	Orbital transition	Type of transition	CI transition energy	Experiment[b]
$^3E_{1g}$, $^3E_{2g}$	$4e_{2g} \rightarrow 5e_{1g}$	d → d	14.6-15.1	18.9
$^1E_{2g}$			21.2	21.8
$^1E_{1g}$			26.7	24.0
$^3E_{1g}$	$8a_{1g} \rightarrow 5e_{1g}$	d → d	36.8	...
$^1E_{1g}$			36.3	30.8
$^1A_{1u}$, $^1E_{2u}$ / $^1A_{1g}$, $^1E_{2g}$	$6e_{1u} \rightarrow 5e_{1g}$ / $4e_{1g} \rightarrow 5e_{1g}$	L → M	~73	37.7
$^1A_{1g}$, $^1E_{1g}$	$4e_{2g} \rightarrow 5e_{2g}$	M → L	~82	41.7
$^1A_{2u}$	$6e_{1u} \rightarrow 5e_{1g}$	L → M	83.8	50

[a] Adapted from ref. 135.
[b] Ref. 130c.

multiplet averages are calculated. A decision between the two assignments of the optical absorption spectrum of ferrocene (Xα-SW vs. CI) on the basis of the data presented in Tables IX to XI would be premature although the Xα-SW assignment is in better overall agreement for the whole energy range (see the discussion below).

A more detailed comparison of the different calculations on ferrocene may be based on the interpretation of the ionization spectrum /134, 136-138/. The HF calculations are of quite different quality. In the first one /134/, a basis set of 293 gaussian functions was reduced to 85 contracted functions. This represents a minimal

Table XII Comparison of calculated ionization potentials of ferrocene with those measured by He(I) photoelectron spectroscopy (energies in eV)

	HF - LCAO				Xα transition state			
Orbital	KT[a]		RO[c]	RO[e]	SW[f]		DV[j]	Experiment[k]
$4e_{2g}$	11.8		5.69	8.3	8.5		6.7	6.88
$8a_{1g}$	14.3		7.46	10.1	7.9		6.7	7.23
$6e_{1u}$	9.45		8.85[d]	11.1	9.3		8.1	8.72
$6e_{1g}$	9.54		8.79[d]	11.2	9.7		8.6	9.39
$6a_{2u}$	13.6		13.03	15.5	11.7	g	10.9	
$3e_{2u}$	14.1	b	13.60	...	11.6	g	11.2	12.3
$3e_{2g}$	14.4	b	...	...	11.7	g	11.3	(13.0)
$5e_{1u}$	15.3	b	...	...	11.5	g	11.8	
$3e_{1g}$	15.2	b	...	...	11.6	g	12.0	13.5
$7a_{1g}$	15.7	b	...	...	12.8	h	12.1	
$5a_{2u}$	19.0		...	...	14.2	i	14.8	
$6a_{1g}$	19.8		...	...	14.6	i	15.1	16.5
$2e_{2u}$	19.9		...	...	16.0		15.2	
$2e_{2g}$	20.3		...	...	16.1		15.4	

Comments to Table XII:

[a] Koopmans' theorem (negative HF eigenvalues); ref. 136 (large basis set).
[b] In ref. 136 assigned to the I.P. at 13.5 eV.
[c] Relaxed orbital (RO) method by taking differences of SCF total energies, large basis set of ref. 136.
[d] In ref. 136 and ref. 132 assigned in reverse order.
[e] Relaxed orbital method for a minimal basis set, ref. 134.
[f] Ref. 138; first two I.P.'s assigned in reverse order.
[g] In ref. 138 assigned to the I.P. at 12.3 eV.
[h] In ref. 138 assigned to the peak at 13.0 eV.
[i] In ref. 138 assigned to the I.P. at 13.5 eV.
[j] Ref. 137. Only the first four I.P.'s have been calculated explicitly by the transition state procedure. The remaining I.P.'s have been estimated on the basis of the data given in ref. 137. Their assignment is tentative.
[k] Ref. 133.

basis set except for the 3d orbitals of Fe which are described by split functions. The basis set of a recent HF calculation /136/ was of better than "double zeta" quality (188 contracted gaussians) in that it included three d-type functions on Fe and a p-type polarization function on the H's. However, no Fe 4p "polarization" function were included in contrast to the smaller basis set. The basis set used in the DV-Xα calculation /137/ consisted of two Slater-type orbitals per atomic orbital (double zeta basis); the total number of sampling points was 5200 (312 on C, 130 on H, and 780 on Fe). The same geometry (D_{5h}) was used in all four calculations except for slightly different bond lengths in one of the HF calculations /137/. For instance, the Fe-C bond distance is 2% shorter: 2.057 Å instead of 2.106 Å. But this small difference should not hamper a comparison with the other calculations.

In Fig. 18 we reproduce the He(I) photoelectron spectrum of ferrocene for the vapour phase /133/. In Table XII we have collected the ionization potentials calculated by the various methods and the assignments for the bands in the experimental spectrum deduced from these results /134, 136-138/.

The most remarkable single point to be made from these calculations is the so-called "break-down of Koop-

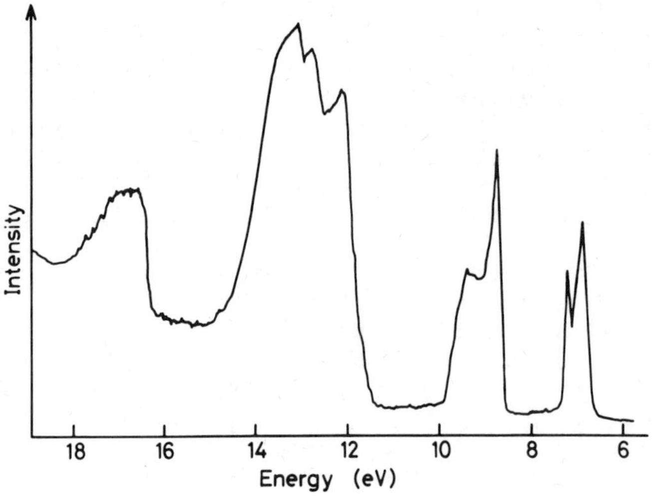

Fig. 18 The He(I) photoelectron spectrum of ferrocene (adapted from ref. 133).

mans' theorem" /134, 143/. This term denotes the phenomenon that the order of HF one-particle energies is different from the total energies of the corresponding states of the positive ion. The relaxation energy, i. e. the energy difference between the total energy of the ion in the frozen orbital approximation (Koopmans' theorem) and the SCF total energy for that state, depends very much on the character of the orbital to be ionized. For the orbitals with predominant Fe 3d character the relaxation energy is 6 - 7 eV whereas for Cp ring orbitals it is only 0.5 - 0.7 eV. These differences in the relaxation energies are not an artefact of the basis set as they appear for all HF calculations of ferrocene /134, 136/. Two important consequences follow from this observation. First, in order to interpret the photoelectron spectrum with HF calculations, it is indispensable in such transition metal complexes to carry out separate self-consistent calculations for every state of the ion to account properly for the relaxation. Ionization potentials are then obtained by taking differences of two SCF total energies. Therefore, the necessary computational effort increases drastically, also because the total energies have to be known with high precision. Compare the influence of the size of the basis set as exhibited by columns IV and V of Table XII. Second, for transition metal complexes

HF theory can no longer be taken as the basis for the one-electron picture underlying qualitative MO theories which are so successful in rationalizing many aspects of chemistry. A rationalization a posteriori assuming an extremely large ligand field splitting of the metal d orbitals beforehand does not seem to be of any help in this context /141/. In these MO treatments heavy use is made not only of the shape of the various orbitals, but also of their relative ordering in energy. As a prototype for such MO theories one may take the Woodward-Hoffmann rules of organic chemistry /144/. But rather similar MO treatments have been successfully applied to inorganic and organometallic complexes quite recently /145/.

Differences in relaxation energies between Fe 3d orbitals and Cp orbitals have also been found in the Xα calculations, but they amount to only about 0.5 eV, both in the SW and the DV method /137, 138/. Otherwise we find a rather uniform shift of all levels when going to the transition state and no change in the level ordering.

A further point warranting closer inspection is the effect of the muffin-tin approximation which may be deduced by comparing columns VI and VII of Table XII. The values and the ordering of the orbital energies are rather similar with two important exceptions. The Xα-SW method predicts the wrong ground state of the ferricenium ion: $^2A_{1g}$ instead of $^2E_{2g}$ as deduced from experiment /131-133/ and found in the HF calculations. The $8a_{1g}$ level of the Xα-SW calculation is associated with the lowest ionization potential /138/. The DV method gives the two Fe 3d levels virtually degenerate so that no conclusion on the ground state of the ion can be made from this calculation.

The clustering of Cp σ and π levels between 11.5 and 12.8 eV in the SW calculation appears also in the DV method, between 10.9 and 12.1 eV. No significance should be attributed to the relative ordering of the top five SW orbitals in this energy range. Comparing the distance in energy from the $6e_{1u}$ to the $6a_{1g}$ level both for the SW and the DV method one realizes a "compression" of the SW levels by about 1.7 eV, an effect which has been found for valence levels in other Xα-SW calculations as well. Taking into account these differences to the Xα-DVM results the assignments of the ferrocene ionization spectrum given originally /138/ should probably be revised on the basis of the more accurate Xα-DV calculation.

Let us finally turn to the interpretation of the experimental ionization spectrum (see Fig. 18) which emerges from the results of all calculations. There is general agreement that the first band is to be associated with the Fe 3d-like orbitals $4e_{2g}$ and $8a_{1g}$, the former having the lowest ionization potential. The next band corresponds to ionization of the $6e_{1u}$ and $6e_{1g}$ Cp-π levels. On account of an energy difference of only 0.06 eV, Bagus et al. /136/ follow Rabalais et al. /132/ in assigning the lower energy peak to the $6e_{1g}$ level. This assignment contradicts all other calculations and does not seem to be justified on the basis of such a small energy difference. The next band in the experimental spectrum is very broad and intense. It has therefore been attributed to ionization of several Cp σ and π levels. Any assignment on the basis of all present calculations can only be tentative. Relaxed orbital calculations are available for only two ionization potentials of this band /136/. Furthermore, no transition state calculations with the Xα-DV method have been published. Rather, the values presented in Table XII have been taken from the orbital energies found in $6e_{1g}$ transition state /137/. Both methods, HF-LCAO (KT) and Xα-DVM, agree in the $6a_{2u}$ and $7a_{1g}$ levels as extremes of this intensive band and in the level ordering. From the clustering of the levels found in both calculations, one might assign the levels $5e_{1u}$, $3e_{1g}$ and $7a_{1g}$ to the peak at 13.5 eV. Also, both methods lead also to the same assignment for the peak at 16.5 eV, although the Xα transition state energies are in better agreement with the experimental value than those derived from HF eigenvalues.

The discussion of this section may be summarized as follows. Despite the gross features of the electronic structure of ferrocene being well understood, there remain several subtle points to be clarified. Even the large HF-LCAO calculation /137/ cannot answer all questions beyond doubt as the basis set is still far from the HF limit. On the other hand, the parametrization used in the Xα-SW calculation (one of the first to use overlapping spheres) may possibly be improved drawing from the increased experience with this method /120, 121/. In any case, one should be cautioned against applying the Xα-SW method without awareness of the approximations involved, even in the "overlapping sphere" version - a truism applicable to any method used in electronic structure calculations.

4.5 Summary

After the presentation of the Xα-SW method and its application to selected problems it might be appropriate to summarize briefly the problems and the classes of compounds to which it has been applied most frequently and most successfully /23, 146/.

The SCF-Xα-SW method is an electron density formalism starting from approximate expression for the total energy. The theory does not deal with the hamilton operator nor does it afford a total wave function of the system, but rather describes the electronic structure _via_ an effective potential, its one-particle energies and the corresponding orbitals.

Certainly, it must be considered a serious disadvantage of the method that no reliable value for the total energy of a system is readily available. No quantitative answers can be given to questions concerning the stability or preferred geometry of a molecular cluster. Although means to amend this deficiency are in sight, no routine procedure has yet been developed. However, the electronic structure of a system may very well be interpreted using energies and character of the Xα-SW orbitals and the charge density, the primary quantity of the method. Additional insight may be gained from distribution of charge over the various regions of space and the SW population analysis. The method is ideally suited to study trends in a series of chemically related compounds, especially if the muffin-tin errors may be assumed to stay approximately constant /147/. Replacing an atom by an heavier one of the same group hardly increases the computational efforts and therefore facilitates further trend studies /92, 93/.

The method probably works best for the interpretation of photoelectron and optical absorption spectra due to the Xα transition state procedure. A consistent one-particle picture is provided even for systems with heavy atoms. Virtual orbitals may be used to characterize excited states. Comparisons of Xα-SW transition state energies with experimental ionization potentials /148/ show in general better agreement than results of semiempirical orbital energies (CNDO etc.) and about the same as _ab initio_ HF-LCAO orbital energies.

One-electron energies are not a very sensitive criterion for the quality of orbital wave functions and charge distributions derived from them. However, the calculation of other observable quantities is somewhat hampered by the special form of SW orbitals. One-particle observables, like dipole momentum or polarizabilities, can be calculated. The Fourier transform of SW orbitals provides information about the momentum distribution of the electrons. It allows the calculation of Compton profiles, a further criterion to evaluate the quality of the wave function /149/.

The coupling constants for the isotropic and anisotropic hyperfine interaction as measured by NMR and ESR, probe spin densities at the nucleus. Mössbauer isomer shifts furnish information about the value of the wave function at the nucleus. The results for the relevant coupling constants, as calculated by the $X\alpha$-SW model, may be considered fair to satisfactory /147, 150/, especially if one takes into account that they are very sensitive tests of the wave function.

Very recently intensities of optical transitions ("oscillator strengths") /151/ and cross sections and angular distributions for photoemission /152, 153/ have been successfully calculated using the $X\alpha$-SW method. All these quantities invoke dipole matrix elements which may conveniently be calculated in the so-called acceleration form due to the local character of the $X\alpha$ potential. The continuum wave functions needed for the study of photoemission processes, can be generated quite easily within the SW formalism. Instead of the secular problem (104), one solves the system of linear equations (103) with inhomogeneities on the right hand side which are caused by the boundary conditions at infinity characterizing scattering evens /152/.

Due to its greater computational simplicity as compared to the HF-LCAO method and due to a higher accuracy than that of semiempirical methods, the SCF-$X\alpha$-SW method seems to be optimal up to now for the study of systems containing one or several transition metal atoms, like coordination compounds, organometallic compounds, metal clusters, impurities and defects in solids. Let us mention here a few representatives for the various systems: the permanganate ion MnO_4^- /20/, platinum tetrachloride $PtCl_4^{2-}$ /154/, Zeise's anion $Pt(C_2H_4)Cl_3^-$ /155/, nickel tetracarbonyl $Ni(CO)_4$ /156/, dirhenium octachloride $Re_2Cl_8^{2-}$ /157/, the cluster $[Fe_4S_4(SCH_3)_4]^{2-}$ as a model for the active center of ferredoxin /158/, the cluster $\square Nb_6C_{12}Nb_8$ as a model to study the possible effect of a carbon vacancy (□) on the superconductivity of NbC /159/ etc. There is also hope to enlarge our understanding of the electronic factors influencing chemisorptive and catalytic processes with the help of SCF-Xα-SW model calculations, especially the role played by transition metals /146/. In the next chapter, we shall take up some topics related to these interesting phenomena.

5. APPLICATIONS TO METAL CLUSTERS AND SURFACE PROBLEMS

Most industrial chemical processes are promoted by catalysts (mainly of heterogeneous type), so that intense scientific interest in their functioning does not come as a surprise. Yet, the diversity of the observed phenomena and their complex nature have so far prevented a detailed understanding of the fundamental processes involved. Large efforts are currently undertaken to uncover the physical basis of heterogeneous catalysis, especially with the help of recently developed methods of surface science /160, 161/. Also, great hopes have been expressed that quantitative information on catalytic systems and surface phenomena in general can be obtained from SCF-Xα-SW cluster calculations /146/. But before we delve into the details of such calculations, it seems appropriate to give a cursory characterization of the underlying problems and to evaluate from a general point of view the possible contributions of cluster calculations to their solution.

5.1 Clusters, Surfaces and Chemisorption

An atom at the surface of a solid may be considered intermediate between a free atom and an atom in the bulk because only part of its chemical valence is saturated. But its interaction with atoms or molecules impinging on the surface from the gas phase is not fundamentally different from that normally found between atoms or molecules. Some of these forces may be weak, like the van der Waals forces, causing only physical adsorption of the adatom. Due to the free valencies, surface atoms frequently form chemical bonds to incoming particles with binding energies in excess of 1 eV. In many cases energies as high as 3 to 4 eV are observed. This phenomenon, called chemisorption, may obviously be taken as the primary step of all surface processes.

However, a catalytic process is much more complex

in that it consists of many elementary steps, such as adsorption, interaction between and surface migration of adsorbates, breaking of bonds, forming of the reaction products and finally their desorption from the surface. Besides the electronic factors influencing the chemisorptive bonds and the structure of reaction intermediates, questions concerning the kinetics of the various elementary reaction steps and their interplay are of central importance for the understanding of a catalytic system. But here, the situation becomes truly discouraging if one considers the present status of the first-principle determination of the kinetics for simple gas-phase reactions /162/. Only atom-diatom reactions seem to be tractable in a quantitative fashion for many years to come. However, as one is not interested in a fully detailed description of an elementary step, statistical theories based on limited information which are applied to reactions in the gas and liquid phase with increasing success /163/, may prove equally useful for reactions on surfaces.

Even if one concentrates on the heterogeneous catalysis of metal surfaces and small metal particles dispersed on support, as we shall do in the following, the problems are enormous because the chemical behaviour of these "real" surfaces depends on numerous parameters which are hard to control, such as crystallographic orientation, structure defects, contaminations, influence of the supporting material etc.

Regarding the almost complete lack of structural information on catalytic systems, it is not surprising that the recent progress in surface science has come through the study of much simpler surfaces, mostly of metals /164-166/. Single-crystal surfaces with definite orientation, almost free from contaminating atoms, have been prepared under ultra-high vaccuum conditions with pressures of 10^{-10} torr and below. New experimental techniques have been developed to study these model systems providing detailed and reliable information. These experimental advances have in turn stimulated the formulation of new theoretical approaches to surface problems. Of course, there is a huge gap between these model systems and industrial catalysts, as is illustrated best by the pressure under which the latter are normally operated (10^3 torr!). Also, single-crystals are not very good catalysts. It is therefore not immediately clear how much the knowledge of well defined metal surfaces will contribute to our understanding of catalytic systems /161/. For some elementary processes, such as the chemisorption of atoms and simple molecules, certain features of the electronic structure are already becoming quite clear.

To describe a surface, we need the same type of information as for any solid or molecule. We must determine the identity of the atoms present, the geometrical arrangement of these atoms and the distribution of electrons surrounding these atoms both in energy and in space. We mention a few spectroscopic techniques /164, 166/ which give access to this information: geometry-sensitive experiments /167/, such as field-ion microscopy and low-energy electron diffraction (LEED), photoemission spectroscopy both at ultraviolet (UPS) and x-ray photon energies (XPS) /168/, core level spectroscopies, like Auger electron spectroscopy, appearance potential spectroscopy or ESCA, vacuum-tunneling spectroscopies based on field emission or ion neutralization, and desorption spectroscopies (thermal and electron stimulated). Very often it is quite difficult to give a quantitative interpretation of the observed spectrum because of the complexity of the system. For instance, the theoretical reconstruction of LEED spectra of overlayer systems have provided information on adsorption sites and bonding distances only recently /169/. The necessary computational efforts are quite substantial. We shall give several examples for the help afforded from theory in interpreting various photoemission spectra of chemisorption systems (see secs. 5.3 and 5.4). However, valuable insight into the structure of a surface system may be obtained even if these techniques (possibly in combination) are only used in a qualitative way as "fingerprinting" /170/.

It is the role of theory to design models and computational techniques in order to asses the main factors influencing the electronic and geometric structure of surface phenomena. Due to the interdisciplinary character of surface science involving both solid state physics and chemistry, the intention of the different theoretical approaches varies from models illustrating principal effects under restrictive simplifying assumptions to those describing rather specific situations.

The following discussion is not meant as a comprehensive review of recent, dramatic developments in the theory of chemisorption on metal surfaces /164-166, 171/. Rather we try to underline the basic physical concepts of the chemisorptive bond. Consider a neutral adsorbate atom /171a/, e.g. a hydrogen atom, with an occupied

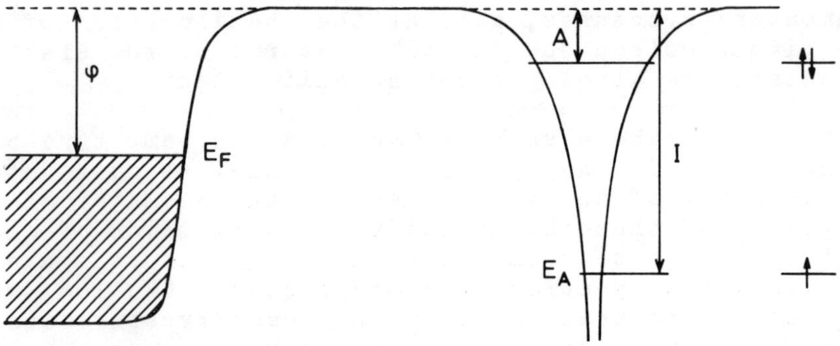

Fig. 19 Energy level diagram for the interaction of an atom with a metal surface.

level E_A as it approaches a metal surface (see Fig. 19). The metal may be characterized by a band of energy levels filled up to the Fermi level E_F. The work function φ is the amount of energy necessary to transfer one electron from the metal to infinity. Since the surface region is only a small fraction of the total solid, the energy bands are unaffected as far as their position in energy is considered. However, the wave functions of extended states are no longer Bloch waves because of the breaking of the translational symmetry due to the presence of a surface. On top of that, so-called surface states split off from the bulk energy bands. They show Bloch-like behaviour in the two dimensions parallel to the surface and decay towards the interior of the solid /178, 179a/.

The ionization potential I of the adatom is approximately given by $-E_A$. If its value is less than the work function φ of the metal, an electron will be transferred from the atom to the solid. Conversely, if the electron affinity A of the atom is greater than φ, an electron will move in the opposite direction. Only if $I > \varphi$ and $A < \varphi$, the atom is stable in its neutral state and the bond will be covalent, otherwise ionic. As a side remark, note that the electron affinity A of a system may be calculated within the Xα model by using a transition state where half an electron has been added to the highest energy level which is not completely filled. For an open shell system, the difference between the ionization and affinity energy is due to the Coulomb repulsion U between the valence electrons and the added electron. Hydrogen presents an extreme case with I = 13.6 eV and A = 0.7 eV so that $U \simeq I - A = 12.9$ eV. Since φ is typically 4 - 6 eV, hydrogen is expected to form a neutral bond.

As the atom approaches the metal surface, the atomic states are no longer exact eigenstates of the system, but acquire a finite lifetime τ as the electrons may tunnel between the adatom and the surface. The atomic level will consequently broaden with a halfwidth $\Delta \simeq \hbar/2\tau$. To put it in mathematical terms, this broadening of a discrete state of the adsorbate results from its interaction with the continuum of delocalized substrate states. The resulting orbitals will be delocalized over the whole system with energy shifts depending on the strength of the interaction and the relative position of the adsorbate levels with respect to the energy bands. A complicated self-consistent problem arises: the orbital mixing determines the charge on the adsorbate which through the intra-atomic Coulomb repulsion U will influence the effective orbital energy.

From these considerations we expect electron-electron correlation effects to be important. However, it turns out /179a/ that the HF approximation is valid if $\pi\Delta > U$ or if the separation of bonding and antibonding orbitals exceeds U. Fortunately, the value of U applicable to the free atom is considerably reduced near a metal surface by screening effects. These may be understood in terms of the image potential V_{im}. Neglecting other effects, the ionization potential $-E_A$ is decreased by the amount V_{im} (classically given by 1/4x where x is the adsorbate-surface distance) since the resulting ion interacts attractively with the metal through its image charge. Similarly, the electron affinity A is increased by V_{im} through the attractive interaction of the negative ion with the metal. The effective Coulomb repulsion therefore becomes

$$U_{eff} = U - 2V_{im}. \tag{139}$$

One should note, however, that a quantum treatment of the image interaction gives results which differ significantly from the classical value at small metal-ion separations /180/. From this discussion, one may conclude that hydrogen adsorption may be rather difficult to treat due to the large Coulomb repulsion U. For other atoms, and even more for molecules, U is much reduced and the selfconsistent field approximation is expected to work quite well.

The first quantitative theory for the chemisorption of hydrogen was proposed by Newns /173a/ using the so-called Anderson hamiltonian. This model is quite simple-non-interacting electrons in energy bands of the metal, interacting electrons in localized states on the adsorbate with suitably chosen parameters. Therefore, only qualitative investigations can be carried out. The model has been extended to include several adsorbate levels and energy bands to describe the chemisorption of molecules on transition metals (e.g. Ni/CO /173b/). But the uncertainty of how to choose the parameters remains in this semi-empirical method. Nevertheless it serves as an important test case for calculations beyond the HF approximation /173c/.

The model has further been used to answer the important question how localized in space a chemisorption bond is. For the case of a strong adsorbate-substrate interaction the results are actually quite simple to interpret /179a,b/. Sharp states are split off from the top and the bottom of the band corresponding to the bonding and antibonding states of a "surface complex". Einstein /174/ has carried out calculations for the change of the local density of states for the case of an atom adsorbed on top of an atom of the (100) face of cubium, a simple cubic s-band solid. These calculations show that for large interaction, a surface complex is essentially decoupled from the remaining "indented solid". The changes in the local density of states decay rapidly as one moves away from the adsorbate (say to the next-nearest neighbours), not only in the surface plane, but also into the solid.

Let us illustrate the resulting main features of the electronic interactions for the chemisorption at a transition metal surface as they will be the systems of our main interest (see Fig. 20). The density of states (DOS) characterizing the transition metal has two rather different contributions. The highly peaked DOS of the localized d orbitals is overlapped on both sides by the rather wide s-band. The Fermi level falls within the d-band if we take Ni as an example. The adsorbate-substrate interaction may then be switched on consecutively, for the sake of the argument. Within the "surface molecule", the adsorbate level at energy E_A interacts strongly with the localized d orbitals of the neighbouring metal atoms producing a (filled) bonding level below the d-band and an (empty) antibonding level above. These "surface molecule" levels are in turn broadened (and slightly shifted)

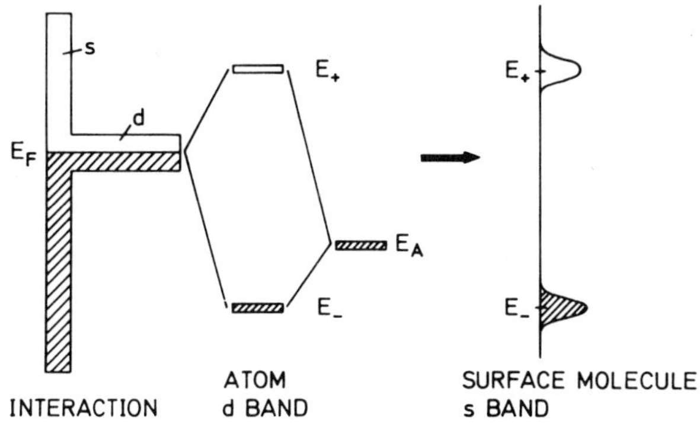

Fig. 20 Energy diagram for an adsorbate orbital E_A interacting strongly with a transition metal surface (schematic).

through their interaction with the continuum of the s-band.

The simple model calculations mentioned above justify the "surface molecule" approach to chemisorption which has been advanced in the chemical literature since long /175/. This approach is very attractive in that it reduces chemisorption to the electronic structure problem of an adsorbate-substrate cluster of sufficient size permitting the application of conventional quantum chemical methods. This cluster approach is becoming very popular in some quarters of quantum chemistry and virtually all methods have been tried: Extended Hückel theory (EHT) /176/, CNDO /177/, HF-LCAO /178/ and the Xα model implemented with the DV method /179/ and the SW formalism /180/.

In the remaining sections, we shall have more to say about how useful SCF-Xα-SW cluster calculations are in understanding surface phenomena. Certainly, if a cluster included enough metal atoms - how many there have to be is an open question - one would get an accurate treatment of the chemisorption problem. Actual clusters, however, include only the nearest-neighbour metal atoms of the adsorbate. In this way, the adatom "sees" its proper surrounding, but this is not true for any of the metal atoms of the cluster. Furthermore, as we have seen from the discussion of Fig. 20, even if we treat the short-range bonding forces properly, there remains the problem that the computed cluster states are not stationary states of the extended system. Charge transfer between the surface cluster and the infinite reservoir of the metal will occur, but is inhibited in a cluster calculation. This may lead to serious deviations from selfconsistency.

Of course, these deficiencies are amended if we attach the "surface molecule" to a properly indented metal surface. This process destroys the selfconsistency already existing in the isolated cluster, and several methods have been proposed to solve this "embedded cluster" problem /181/. Most of them use the LCAO tight-binding formalism to describe the metal. In practice, however, cluster calculations are scarce up to now /181a,e/ and of highly approximate character. For instance, only one metal atom had been included in the surface molecule.

To be sure, an accurate treatment of chemisorption would best combine the local bonding approach with the delocalization effects provided by the metal. As long as no accurate procedure applicable to this effect is available, however, we feel that it is worthwhile to use calculations based on the local bonding approach in a small cluster to seek and gain additional insight into the bonding properties. While errors will be inherent in such a treatment - especially when the collective participation of the metal in bonding and hole creation (level broadening and relaxation) are important - these errors should not be prominent if the trends in a series of systems are investigated instead of trying to make absolute comparisons for a single system. In this way, the "metal" effects which will not change too much from system to system will essentially drop out, the remaining differences should be interpretable in a local picture. For an example of such a study, see sec. 5.4. In any case it appears to us that cluster calculations with a larger number of metal atoms treated for instance with the Xα-SW method may give a more realistic picture of the local electronic structure than simplistic models relying on a sophisticated formalism (see secs. 5.2 and 5.4). This might be true all the more if one wants to describe the electronic structure of the active site on the "rough" surface of a catalyst.

For single crystal surfaces with a periodic overlayer of adsorbates, the whole problem may be circumvented by applying the recently formulated SCF-Xα-KKR band structure formalism /182/ for a film geometry. No applications to the chemisorption problem have yet appeared.

Let us conclude this review by mentioning two other approaches to chemisorption. In cases of a large intra-atomic Coulomb repulsion U, correlation of the electronic motion will play a dominant role. An analogue of the valence-bond theory, where bonding is brought about by spin-pairing, might be more appropriate /183/. A disadvantage of the resulting, rather involved formalism is that it abandons the concept of one-electron levels which has proved so useful in the interpretation of many spectroscopic data.

The density functional theory /14/ has also been applied to surface problems and chemisorption /184/. The metal is represented by a half space with uniform positive background. This may serve as a fair model for free-electron metals, but certainly will not reproduce the characteristics of transition metal surfaces. The first calculation was performed by Smith, Ying and Kolm /194/ for the case of H chemisorption using linear response theory to treat the proton-substrate interaction. The most important point to note in connection with this calculation is that, as the proton is moved away from the metal surface, the electron stays on the metal. However, from experiment it is known that a neutral hydrogen atom is desorbed. This spurious result has been attributed to the inability of linear response theory to provide any bound state for the electron in the hydrogen atom far from the surface /181b/. An exact calculation for the chemisorption of H, Li and O /184c/ is free from this defect. Quite resonable agreement between measured and calculated dipole moments and binding energies for Na, K and Cs have been obtained in the linear response formalism /184b/. In principle, the density functional formalism can be extended to other cases, although actual calculations will undoubtedly be quite difficult.

5.2 The Electronic Structure of Transition Metal Clusters

The electronic structure of small metal clusters are intrinsically interesting in the way they relate to the bulk band structure and surface states of the corresponding solid. Considerable current interest derives from the fact that the active centers of commercial heterogeneous catalysts often consist of small metallic clusters, supported on porous material such as silica /146/. However, these cluster, as small as 10 Å in size, are still quite large in that they contain typically 100 - 1000 metal atoms. The transition metal clusters studied so far with

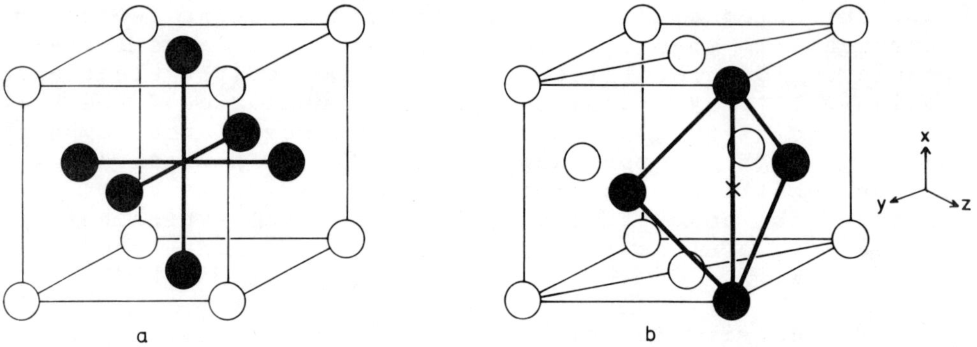

Fig. 21 Schematic representation of an fcc lattice.
a) The octahedral model cluster (shaded spheres).
b) The four atom cluster modeling a (110) substrate surface. (Also indicated is the coordinate system used in the Ag_4O cluster calculations, see sec. 5.4).

with the Xα-SW method consisted of up to 13 atoms. Here, we are interested in such clusters mainly because they can form models for the various substrate surfaces in local chemisorption studies.

Let us begin our MO studies of transition metal clusters by the Xα-SW method with Ni_6, Cu_6 and Ag_6 clusters of octahedral symmetry /195/. The atoms sitting in the middle of the cube faces of an fcc unit cell form an octahedron with the bond length equal to the nearest neighbour distance (see Fig. 21a). The model cluster to be used later on for the (100) surface (see sec. 5.4) may be derived from this octahedron by removing the "top" atom, whereas the model cluster for the (110) surface originates by removing two adjacent atoms (Fig. 21b). The bond distances were chosen as in the corresponding bulk metals: 2.49 Å, 2.54 Å and 2.89 Å for Ni, Cu and Ag, respectively. The sphere radii of the various spheres are completely determined by symmetry and the condition of touching spheres. For close-packed clusters of the type investigated in this section, the muffin-tin approximation has proven fully satisfactory, and the orbital energies do not change appreciably with moderate amounts of sphere overlap. This is consistent with the criteria for using overlapping spheres /116, 120/.

The resulting electronic energy levels for the M_6

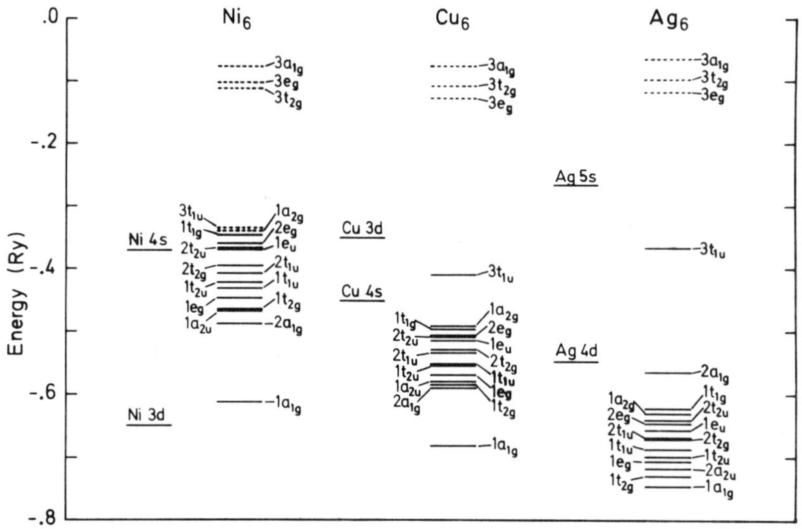

Fig. 22 Comparison of SCF-Xα-SW orbital energies for different octahedral metal cluster M_6 (M = Ni, Cu, Ag). Full lines represent occupied orbitals, dashed lines empty ones. The SCF-Xα orbitals for the various free atoms are also shown.

clusters (M = Ni, Cu, Ag) are shown in Fig. 22 together with the Xα atomic levels of Ni $3d^8\,4s^2$, Cu $3d^9\,4s^2$ and Ag $4d^{10}\,5s^1$. The d orbitals of six atoms at the corner of an octahedron generate the irreducible representations $a_{1g} + a_{2g} + a_{2u} + 2e_g + e_u + t_{1g} + 2t_{2g} + t_{2u}$ of the group O_h, whereas the s orbitals generate levels of $a_{1g} + e_g + t_{1u}$ symmetry. For Ni_6 we find that the d levels form a dense manifold ($2a_{1g}$ - $1a_{2g}$), bounded below and above by the $1a_g$ level at -0.613 Ry and the $3t_{1u}$ level at -0.335 Ry which are predominantly s- and p-like, but also contain some d-character. Of the high lying levels, the $3e_g$ level at -0.103 Ry may be formally counted as the one generated by the Ni 4s orbitals. These levels are filled with 10 valence electrons per Ni atom up to the $1t_{1g}$ level at -0.346 Ry which we may therefore identify as the Fermi level of the cluster. The results for the Cu_6 cluster are very similar to Ni_6 except that the d level cluster is strongly shifted downwards and that the $3t_{1u}$ Fermi level (filled with four electrons) lies well above at -0.411 Ry, due to the increased number of electrons. We refrained from performing spin-polarized

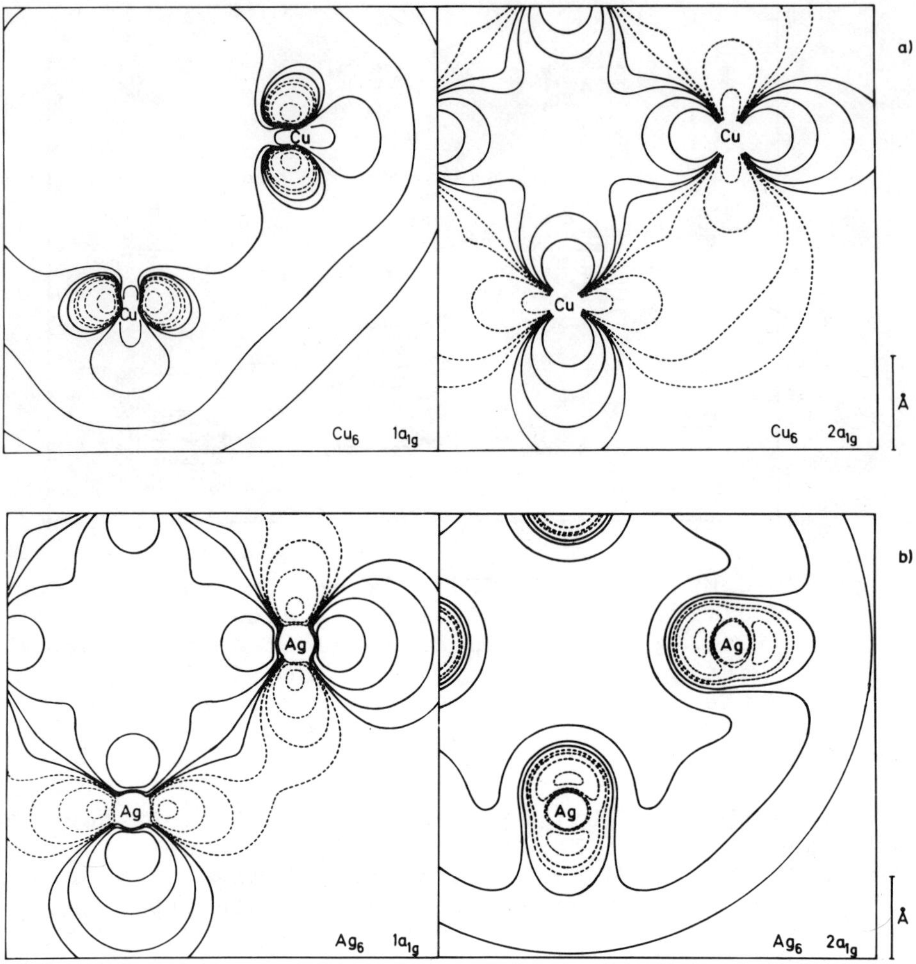

Fig. 23 Contour plots of the $1a_{1g}$ and $2a_{1g}$ orbital wave functions of a) Cu_6 and b) Ag_6. Different signs of the orbitals are represented by full and dashed lines. Note the different scales!

calculations because the consequences for the bonding picture are negligible in this simple model.

When going from Cu_6 to Ag_6 the energy level spectrum changes significantly in that the cluster of d-like levels drops below the lowest s-p level $2a_{1g}$ (see Fig. 22). The most strongly bound level is again of a_{1g} symmetry, but now it belongs to the "bottom of the d-band". To illustrate this change in electronic structure, we show contour plots of the $1a_{1g}$ and $2a_{1g}$ levels, both for Cu_6 and Ag_6, in a plane containing four atoms and the center of the octahedron (Fig. 23). The change in the $1a_{1g}$ level from an s-type level with d admixture in Cu_6 to an almost pure d level in Ag_6 shows up clearly. Both orbitals, the $1a_{1g}$ level in Cu_6 and the $2a_{1g}$ level in Ag_6, show small contributions from d-waves, but their dominant s-character is obvious, as one does not find a nodal plane when going around one atom at a sufficient distance, say the muffin-tin radius.

Although this qualitative discussion of the energy level spectrum furnishes valuable insight into the electronic structure of these transition metal clusters, it is desirable to have a quantitative measure for the s-d interaction at hand when comparing with bulk electronic band structures of these materials. The SW population analysis as mentioned in sec. 4.1 is ideally suited for this purpose. As a measure of the contribution of a specific atomic orbital to the molecular orbital under consideration we define

$$Q_p(\ell) = \sum_m Q_p(\ell,m) \tag{140.a}$$

where (cf. eq. (68)):

$$Q_p(\ell,m) = 4\pi\left[C^p_{\ell m}\right]^2 \int_0^{b_p} \left[R^p_\ell(r)\right]^2 r^2 dr \tag{140.b}$$

The "partial wave" charges $Q_p(l)$ add up to the total charge Q_p inside an atomic sphere, as they should (cf. eq. (120)).

For clusters with a large number of energy levels it is very convenient to present information like the SW populations as density-of-states (DOS) profiles which permit a close comparison with bulk DOS without sacrificing the MO interpretation. If ψ_i is an energy eigenfunction with eigenvalue ε_i and $\varphi_{p,\ell}$ a localized atomic-like wave function of symmetry type l centered on the site p, then the projected or local density of states is defined by

$$N_p(\varepsilon,\ell) = \sum_i |\langle\varphi_{p,\ell}|\psi_i\rangle|^2\, \delta(\varepsilon-\varepsilon_i) \tag{141}$$

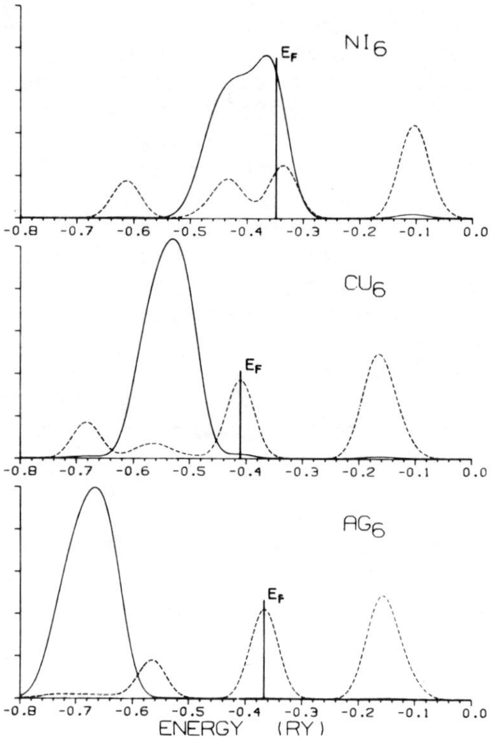

Fig. 24 Population densities for octahedral clusters M_6 (M = Ni, Cu, Ag). Dashed lines: sum of s and p population (x5); solid lines: d population.

By "broadening" the discrete eigenvalue spectrum of a cluster with a set of gaussians we obtain an analogous "population" density of states of the form

$$N_p(\varepsilon,\ell) = \sum_i \frac{n_i}{\sqrt{2\pi\sigma}} Q_p(\ell,i) \exp\left(-\frac{(\varepsilon-\varepsilon_i)^2}{2\sigma^2}\right). \qquad (142)$$

In this expression, ε_i is the orbital eigenvalue, n_i the occupation number, $Q_p(l, i)$ the SW population defined in (140.a) (with a suitable index referring to the orbital i) and σ is an appropriately chosen broadening parameter. One may interpret this procedure as simulating the effect of adding more shells of neighbour atoms to the cluster thereby generating new energy levels between the already closely spaced eigenvalues. Of course, for small clusters like the present ones, no meaning should be attached to the finestructure of the spectrum and therefore a broadening parameter $\sigma = 0.025$ Ry was chosen. For completeness, let us mention that the expression for the total DOS, N(ε), of a cluster to be used later on may be derived from (142) by formally putting $Q_p(1,i) \equiv 1$.

The resulting population densities for the octahedral clusters are displayed in Fig. 24, separately for the sum of the s- and p-orbital populations and the d-orbital population. The similarity of the spectra to a density of states is quite striking: a high and narrow peak due to the strongly localized d-like states and several small contributions of s- and p-like states spread over the energy scale. Continuing the analogy with the band structures of bulk metals, one may therefore regard the levels from $2a_{1g}$ to $1a_{2g}$, taking the Ni cluster as an example, as a precursor of the d-band and may attribute the remaining levels to the "s,p-band". As in the bulk, one finds the s,p-band overlapping the d-band and a strong interaction between the two kinds of bands (see Fig. 24).

For ease of reference, we include the energy band structure of Ni, Cu and Ag along one direction of k-space, as obtained from APW and KKR calculations /186/ which also employ the Xα exchange approximation (see Fig. 25).

The same conclusions hold for the Cu_6 cluster and bulk cupper; however, both the interaction and the overlap of the two bands are clearly reduced and the d-band

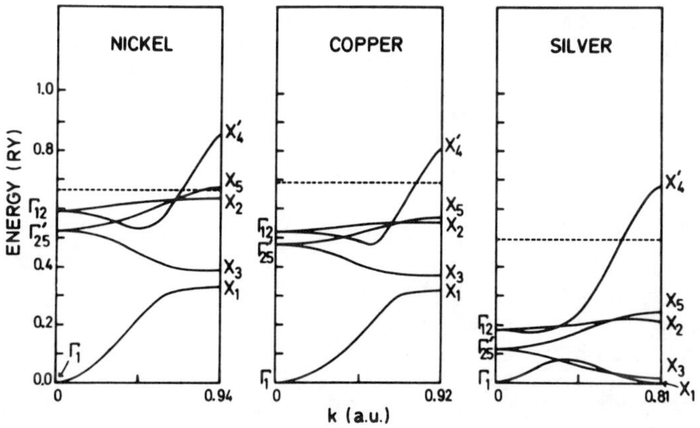

Fig. 25 Bulk energy band structure of Ni, Cu and Ag along one direction kf "k-space" from APW and KKR calculations. The bands have been adjusted such that the lowest band at Γ_1 has the same energy value.

Table XIII Comparison of energy differences for octahedral clusters and bulk energy bands (all energies in eV)

	d-band width	s,p-band width	bottom s,p- to top d-band	top d-band to E_F
cluster				
Ni	2.0	6.9	3.7	-0.1
Cu	1.3	6.9	2.6	1.1
Ag	1.6	5.4	-0.9	3.6
bulk				
Ni[a,h]	4.9	11.4	7.4	-0.02
Cu[b]	3.3	10.8	5.7	1.8
Ag[c]	4.1	7.9	1.9	4.1
expt.				
Ni[i]	3.0[d];3.3[e]	...	...	-
Cu	3.0[f];3.1[d]; 3.2[g]	...	...	1.9[d,g];2.0[f]
Ag	3.2[f,g];3.5[d]	...	...	3.9[d,f];4.0[g]

a-g: References 186a,b,c, 187, 188a,b,c.
h: Values for the minority spin.
i: Only the filled part of the d-band is observed by XPS and UPS.

is completely filled. This semiquantitative agreement between the electronic structure of the bulk and a finite cluster may also be carried over to silver. In both systems, the d-band is found entirely below the s-p-band, with the distance to the Fermi level significantly increased as compared to copper.

In Table XIII we compare the widths of the d- (X_5-X_1) and the s,p-band (X_4'-Γ_1) and the distances from the top of the d-band to the bottom of the s-p-band (Γ_{25}'-Γ_1) and to the Fermi level (X_5-E_F) for the bulk and the corresponding quantities for the cluster of the three materials, Ni, Cu and Ag /186/. For Ni, both in Table XIII and Fig. 25 the values for the bands of minority spin had been taken /186a/. Also given in Table XIII are experimental values where available. They were obtained from XPS /187/ and UPS /188/ measurements, the latter in the energy range from 26.9 eV to 40.8 eV to exclude the difficulties inherent in the interpretation of He(I) spectra. The energy differences for the clusters are, as expected, much smaller than those from band structure calculations. The only exception is the distance from the top of the d-band to the Fermi level for Cu and Ag. A direct comparison of bulk energy bands with the discrete energy spectrum of a finite cluster is only possible in a limited sense.

Firstly, clusters of six atoms are certainly too small for any quantitative agreement with bulk values to be expected; we shall turn to larger clusters in a moment. Furthermore, the DOS may be taken to represent the photoelectron spectrum only to first order as matrix elements are completely neglected. This may be seen from the dependence of the spectrum of bulk Ni on the photon energy /186b/. Dipole matrix element effects may also be responsible for the difference between the calculated width of the Ni d-band of ~ 4.5 eV (occupied part only) and the measured value of ~ 3 eV /188b/. On the other hand, Messmer, Johnson et al. /189/ suggest that the narrow optical d-band DOS for Ni can be explained by the fact that in UPS measurements only two to three surface layers are sampled. They note that the d bandwidth of the Ni_{13} cluster is in relatively good agreement with the UPS data and conclude that a "localized" cluster description might be a more realistic representation of the initial state of d-band photoemission from surfaces than the delocalized Bloch representation of bulk band structure theory.

Despite these restrictions, the good correlation of all energy differences, from Ni to Cu to Ag in Table XIII is quite striking. For instance, the width of the d-band is largest for Ni, smallest for Cu, both for the cluster and for the bulk band structure. This important result shows that already a cluster of six atoms reflects in essence the electronic properties of the corresponding metal. We are therefore confident that clusters of this size are sufficiently large to perform a reliable comparative study of the oxygen chemisorption on surfaces of these metals. This will be the subject of sec. 5.4.

Similar SCF-Xα-SW calculations have been carried out by Messmer, Johnson et al. for 13-atom clusters of copper, nickel, palladium and platinum having cubo-octahedral geometry /189/. This structure as shown in Fig. 26 corresponds to the local arrangement of atoms in the fcc metals. Note that the central atom of this cluster has the same nearest-neighbour environment as a metal atom in the bulk

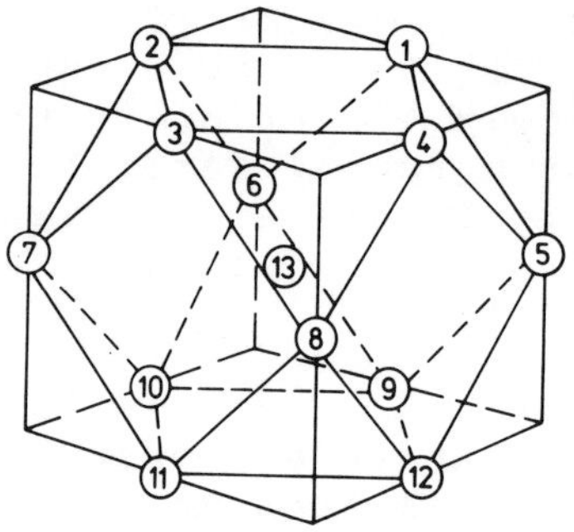

Fig. 26 Cubo-octahedral cluster containing 13 atoms.

and that the surface planes of the cluster may be classified as (100) and (111).

Let us look more closely at the series of copper clusters with two, six and thirteen atoms, all with nearest-neighbour distances equal to the bulk crystalline value. In Fig. 27, the resulting energy level spectra are compared to the DOS for bulk Cu, as calculated by the SCF-Xα-KKR band structure method /186b/. Since the approximations both in the KKR and the SW method are very similar, the comparison is truly consistent, except for the fact that the top of the d-band had to be roughly aligned with the corresponding levels of the Cu_{13} cluster because of the arbitrariness in the zero of energy in a band structure calculation. At first sight, the overall downward shift of the d-like orbitals from Cu_2 to Cu_{13} seems to prevent a close agreement with photoelectron spectra. However, this discrepancy is significantly reduced when comparing transition states which include the proper relaxation. This shift decreases with increasing cluster size, approaching zero in the limit of the infinite crystal. The ionization potential for the highest occupied orbital of a Cu_{13} cluster as calculated by the transition state procedure is 7.3 eV which may be compared to the work function, 4.7 eV, of bulk Cu. Comparing the cluster energy levels with those for the isolated Cu atom in the two configurations $3d^9 4s^2$ and $3d^{10} 4s$ as shown in Fig. 27, a rationalization of the changing position in energy of the d-like ground state levels may be given in terms of the relative occupancy of the d orbitals. The higher the occupation of an orbital is, the higher its energy will be (see Fig. 3). As a consequence of the

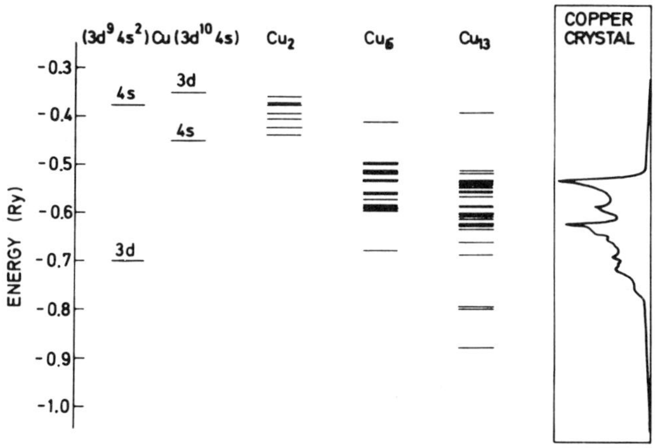

Fig. 27 Comparison of occupied SCF-Xα-SW electronic energy levels for Cu_2, Cu_6 and Cu_{13} with the electronic DOS of bulk Cu calculated by KKR band-structure theory (Ref. 186b). Also shown are the SCF-Xα energy levels for the isolated Cu atom in two configurations.

increasing overlap of the s- and the d-"band" when going from Cu_2 to Cu_{13}, one will find an increasing occupation of the 4s orbital of Cu and concomitantly a decreasing occupancy of the d orbitals. Inspection of the atomic energy levels for the two different configurations as well as similar trends in nickel clusters support this interpretation.

The clustering of the energy levels and the band widths of the d orbitals in Cu_{13} closely match the main features of the d-band DOS for bulk Cu already much better than the Cu_6 cluster considered previously (see Fig. 27). Delocalized s,p-like Bloch conduction-band states of crystalline Cu are responsible for the low-density tails in the DOS for bulk Cu at both sides of the d-band peaks. There is a close correspondence of these states to delocalized s,p-like cluster levels of Cu_{13} in the same energy range. The lowest unoccupied s,p levels occur just above the Fermi energy (at ~ 0.350 R_y). The energy for electronic transitions to these levels from the top of the d band in Cu_{13} occurs in the same energy range (2.0-2.4 eV) as interband transitions which are responsible for the characteristic color of solid copper. The analogy of the electronic structure of Cu_{13} to that of bulk, however, should not be pushed too far as significant cluster effects show up both in the level spectrum and the shape of the orbitals. The three lowest levels well below the manifold of closely spaced d levels (see Fig. 27) which are of symmetry a_{1g} (-0.879 Ry), t_{2g} (-0.800 Ry) and e_g (-0.793 Ry), are to a very large extent localized on the central atom. The charge fractions in the central sphere are 24%, 84% and 85%, respectively. This localization shows that the electrons experience a potential well near the central atom which is much deeper due to the overlap effect from the neighbour atoms. All three orbitals are strongly bonding between the central atom and the outer atoms of the cubo-octahedron. The t_{2g} and the e_g level may accordingly be considered as the d orbitals of the central atom. They have therefore no counterpart in bulk Cu, but are artefacts of the cluster.

Interesting magnetic effects may be observed for the series of nickel cluster /189/: Ni_6, Ni_8 and Ni_{13}. All levels up to $1t_{2g}$ (cf. Fig. 22) are completely filled in Ni_6 with no net spin. For a Ni_8 cluster in the geometry of a simple cube, a "paramagnetic" electronic structure with four net spins results from a spin-unrestricted SCF-Xα-SW calculation. Similarly, a paramagnetic ground state is found for Ni_{13} with six unpaired spins in the two topmost occupied levels. The exchange splitting for d-like orbitals of this cluster was found to be approximately 0.015 Ry which may be compared to values between 0.04 and 0.06 Ry determined at corresponding symmetry points of the crystalline energy bands. The corresponding spin magneton numbers per atom are 0, 0.25 and 0.46, respectively. These values may be compared to those characteristic of bulk ferromagnetic nickel (experiment: 0.56, band structure calculation: 0.58 /190/). With regard to these results, it is interesting to note that paramagnetic magneton numbers for small nickel crystallites have been found to decrease continuously from the bulk value as the crystallite size is decreased from 100 to approximately 10 Å /191/.

Xα-SW calculations have also been carried out on 13-atom cluster of Fe, Ni and Cu with an arrangement of the atoms chosen to represent two layers of a (100) surface /192/. However, no comparison is possible with the results discussed above as these calculations are not selfconsistent.

Besides the SCF-Xα-SW method, the extended Hückel (EH) /176, 193/ and the CNDO /194/ method have been used to study the electronic structure of transition-metal clusters and the interaction of these clusters with adsorbates. In this context the question arises how well the various methods succeed in representing the substrates for chemisorption. A further point is how many atoms should be included in the cluster model. Messmer, Johnson et al. /189/ have carried out a very informative comparison of the various computational methods for the cubo-octahedral cluster Ni_{13}. They chose the total DOS as criterion for their comparison generated from the discrete spectrum of the cluster with a gaussian broadening σ = 0.01 Ry, to give more resolution than in Fig. 24. The DOS profile resulting from a non-spinpolarized Xα-SW calculation is compared in Fig. 28 to majority-spin DOS for crystalline nickel, as calculated with an SCF-Xα-LCAO band-structure method by Callaway and Wang /190/. There is rather satisfactory agreement between both profiles, especially regarding the fact that the Fermi energy intersects in both cases the highly peaked top of the d band. In the cluster Ni_{13}, the sharp peak in the DOS near the Fermi level is due to antibonding and nonbonding d orbitals whereas the broad peak around 0.1 Ry corresponds to largely bonding d orbitals. The peaks near 0.25 and 0.4 Ry are caused by the strongly bonding MO's which have a dominant localization on the central atom (cf. the above discussion of Cu_{13}). These results strongly suggest that the principal features of the band structure of crystalline nickel are determined by short-range order which may be well represented by the 13-atom cubo-octahedral cluster. Convergence to bulk data may probably be reached with clusters containing between 25 and 50 atoms.

Also shown in Fig. 28 are the DOS profiles for the same cluster generated from the results of EH and CNDO calculations. In the EH calculation a double zeta Slater-type basis was used to represent the d orbitals /176c/. The CNDO parameters /194b/ were chosen by trying to match bulk quantities such as d bandwidth, Fermi energy and binding energy per atom for an octahedral Ni_6 cluster. The energy scales of the various calculations have been shifted so as to line up the bulk Fermi level with the highest occupied orbital of the cluster. For the EH case, one finds a highly peaked DOS near the Fermi energy but the d bandwidth is much narrower than in the bulk. The s-d hybridization which is responsible for the proper configuration of the Ni atom in bulk material and for the long tail of the DOS towards lower energies is practically absent in the EH cluster results. For the CNDO calculation, virtually no agreement is found with

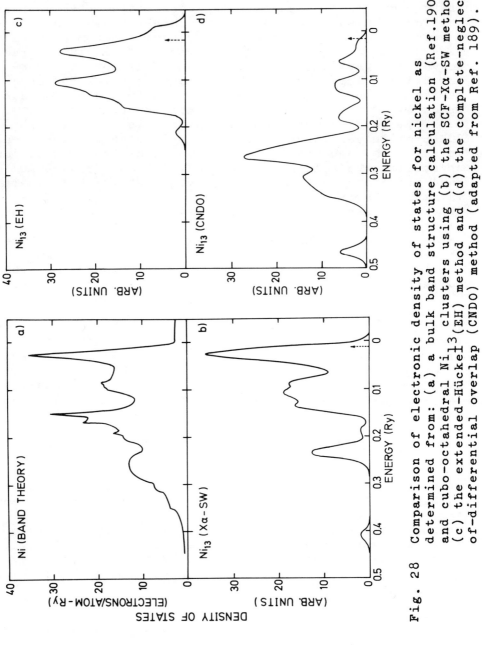

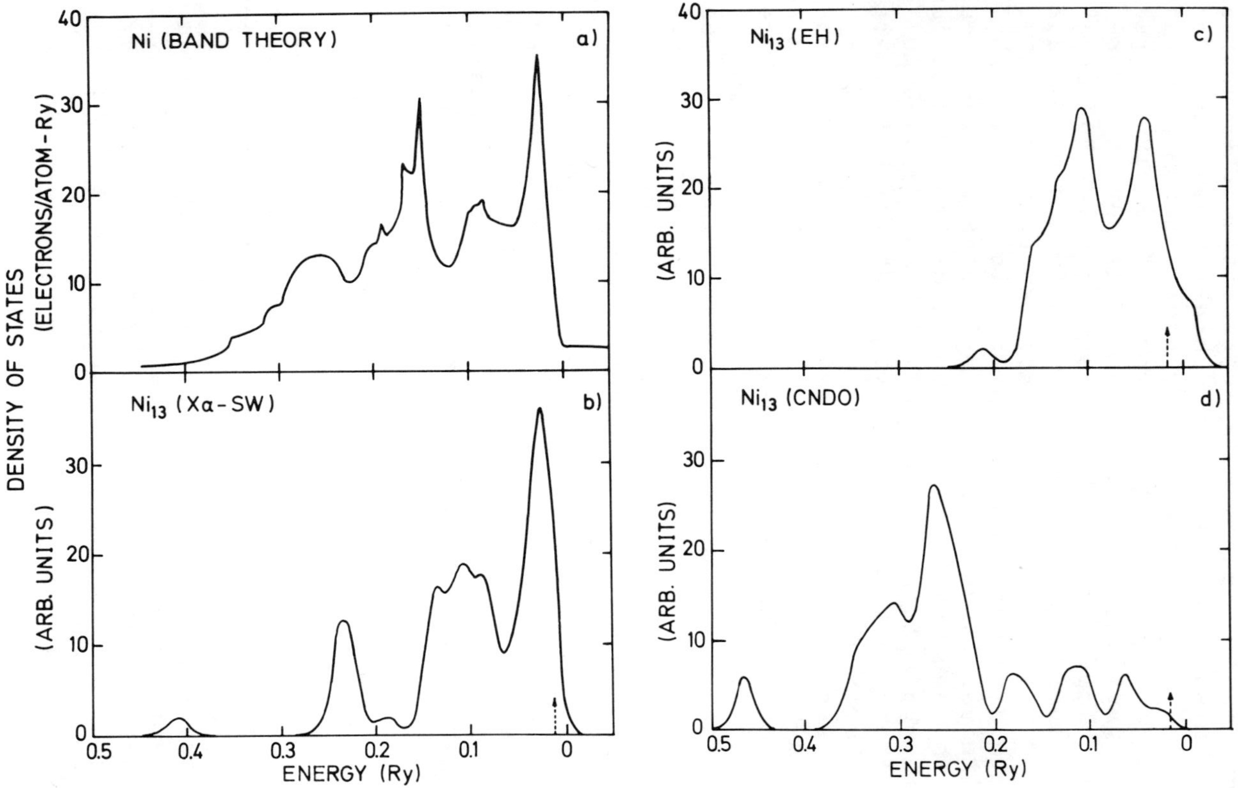

Fig. 28 Comparison of electronic density of states for nickel as determined from: (a) a bulk band structure calculation (Ref.190), and cubo-octahedral Ni_{13} clusters using (b) the SCF-Xα-SW method, (c) the extended-Hückel (EH) method and (d) the complete-neglect-of-differential overlap (CNDO) method (adapted from Ref. 189).

the results for the bulk. The high peaks near the Fermi energy are missing in the DOS and appear in an energy range where the bulk DOS is strongly decaying.

From this discussion, we see that the SCF-Xα-SW method seems to provide by far the best results for the electronic structure of transition metal clusters. This success is not accidental as can be seen from the other EH calculations on clusters of Cu, Ag and Pd with up to 55 atoms /193/. Even for clusters of 19 atoms they calculate d bandwidths of 0.2 to 0.3 eV whereas Xα-SW clusters of six atoms give values of 1.3 to 1.6 eV (see Table XIII) and clusters with thirteen atoms reach values of the order of several eV, close to experimental d bandwidths. Messmer, Johnson <u>et al.</u> trace the discrepancies in EH calculations to the arbitrariness of the parametrizations used. They point out that the EH method has not been successful in the determination of band structures for transition metals in contrast to the SCF-Xα-KKR method. It should be noted, however, that the success of the EH method is established for investigations on the band structure and finite cluster models of certain semiconductors /195/.

Let us mention here several other Xα-SW calculations on the electronic structure of related clusters. An extensive study on clusters containing up to 13 Li atoms was carried out by Fripiat <u>et al.</u> employing the spin-polarized version of the Xα-SW model /103/. They determined binding energies per atom and equilibrium bond lengths and compared the results to the corresponding bulk quantities. (We recall from sec. 4.2, that fair agreement with experiment had been obtained for these quantities in Li_2). They compared two Li_{13} cluster of different structure: an icosahedron with an atom at the central position and a cubo-octahedron. The former structure is more stable (by 1.81 eV) and has slightly shorter bonds (3.23 Å; bulk: 3.04 Å). No DOS profiles have been given.

One might wonder, in fact, how well the electronic structure, and correspondingly the DOS of free-electron metals might be approximated by cluster models. Calculations on aluminum clusters containing 5, 9 and 25 atoms have been performed recently /196/. The energy difference between the Fermi level and the lowest orbital is 0.50, 0.64 and 0.77 Ry, respectively. This may be compared to the bandwidth of 0.82 Ry from an Xα-LCAO band structure calculation. The overall shape of the DOS profile for the 25-atom cluster follows quite closely a curve proportional to $\varepsilon^{1/2}$ as appropriate for free-electron metals.

Recently, a cluster of NbC, a partially covalent material with sodium chloride structure, has been studied /159/. The cluster, $CNb_6C_{12}Nb_8$, contains a carbon atom at the center with three complete shells of neighbour atoms. The C-C_{12} substructure is identical to the cubo-octahedron studied above. The resulting DOS profile agrees remarkably well with the corresponding results of an APW band structure calculation where identical parameters (exchange parameters α, muffin-tin sphere radii) had been used.

5.3 Chemisorption of Carbon Monoxide

We now take up the discussion of Xα-SW cluster calculations relevant to the chemisorption problem. From considerations in sec. 5.1 we know that the surface molecule approach should provide a reasonable description in cases of strong adsorbate-substrate interactions. Drawing from the experience with applications of the Xα-SW method to molecular systems, which surface problems do we expect to be amenable to Xα-SW cluster calculations? Questions concerning the energetically most favourable site of adsorption, the equilibrium distance of an adsorbate to the topmost atomic layer of the surface or the heat of adsorption will not receive a direct quantitative answer, as they would imply very accurate values for the total energy. Furthermore, models with fairly small metal clusters will not be able to reproduce (or predict) trends of the above quantities with the orientation of the surface plane ((100) vs. (111) etc.). We shall have to rely on the information provided by the cluster level spectrum and the orbital wave functions, and we expect useful contributions of Xα-SW cluster calculations to the interpretation of photoelectron (PE) spectra of chemisorbed species.

Since Eastman and Cashion /197/ in 1971 showed for the first time conclusively that information on electronic energy levels of chemisorbed atoms and molecules (oxygen, CO on polycrystalline nickel films) could be obtained from UPS experiments, a large amount of experimental and theoretical work has been spent to relate these spectra to the chemisorption problem. In such an experiment, one typically compares the PE spectrum of a clean surface to that recorded after increasing exposure to the adsorbate gas and one tries to interpret the difference spectrum. This is certainly a more difficult task than for spectra obtained from the adsorbate gas alone, as a number of interferences between substrate and adsorbate emission can take place. For higher photon energies (e.g. HeII), the assumption of emission into a free electron state can be expected to be quite good. Otherwise, for upper states too close to the vacuum level, the density of the final states might introduce additional structure. In Fig. 29, the HeII PE spectrum for a Ni (100) surface is shown after exposure to 0.6 L of CO (1 L = 1 Langmuir = 1 x 10^{-6} Torr sec) and besides the main emission peak due to the Ni d-bands two additional peaks P_1 and P_2 show up, approximately 8 eV and 11 eV below the Fermi energy E_F. It is therefore quite obvious to interpret these additional features as adsorbate energy levels. A quantitative discussion of this adsorbate density of states, as obtained from the corresponding difference spectra, however, is not straightforward because frequently negative peaks are observed (see for example sec.5.4). Furthermore, the intensity of photoemission from any one peak is strongly affected by the various angular variables of the experiment (electron take-off angle, angle between incoming light and outgoing electrons, between incoming light and surface, polarization of the light).

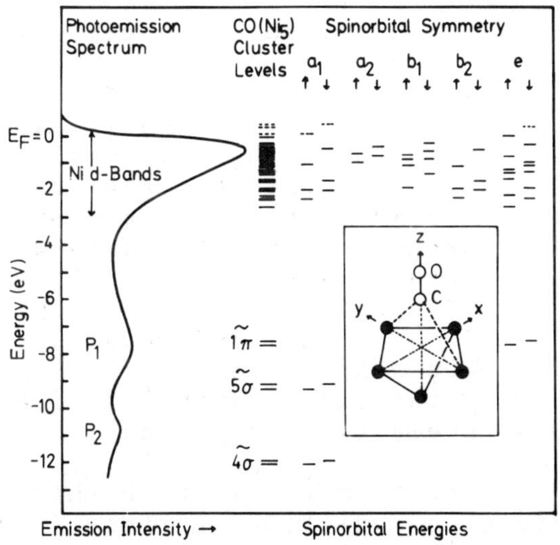

Fig. 29 Comparison of theoretical spin-unrestricted energy levels for the $CO(Ni)_5$ cluster and experimental HeII photoemission spectrum for CO adsorbed on Ni (100). Occupied levels are denoted by solid lines, unoccupied levels by dashed lines. Insert shows the model cluster (adapted from Ref. 203).

The negative features in the difference spectra will not be explicable in a cluster model alone as they may have various reasons. Apart from the quenching of true surface states of the clean surface, they may be caused by the decreased effective source depth of the substrate due to the deposition of an adsorbate layer; by changes in scattering of the outgoing electrons due to the changed potential variations in the surface region and by a redistribution of substrate electrons due to the formation of the chemisorptive bond.

The situation might be further complicated by broad and complex emission peaks from the metal substrate overlapping the energy region of adsorbate levels, by the presence of various chemisorption states of the adsorbate and by the possibility of its decomposition upon chemisorption (e.g. the so-called α, β, γ and virgin states of CO on W /165, 168/).

The spectrum of CO chemisorbed on Ni with two broad, but well-defined peaks below the metal band seems to be simple enough to be treated with a surface molecule approach, especially as the heat of adsorption is quite high. The values of 26.5 kcal/mole for the (110) surface and 30 kcal/mole for the (100) and (110) surface /198a/ are comparable to the dissociation energy of $Ni(CO)_4$, 35 kcal/mole /198b/. The first 'tentative' assignment of PE spectrum /197/ was based on a comparison with the gas phase results for CO /99a, 100/ identifying peak P_1 at lower binding energy with the highest occupied 5σ MO of CO and P_2 with the 1π orbital (see Fig. 30). However, as the interaction between the adsorbate and substrate is accompanied by large relaxation and bonding shifts, such a straightforward procedure may lead to erroneous assignments. The interpretation of PE spectra for CO chemisorbed on a number of transition metal surfaces - all showing two adsorbate peaks - has been a highly controversial matter over the past years, but finally a definitive assignment seems to be at hand.

From LEED experiments it has been shown that CO forms a c(2x2) structure on a Ni (100) surface /199/. This implies a CO/Ni ratio of 1:2 which coupled with infrared studies suggests a bridged structure. It is also common belief that CO bonds to the surface upright through the carbon end resulting in a structure with a local fourfold symmetry axis (see insert of Fig. 29). In analogy to transition metal carbonyl complexes /141/, the chemisorption bond has been described by a donor-acceptor

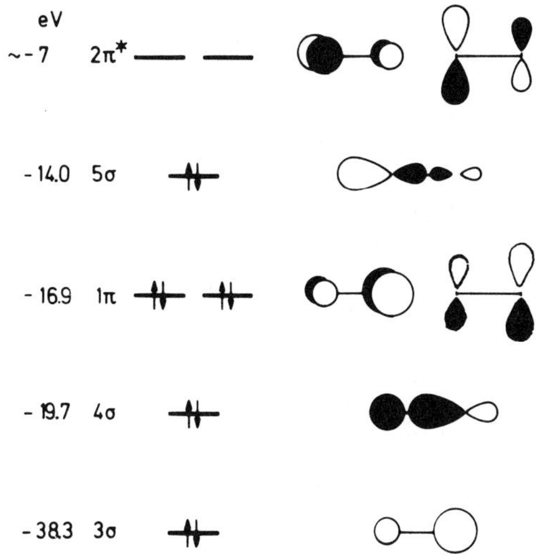

Fig. 30 Binding energies and schematic orbital shapes for the valence electrons of carbon monoxide.

mechanism /200/. From the 5σ orbital which may be characterized as a carbon lone pair orbital electrons are transfered towards the metal. On the other hand, one has a back-donation into the antibonding 2π* MO unoccupied in the free CO molecule. The bonding interaction is expected to lower the energy of the 5σ MO relative to the 1π orbital which is mostly located on the oxygen atom.

Several rather different types of calculations have been performed to find an interpretation of the experimental spectrum. The first is due to Doyen and Ertl /173b/ who use an Anderson model including the coupling of the 2π* and 5σ orbitals of CO to the metal d-band. However, from all other calculations which do not have these restrictions it has been concluded that the coupling of the 1π and 4σ levels to the metal (including the s,p-band) is essential for an understanding of the bonding. Blyholder /177b/ performed a CNDO calculation on CO chemisorbed on a (111) plane group of ten Ni atoms and concluded that the bonding of CO to Ni largeley involves Ni s and p orbitals, with little contribution from the d orbitals. He assignes peak P_1 both to the $1\tilde{\pi}$ and $5\tilde{\sigma}$ orbitals and P_2 to the $4\tilde{\sigma}$ orbital. (Concurring with common practice, the symbol tilda is used to stress that these levels have been modified due to interaction with the substrate). From the discussion in the previous section, we know that CNDO gives a very bad description of the metal cluster DOS, and therefore these results have to be viewed with great caution. The same assignment was given by Cederbaum et al. /178a/ on the basis of a HF-LCAO calculation on the linear NiCO "molecule". The latter is certainly a very crude model for a surface molecule on account of the number of metal atoms included, the lack of what is believed to be the proper local coordination of the CO molecule and the rather limited basis set employed to describe the metal atom (two d-type contracted GTO's out of five gaussians). Anderson and Hoffmann, on the basis of their EH calculations /176c/, follow the original assignment of the spectrum ($5\tilde{\sigma}$, $1\tilde{\pi}$).

Batra and Robaux /201/ carried out an Xα-SW calculation on the cluster CO Ni_5. The Ni_5 cluster was chosen to represent a (100) surface; it may be derived from an octahedron by removing one atom (see Fig. 21a). From the discussion in sec. 5.2 one may expect that this cluster will provide a fair description of the metal substrate. The CO molecule was assumed in its experimentally inferred position along the fourfold axis of the Ni_5 cluster with the C atom at a distance of 2.0 a.u. above the plane of the four Ni atoms. For the other distances equilibrium values of the bulk and the free molecules were chosen: $d_{Ni-Ni} = 4.70$ a.u., $d_{CO} = 2.13$ a.u. The overlapping sphere model was used with the radii $r_{Ni} = 2.85$ a.u., $r_C = 1.60$ a.u., $r_O = 1.37$ a.u. This choice of radii gives transition state energies for the free CO molecule that compare favourable with experimental ionization potentials. The cluster has C_{4v} symmetry. The exchange parameter was set uniformly to $\alpha = 0.75$. The calculation predicted three main peaks: one due to the emission from the Ni d-bands (about 3.5 eV wide) and two levels, 12e at -8.5 eV and $17a_1$ at -11 eV measured with respect to the highest d-like level of the cluster. Therefore, Batra and Robaux assigned the peaks P_1 and P_2 to the levels $1\tilde{\pi}$ and $5\tilde{\sigma}$, respectively, disagreeing with all previous assignments.

From additional PE experiments /202/, there is now sufficient evidence that this interpretation is not correct. The levels $5\tilde{\sigma}$ and $1\tilde{\pi}$ have to be associated with P_1 and the level $4\tilde{\sigma}$ with P_2. The best indication is an angular resolved measurement /202a/ on Ni (111) which shows additional structure near ~6.5 eV in addition to the peak near 7.6 eV. The variation of the relative intensity of the peaks P_1 and P_2 with photon energy /202b/ and electron take-off angle /202c/ was found to be compatible only with this assignment. Very recently, the $3\tilde{\sigma}$ level was detected approximately 29 eV below the Fermi level with a full width at half maximum of 10 to 12 eV /202e/. If one assumes the relaxation of the valence levels to be approximately constant, the stabilization of the $5\tilde{\sigma}$ orbital due to bonding may be inferred from the reduced distance to the $1\tilde{\pi}$ orbitals to be 1.5 eV.

In a spin-unrestricted Xα-SW calculation in the same CO Ni_5 cluster used previously, Batra and Bagus /203/ concluded that their results are compatible with both assignments, the former one /201/ and that confirmed by experiment as discussed above. They used somewhat smaller overlapping spheres: r_C = 1.50 a.u., r_O = 1.29 a.u. The radius for the Ni spheres is not reported. The resulting level spectrum is shown in Fig. 29. Note that the levels $1\tilde{\pi}$ and $5\tilde{\sigma}$ are inverted as compared to the free CO molecule. A similar inversion has been concluded recently /202d/ from angular resolved PE spectra of CO on Ni (111) with orbital energies rather close to these cluster results ($1\tilde{\pi}$: 7.1 eV; $5\tilde{\sigma}$: 8.7 eV).

In Fig. 29 all cluster levels have been rigidly shifted to line up the highest e↑ level with the experimental Fermi level. This highest occupied cluster level shows a significant contribution of the CO $2\pi^*$ orbital and is therefore involved in backbonding. The corresponding e level is unoccupied. The affinity level $2\pi^*$ of nitric oxide, NO, lies energetically much lower and consequently, in a similar cluster calculation on NO Ni_5 /204/ a backbonding level is found at 2.2 eV below the Fermi level. This corresponds rather well to the increased photoemission at about 2 eV below the Fermi level found for NO chemisorbed on polycrystalline Ni. From their calculation on the CO Ni_5 cluster, Batra and Bagus noted that the gas phase orbitals of CO have been significantly perturbed due to the interaction with the metal which involves Ni s,p as well as d orbitals.

Transition state energies from surface clusters invariably turn out too large by 2 to 3 eV after an approximate correction for the work function. For comparison with experiment, one therefore has to line up the experimental Fermi level with one of the topmost occupied cluster levels by a rigid shift of the energy scale. Apparently, relaxation shifts (towards smaller ionization potentials), as observed in the photoemission from adsorbate levels, are not reproduced quantitatively in model clusters employing only a few metal atoms. To put it differently, the relaxation of the levels during the transition state procedure comes out too large for clusters of that size /189/. However, there is also an experimental reason justifying this procedure. Adsorbate energy levels are referred to the Fermi energy, but not to the vacuum energy. The difference between these two quantities, the work function, is not known with certainty, as it changes under chemisorption /170/.

5.4 Chemisorption of Oxygen

The chemisorption layers of oxygen on transition and noble metal surfaces belong to the systems most widely studied with PE spectroscopy /197, 205/. In order to interpret these PE spectra and to gain a better understanding of how oxygen binds to different metal surfaces, we have carried out a series of Xα-SW calculations on metal clusters interacting with an oxygen atom /185/.

We first discuss the system O/Ag (110) for which extensive experimental information is available. The chemisorption of oxygen on Ag (110) surfaces has been characterized by LEED/AES and kinetic measurements /206 a,b/, atom-surface scattering /207/, UPS /205a/ and XPS /208/. UPS difference spectra have been reported for increasing oxygen exposure up to 10^4 Langmuir, both for He(I) and He(II) photon energies; XPS difference spectra are available for the saturated layer only. Oxygen chemisorption on Ag (110) is a slow process /205a/ and the maximum coverage after 10^4L is believed to be only 0.5 monolayers. At this coverage, one finds both for the He(I) and the He(II) difference spectra three distinct peaks above the d band region, i.e. between -4.0 eV and the Fermi level; the XPS difference spectra show increased emission in the same range. The identification of definite positive features in the d-band region on top of the general attenuation is difficult (see the discussion below). From LEED, kinetic and atom-surface scattering experiments it has been concluded that the oxygen atoms most probably sit in the troughs of the (110) surface between the silver atoms of the top layer at approximately the same height as these atoms. The oxygen atom will certainly not be expected below this plane as the distance to the four nearest neighbours is 2.04 Å (assuming an undistorted Ag (110) surface with a bulk metal-metal distance of 2.89 Å). This is just the silver-oxygen distance in Ag_2O.

The cluster chosen to represent the interaction with the (110) surface consists of the four nearest neighbour silver atoms, two from the top rows and two from the trough between them (cf. Fig. 21b). The oxygen atom is put into the position inferred from experiments. One may think of this silver-oxygen cluster as being derived from an octahedron of six silver atoms with an oxygen atom in the center and two adjacent silver atoms taken away. Quantitative agreement with experimental results may certainly not be expected from clusters of this size, although the main features of PE spectra and the chemisorption bond are expected to evolve. The effect of adding one atom to such a small metal cluster might, however, be undesirably large, as it changes the coordination of most of the atoms. Although certain differences may be noticed when comparing the results of the Ag_6 cluster (see sec. 5.2) to the Ag_4 cluster to be discussed shortly or the Ag_4O cluster to the Ag_5O cluster (see below), the main features of bonding and level spectra stay unchanged. Therefore clusters of 4 or 5 Ag atoms seem to be large enough at the intended level of sophistication.

To study the Ag_4O cluster, several SCF-Xα-SW calculations have been performed with varying parameters and configurations. In the calculation B non-overlapping muffin-tin spheres were used both for silver and oxygen. For the cluster derived from an octahedron as mentioned above, all sphere radii are determined by demanding touching spheres. The various radii are: r_{Ag} = 2.7307 a.u., r_O = 1.1311 a.u. and r_{OUT} = 6.5925 a.u. Calculation A was just the four Ag atoms in the same geometry with oxygen replaced by an empty sphere of the same size. As a considerable portion of the valence electron charge is found in the intersphere region, we decided to use overlapping spheres. Because of the very satisfactory results for the metal clusters (see sec. 5.2) and the charge distribution in the chemisorption case to be discussed shortly, we decided to increase only the oxygen sphere to r_O = 1.37 a.u., a value which has been used in previous calculations /201/ (calculation C). As we are not so much concerned with obtaining the best fit to experimental spectra, but more with trends and their explanation, we have not varied the oxygen sphere radius further. The distance of the oxygen atom with respect to the top layer of surface atoms is somewhat uncertain. In order to examine the influence of this parameter on the PE spec-

spectra and the bonding we placed the oxygen atom (r_O = 1.37 a.u.) on the twofold axis of the cluster at a position which corresponds to a distance of 0.5 a.u. to the top layer of the surface (calculation D).

For the sake of clarity and ease of reference, we summarize the salient features of the model clusters used:

(A) Ag_4 cluster of Fig. 21b with an empty sphere in place of the oxygen atom.
(B) Ag_4O cluster of Fig. 21b with touching spheres.
(C) Ag_4O cluster as in (B), but with increased oxygen sphere (radius: 1.37 a.u.; overlapping spheres).
(D) Ag_4O cluster with sphere radii as in (C), but the oxygen atom lifted above the site of (B) and (C) by 0.5 a.u. along the two-fold axis.

The resulting energy level spectra for all four calculations (A-D) are shown in Fig. 31. The cluster has C_{2v}

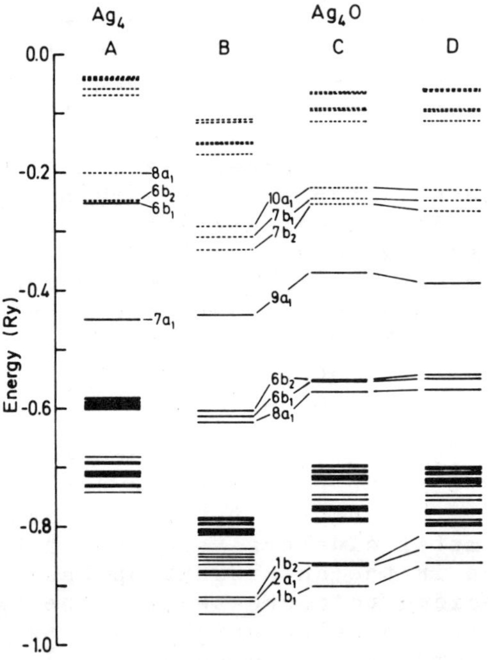

Fig. 31 Comparison of the SCF-Xα-SW orbital energies for the different model clusters Ag_4 (A) and Ag_4O (B-D). Only those levels are labeled which are discussed in the text. For the different model parameters, see the text. Dashed lines represent empty levels.

symmetry and the levels have been labeled according to the irreducible representations a_1, a_2, b_1 and b_2 of this group. The y-axis has been taken along the troughs (see Fig. 21b). The energy level spectrum for the four-atom cluster (A) shows the four levels $7a_1$, $6b_1$, $6b_2$ and $8a_1$, which are derived from Ag 5s atomic orbitals, well above the d levels (from -0.743 to -0.583 Ry). They are filled up to the $6b_1$ level at -0.248 Ry. One is able to recognize a qualitative similarity with the energy level spectrum of the six atom clusters discussed previously in that the levels derived from Ag 5s again lie clearly above the "d-band" (cf. Fig. 22). When this cluster interacts with oxygen (calculation B), four additional levels appear in the spectrum. The $1a_1$ level at -1.787 Ry is mostly localized in the oxygen sphere and represents the O 2s atomic orbital. Between the d-like levels localized in the Ag spheres and the levels correlating with the s-type levels in the Ag_4 cluster ($9a_1$, $7b_2$, $7b_1$ and $10a_1$), we find three more levels ($8a_1$, $6b_1$, $6b_2$) close in energy which derive mainly from the O 2p atomic orbitals. A SW population analysis of these levels and the highest occupied orbital confirms that these levels have pure p character within the oxygen sphere, but that both s and d waves as well as some p waves contribute to the appreciable localization in the Ag spheres. Orbital contour maps of these four orbitals reveal that they are all antibonding with respect to silver-oxygen bonds. (This results from the predominance of the antibonding interaction with Ag 4d orbitals over the bonding interaction with Ag 5s.) The corresponding bonding levels are found at the bottom of the "d-band": the levels $1b_1$, $2a_1$ and $1b_2$. Their localization in the oxygen sphere is, however, much smaller than that of the upper levels. We come back to a more quantitative discussion of this point later on. Due to the three O 2p levels $8a_1$, $6b_1$ and $6b_2$ below the "Fermi level" $9a_1$, one finds the $7b_1$ level (formerly $6b_1$) now empty. This suggests a partial charge transfer from Ag_4 to O.

Comparing the energy level spectra A and B one notices a shift of the latter towards more negative energies. This shift is approximately 0.2 Ry for the Ag d levels and is mostly caused by the large amount of charge contributed to the intersphere region by oxygen. (Only 6.32 electrons are located in the oxygen sphere showing that the description of this atom needs improvement.) The radius of its sphere was enlarged in calculation C as a consequence of this observation and the result was a more or less uniform shift of the spectrum by 0.05 to 0.07 Ry.

One now has 7.04 electrons inside the enlarged oxygen sphere. The picture of the oxygen-silver bonding essentially stays the same. Lifting the oxygen atom above the top row silver atoms (calculation D) resulted only in minor changes in the spectrum. The bonding levels, however, are no longer split off from the d band as in B and C.

How well do the energy level spectra B to D correlate with the PES difference spectra for O/Ag (110)? To compare experimental and theoretical level spectra, the Xα-SW orbital energies were shifted uniformly such that the top of the d-bands coincided. The assumption of a uniform relaxation was tested for case C by calculating selected transition states over the whole energy range; it held true within the limits of $\pm$ 0.3 eV which is acceptable for our purpose. In Fig. 32, we compare the He(I), He(II) and XPS difference spectra /205a, 208/ with those levels which show some localization in the oxygen sphere. Also indicated is the position of the d-band.

From these results it is clear that the observed peaks above the d-band correlate with the antibonding oxygen levels of the Ag_4O cluster. An additional level appears in the vicinity of the Fermi level ($9a_1$); in experimental spectra a tailing of the increased emission up to E_F is noted. The calculated bonding levels fall in the range of the d-bands. The experiment here is not so clear-cut. Because of the strong attenuation of the substrate emission in the band region and the difficulties in interpreting these changes /170/ no clear evidence can be derived from UPS spectra. In the XPS spectrum, however, a small positive peak is observed at about -7.5 eV /208/. All three calculations give almost equal agreement with the PES spectra, although there are reasons to prefer model D to the others. Here, the oxygen bonding levels are less split off from the bottom of the d-band, and the contribution of the levels above the d-band to the charge in the oxygen sphere is strongest in analogy to the experimental observation. This charge amounts to 2.148, 2.022 and 2.454 for the models B, C and D, respectively, whereas the corresponding values for the bonding levels are 0.792, 1.444 and 0.964.

In conclusion, we obtain a satisfactory interpretation of the PE difference spectra and for the silver-oxygen chemisorptive bond formed by the interaction of O 2p orbitals both with Ag 4d and 5s orbitals. Contour maps illustrating this statement will be shown later on. The

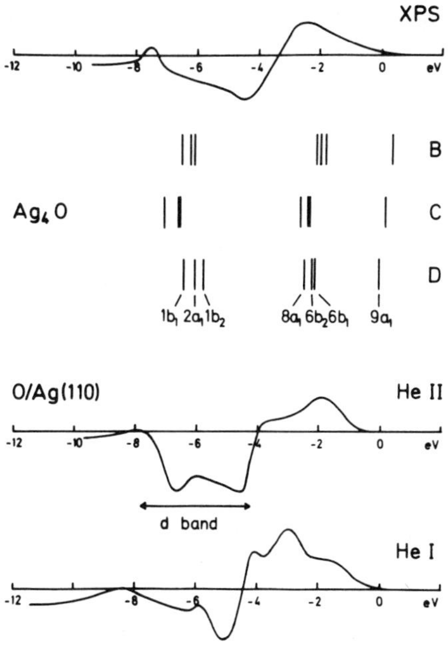

Fig. 32 Comparison of He(I), He(II) and XPS difference spectra of the system O/Ag (110) with the energy level spectra of the Ag_4O model clusters. The level spectra have been shifted such that the energy of the highest d-like orbital coincides with the top of the d-band. Only levels involved in the O-Ag interaction are shown.

oxygen atom may sit slightly above the top plane of the (110) surface. The agreement between the experimental and the theoretical spectrum is not quantitative. One reason may be that the Ag s-band is represented only by very few, energetically separated levels of a Ag_4O cluster.

Besides this calculation on the Ag_4O cluster, Xα-SW calculations on a Ni_5O cluster have been reported previously /209/. An obvious next step is to consider the chemisorption of oxygen on a copper surface and to compare the interaction of oxygen with the three metals Ni, Cu and Ag. We chose the same five-atom cluster representing a (100) surface which has been used in sec. 5.3 to investigate the system CO/Ni (100). This cluster plus the oxygen atom on its fourfold axis has C_{4v} symmetry; it is to be understood as a model. Only for Ni are UPS spectra of oxygen chemisorbed on a (100) surface available /197/, whereas for Cu it is a (111) surface /210/ and for Ag a (110) surface /205a/. However, the UPS difference spectra for adsorption on different metal surfaces are usually found to be similar, especially the split-off peaks due to adsorbed species.

One parameter of the model cluster M_5O (M = Ni, Cu, Ag) is the distance of the oxygen to the plane of the four metal atoms. For the system O/Ni (100) there are extensive LEED studies and there is now general agreement that the oxygen sits in the fourfold site, approximately 0.9 Å above the top layer /169/. There exists experimental evidence from LEED that oxygen prefers a similar geometry on Cu (100), but no distances are known. The lack of information for O/Ag (100) is even more pronounced where oxygen does not form an ordered adsorption layer /206/. The bond distance inferred for O/Ag (110) is very close to that for the oxide Ag_2O. We therefore decided, for Cu and Ag, to use the bond distances of the corresponding oxides. The resulting distances of the oxygen atom to the plane of the four metal atoms are then 0.76 Å and 0.16 Å for Cu and Ag, respectively. The sphere radii were chosen as in the models C and D, i.e. no overlap between the metal atom spheres and a constant sphere radius of 1.37 a.u. for oxygen. The rationale for this procedure was to use as few adjustable parameters as possible.

The resulting energy spectra for the three M_5O clusters are shown in Fig. 33. In order to facilitate a comparison between these spectra, we have started the serial numbers of the levels belonging to the same irreducible representation of C_{4v} point group, in the valence orbital region (i.e. $1a_1$ for O 2s). The highest occupied orbital in the Ni cluster is the 7e level containing two electrons. Five more electrons in the Cu cluster lead to the 8e levels containing one electron as the highest occupied orbital. In the Ag cluster the $8a_1$ and the 8e orbital kept oscillating during the iterations towards selfconsistency. These oscillations could only be stopped by using fractional occupation numbers: 0.75 in the 8e level and 0.25 in the $8a_1$ level resulted in equal energies for these levels. No spin-polarized calculations have been carried out.

In all three energy level spectra we find a cluster of levels representing the d-bands of the bulk and below it the somewhat split-off 1e and $2a_1$ levels. The O 2p atomic orbitals contribute only to levels of the irre-

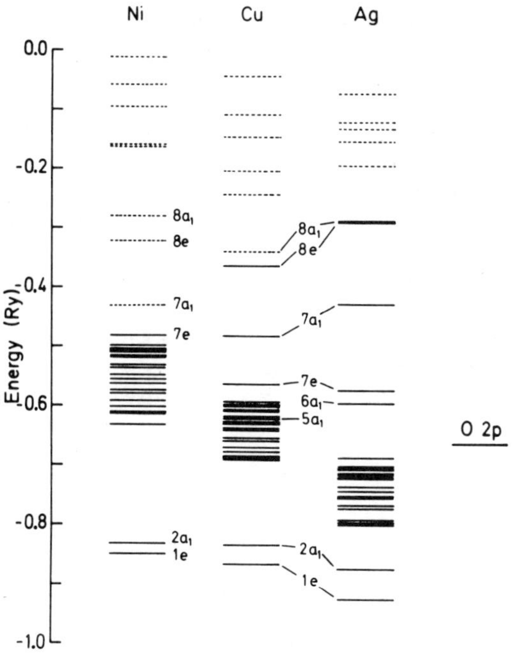

Fig. 33 SCF-Xα-SW orbital energy spectra for the M_5O cluster (M = Ni, Cu, Ag). Dashed lines represent empty levels. For the sake of comparison the SCF-Xα O 2p level is also shown.

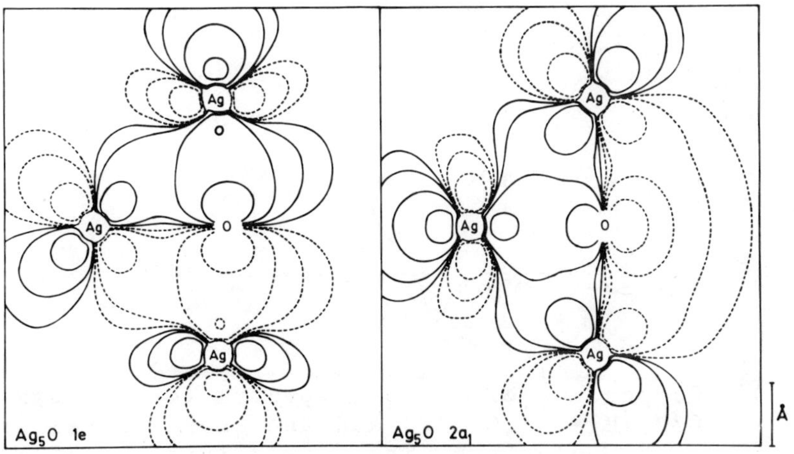

Fig. 34 Contour maps for the oxygen-metal bonding orbitals 1e and $2a_1$ of the Ag_5O cluster.

ducible representations $a_1(p_z)$ and $e(p_x,p_y)$. The levels $2a_1$ and 1e are oxygen-metal bonding. Contour maps of these orbitals are shown in Fig. 34 exhibiting contributions of Ag 4d orbitals of all metal atoms of the cluster. A SW population analysis reveals also small Ag 5s contributions. All other orbitals of the cluster are either nonbonding or antibonding with respect to the oxygen-metal bonds. Filled, strongly antibonding levels occur only for the Cu and the Ag cluster above the "d-band": $6a_1$ (for Ag only), 1e, $1a_1$, 8e, and $8a_1$ (for Ag only).

When comparing the charge distributions and the results of a SW population analysis for these clusters, several trends may be established concerning the oxygen-

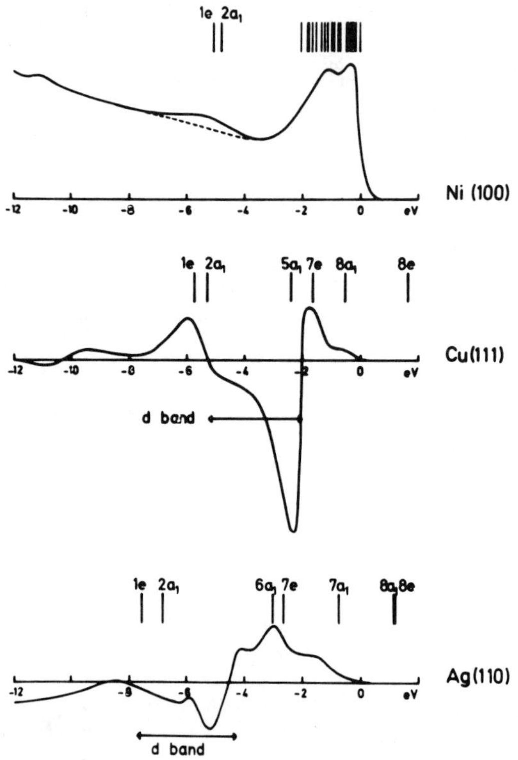

Fig. 35 Comparison of the UPS spectra for the systems O/M (M = Ni,Cu,Ag) and the energy level spectra of M_5O clusters. The energy level spectra have been shifted such that the top of the d-band coincides with the energy of the highest d-like orbital.

metal bonding. Going from Ni to Cu to Ag, one finds:

1) a reduction of the gap between the bonding 1e and $2a_1$ levels and the bottom of the "d-band".

2) a decreasing localization of the bonding levels in the oxygen sphere (only O 2p, average: 49%, 38%, 27%).

3) increasing contributions of the antibonding levels to the population of the O 2p atomic orbitals (10%, 48%, 71%).

4) decreasing contributions of metal s orbitals to the lower split-off levels and increasing contributions to the "oxygen" orbitals above the "d-band".

From these points, especially 1) to 3), a decreasing strength of the oxygen-metal chemisorptive bond may be deduced, in satisfactory agreement with measured adsorption energies /206/.

The results of these cluster calculations are closely paralleled by the UPS spectra for the corresponding systems. In Fig. 35, we compare He(I) spectra for the systems O/Ni (100) /197/, O/Cu (111) /210/ and O/Ag (110) /205a/ with the energy level spectra for the M_5O cluster. As mentioned previously, this has been done by shifting the theoretical spectrum such that the top of the d-bands (for Ni the highest occupied orbital and the experimental Fermi level) coincide. In the cases where difference spectra are available (Cu and Ag) only those levels are shown which have a contribution in the oxygen sphere. The whole spectrum is shown for Ni_5O. We obtain two oxygen levels at -5.0 eV and -4.8 eV (below the Fermi level E_F) falling within the experimental oxygen level at -5.5 eV which has a width of roughly 3 eV. The overall agreement with the experimental spectrum is as good as in the previous calculation where overlapping Ni spheres had been used /209a/.

The UPS difference spectrum for O/Cu shows one peak below (around -6.0 eV) and one peak above (-1.7 eV) the strongly attenuated d-bands. Theoretically, one obtains several bonding and antibonding levels in the expected energy range. The same is true for the system O/Ag. Here, the peaks above the d-band are even more pronounced. No oxygen levels can be identified with certainty in the UPS spectra within or below the d-bands as has been discussed above. The fact that we find experi-

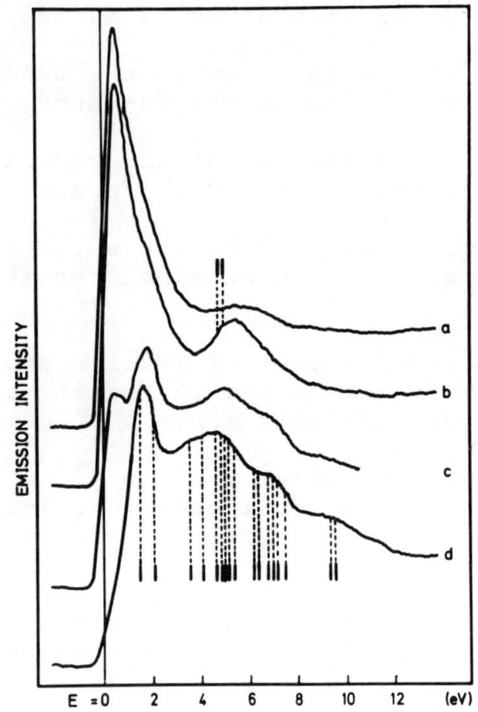

Fig. 36 UV photoelectron spectra (40.8 eV) illustrating the interaction of oxygen with a Ni(III) surface. a) clean surface; b) oxygen chemisorption; c) transition to the oxide; d) after growth of a "surface oxide" layer. Superimposed on the UPS curves are the spin orbital energies for the representative Ni_5O cluster (top) and NiO_6^{6} NiO_6^{10-} cluster (bottom) calculated by the Xα-SW method (adapted from Ref. 212).

mentally a pronounced O 2p peak below the d-band for Ni, two O 2p peaks on either side for Cu and several strong O 2p peaks above the d-band region of Ag, is in very good agreement with the localization of the corresponding levels of the cluster within the oxygen sphere as described in points (2) and (3) above.

In the interaction of oxygen with nickel surfaces, three distinct stages can be discriminated experimentally /211/: rapid chemisorption, surface oxide formation

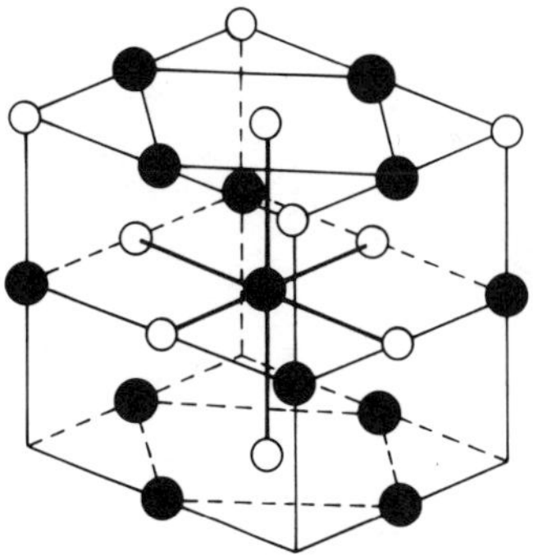

Fig. 37 Schematic diagram illustrating the possible configuration of atoms for incipient surface nickel oxide resulting from moderate to heavy oxygen chemisorption (Ni atoms: dark spheres; O atoms: light spheres). The six-fold coordination of the "central" Ni atom is clearly indicated.

and slow growth of bulk oxide. This process may be followed very nicely by PE spectroscopy /212/. In Fig. 36, He(II) spectra are shown for a Ni(111) surface with increasing coverage by oxygen. Only the PE spectrum of the first stage (Fig.36b) is described properly by the energy levels of the cluster Ni_5O. After an oxygen exposure of approximately 20 L, the main features of the PE spectrum have completely changed (Fig. 36d). The sharp peak near the Fermi level disappeared and a new peak near 1.7 eV evolves, together with a broad maximum centered at approximately 5 eV and a moderate peak at 9 eV.

A completely different cluster model is necessary for the interpretation of the PE spectrum after large oxygen exposure. In Fig. 37, a "monolayer" of nickel oxide, as it may have grown on the nickel surface, is shown schematically. Although this surface oxide will probably have an amorphous structure, the local coordination of the various atoms will be similar to the bulk. It is therefore not surprising to find a rather similar PE spectrum as in bulk NiO, because the nature of the electronic structure of the oxide depends only on the short-range coordination chemical bonding. This observation lead Messmer, Tucker and Johnson /213/ to use a NiO_6^{10-} cluster. This cluster represents the local environment in the bulk oxide, namely a Ni^{2+} ion octahedrally coordinated by six O^{2-} ions (see Fig. 37). Molecular type boundary conditions are not entirely appropriate for a cluster representing a portion of a solid or surface nickel oxide, especially as the clusters carries a negative charge of -2. To prevent accumulation of excess electrons outside the outer sphere on neighbouring equivalent clusters, the model cluster was embedded in a potential well. The depth of this well was determined such that no more than 0.1 electron was allowed in region III. After convergence, the orbital energies were corrected for the well depth by a constant energy. The energy levels resulting from a spin-polarized Xα-SW calculation on the $Ni\ O_6^{10-}$ cluster are also shown in Fig. 36. The agreement with the experimental spectrum for high oxygen coverage is quite striking. Again, the energy scales have been adjusted by using an average transition state relaxation energy of -2 eV and a "work function" of approximately 5.5 eV. The sharp peak near 1.7 eV is associated with photoemission from the Ni 3d-like ligand field levels which are largely antibonding with respect to the O ligands. The broad maximum near 5eV is caused by emission from a wide band of predominantly nonbonding O 2p-like orbitals. The moderately intense peak at 9 eV has been attributed to levels with bonding character at the bottom of the O 2p-band. The bonding is due to an O 2p-Ni 4s interaction. The system O/Ni, described by two different clusters according to the physical situation, serves as a convincing example both for the accuracy and the flexibility of SCF-Xα-SW cluster calculations as tools in the interpretation of PE spectra of surface systems.

ACKNOWLEDGEMENTS

The author is indebted to Prof. G.L. Hofacker for continuous interest and support. He would like to thank Prof. D. Menzel for stimulating discussions and Dr. K.D. Hänsel for valuable help during the preparation of this manuscript. This work was supported in part by the Deutsche Forschungsgemeinschaft through Sonderforschungsbereich 128 and by the Fonds der Chemischen Industrie.

References

1. D. R. Hartree, Proc. Camb. Phil. Soc., 24, 89, 111, 426 (1928).
2. J.C. Slater, Phys. Rev., 32, 339 (1928).
3. C.C.J. Roothaan, Rev. Mod. Phys., 23, 69 (1951).
4. J.C. Slater, Quantum Theory of Atomic Structure, Vol. I and II (McGraw-Hill, New York, 1960).
5. J.A. Pople and D.L. Beveridge, Approximate Molecular Orbital Theory (McGraw-Hill, New York, 1970).
6. C.A. Coulson and I. Fischer, Phil. Mag., 40, 386 (1949).
7. T. C. Collins, in Electronic Structure of Polymers and Molecular Crystals, eds. J.-M. André and J. Ladik (Plenum Press, New York, 1975) p. 405.
8. J.C. Slater, Phys. Rev., 81, 385 (1951).
9. N.M. Hugenholtz, Physica, 23, 481 (1957).
10. F. Herman and S. Skillman, Atomic Structure Calculations (Prentice Hall, Englewood Cliffs, 1963).
11. e.g.: Computational Methods in Band Theory, eds. P.M. Marcus, J.F. Janak and A.R. Williams (Plenum Press, New York, 1971).
12. J.W.D. Connolly, Phys. Rev., 159, 415 (1967).
13. R. Gaspar, Acta Physica Hung., 3, 263 (1954).
14. W. Kohn and L.J. Sham, Phys. Rev., 140, A1133 (1965).
15. J.C. Slater and J.H. Wood, Int. J. Quantum Chem., S4, 3 (1971).
16. J.C. Slater, Adv. Quantum Chem., 6, 1 (1972).
17. J.C. Slater, J. Chem. Phys., 43, S228 (1965).
18. K.H. Johnson, J. Chem. Phys., 45, 3085 (1966); Int. J. Quantum Chem., S1, 361 (1967).
19. a) J. Korringa, Physica, 13, 392 (1947);
    b) W. Kohn and J. Rostocker, Phys. Rev., 94, 1111 (1954).
20. K.H. Johnson, Adv. Quantum Chem., 7, 147 (1973).
21. P. Weinberger and K. Schwarz, in Physical Chemistry, Series Two, Vol. 1, Theoretical Chemistry, eds. A.D. Buckingham and C.A. Coulson (Butterworths, London, 1975) p. 257.

22. J.W.D. Connolly, in Modern Theoretical Chemistry: Vol. IV. Approximate Methods, ed. G.A. Segal (Plenum Press, New York, 1976).

23. K.H. Johnson, Ann. Rev. Phys. Chem., 26, 39 (1975).

24. J.C. Slater, The Self-consistent Field for Molecules and Solids, Vol. IV of: Quantum Theory of Molecules and Solids (McGraw-Hill, New York, 1974).

25. F. Bloch, Z. Physik, 57, 545 (1929); P.A.M. Dirac, Proc. Camb. Phil. Soc., 26, 376 (1930).

26. E. Wigner and F. Seitz, Phys. Rev., 43, 804 (1933).

27. see also ref. 4, Vol. II, app. 22, p. 323.

28. T.C. Koopmans, Physica, 1, 104 (1933).

29. J.C. Slater, J.B. Mann, T.M. Wilson and J.H. Wood, Phys. Rev., 184, 672 (1969).

30. J.C. Slater, J. Chem. Phys., 57, 2389 (1972).

31. M. Ross, Phys. Rev., 179, 612 (1969); W.E. Rudge, Phys. Rev., 181, 1033 (1969); L.J. Sham, Phys. Rev., A1, 169 (1970).

32. J.B. Mann, Los Alamos Sci. Lap. Rep. LA-3690 (1967) and LA-3691 (1968).

33. K. Schwarz, Phys. Rev., B5, 2466 (1972); Theoret. Chim. Acta., 34, 225 (1974).

33a. K. Schwarz and J.W.D. Connolly, J. Chem. Phys., 55, 4710 (1971).

34. I. Lindgren, Ark. Fys., 31, 59 (1966).

35. E.A. Kmetko, Phys. Rev., A1, 37 (1970).

36. J.H. Wood, Int. J. Quantum Chem., S3, 747 (1970).

37. M. Berrondo and O. Goscinski, Int. J. Quantum Chem., S3, 775 (1969).

38. J.C. Slater and J.H. Wood, Int. J. Quantum Chem., S4, 3 (1971).

39. R. Schrieffer, J. Vac. Sci. Technol., 13, 335 (1976).

40. P.S. Bagus and B.I. Bennett, Int. J. Quantum Chem., 9, 143 (1975).

41. O. Goscinski, B.T. Pickup and G. Purvis, Chem. Phys. Lett., 22, 167 (1973).

42. O. Goscinski, G. Howat and T. Åberg, J. Phys. B8, 11 (1975).

43. T.M. Wilson, J.H. Wood and J.C. Slater, Phys. Rev., A2, 620 (1970).

44. P. Hohenberg and W. Kohn, Phys. Rev., 136, B864 (1964).

45. L. Hedin and S. Lundqvist, in Solid State Physics, eds. F. Seitz, D. Turnbull and H. Ehrenreich (Academic Press, New York, 1969) Vol. 23, p. 1; J. Phys. C, 4, 2064 (1971).

46. O. Gunnarsson and P. Johannson, Int. J. Quantum Chem., 10, 299 (1976).

47. e.g.: J.R. Smith, in Interactions on Metal Surfaces, Topics in Applied Physics, ed. R. Gomer (Springer, Berlin, 1975), Vol. 4, p. 1.

48. J.R. Sabin, J.P. Worths, and S.B. Trickey, Phys. Rev., B11, 3658 (1975); J.R. Sabin and S.B. Trickey, J. Phys. B, 8, 2593 (1975).

49. I. Lindgren and K. Schwarz, Phys. Rev., A5, 542 (1972).

50. F. Herman, J.P. Van Dyke and I.B. Ortenburger, Phys. Rev. Letters, 22, 807 (1969).

51. K. Schwarz, Chem. Phys., 7, 94 (1975).

52. P. Lloyd and P.V. Smith, Adv. Phys., 20, 69 (1972).

53. A.R. Williams, Int. J. Quantum Chem., S8, 89 (1974).

54. P.M. Morse and H. Feshbach, Methods of Theoretical Physics (Mc-Graw Hill, New York, 1953), Vol. I, chap. 7.

55. S.M. Blinder, in Physical Chemistry, Series Two, Vol. 1, Theoretical Chemistry, eds. A.D. Buckingham and C.A. Coulson (Butterworths, London, 1975) p.1.

56. L. Scheire and P. Phariseau, Int. J. Quantum Chem., S9, 105 (1975).

57. L. Scheire and P. Phariseau, Int. J. Quantum Chem., S8, 109 (1974).

58. L. Scheire and P. Phariseau, Physica, 74, 596 (1974).

59. e.g. D. Jackson, Classical Electrodynamics (Wiley, New York, 1962) chap. 16.

60. M. Danos and L.C. Maximon, J. Math. Phys., 6, 766 (1965); R. Nozawa, J. Math. Phys., 7, 1841 (1966); E.O. Steinborn and E. Filter, Theor. Chim. Acta, 38, 247, 261 (1975).

61. D.M. Brink and G.R. Satchler, Angular Momentum (Clarendon Press, Oxford, 1968), 2nd. ed.

62. L. Scheire and P. Phariseau, Int. J. Quantum Chem., 9, 887 (1975).

63. R. Kjellander, Chem. Phys., 12, 469 (1975).

64. R. Kjellander, Chem. Phys. Lett., 29, 270 (1974).

65. C.Y. Yang and S. Rabii, Phys. Rev., A12, 362 (1975); B.G. Cartling and D.M. Whitmore, Int. J. Quantum Chem., 10, 393 (1976).

66. L.F.Mattheis, Phys. Rev., 133, A1399 (1964).

67. P.O. Löwdin, Adv. Phys., 5, 1 (1956).

68. T. Loucks, Augmented Plane Wave Method (Benjamin, New York, 1967).

69. J.W.D. Connolly and J.R. Sabin, J. Chem. Phys., 56, 5529 (1972).

70. U. Mitzdorf, Theor. Chim. Acta, 37, 129 (1975).

71. M.L. Sink and G.E. Juras, Chem. Phys. Lett., 20, 474 (1973).

72. R.F.W. Bader, Acc. Chem. Res., 8, 34 (1975).

73. J.C. Slater, Quantum Theory of Molecules and Solids, Vol. II (McGraw-Hill, New York, 1965) chap. 4.

74. S. Larsson and J.W. D. Connolly, Chem. Phys. Lett., 20, 323 (1973).

75. J.W.D. Connolly and K.H. Johnson, Chem. Phys. Lett., 10, 616 (1971).

76. J.B. Diamond, Chem. Phys. Lett., 20, 63 (1973).

77. e.g. M. Tinkham, Group Theory and Quantum Mechanics (McGraw-Hill, New York, 1964).

78. A. Veillard, in: Computational Techniques in Quantum Chemistry and Molecular Physics, eds. G.H.F. Diercksen, B.T. Sutcliffe and A. Veillard (Reidel, Dordrecht, 1975) p. 201.

79. V. Heine, in: Solid State Physics, Vol. 24, eds. H. Ehrenreich, F. Seitz and D. Turnbull (Academic, New York, 1970) p. 1.

80. F.S. Ham and B. Segall, Phys. Rev., 124, 1786 (1961).

81. N. Rösch and J. Ladik, Chem. Phys., 13, 285 (1976).

82. F.C. Smith, Jr. and K.H. Johnson, private communication.

83. J.B. Danese and J.W.D. Connolly, J. Chem. Phys., 61, 3063 (1974).

84. W. von Niessen, L.S. Cederbaum, G.H.F. Diercksen and G. Hohlneicher, Chem. Phys., 11, 399 (1975).

85. W.M. Tolles and W.D. Gwinn, J. Chem. Phys., 36, 1119 (1962); K. Kimura and S.H. Bauer, J. Chem. Phys., 39, 3172 (1963).

86. a) J.I. Musher, Angew. Chem., 81, 68 (1969);
b) J.I. Musher, J. Amer. Chem. Soc., 94, 1370 (1972);
c) V.B. Koutecky and J.I. Musher, Theor. Chim. Acta, 33, 227 (1974).

87. L. Pauling, The Nature of the Chemical Bond (Cornell University Press, Ithaca, 1960), 3rd ed.

88. G. Dogett, Theor. Chim. Acta, 15, 344 (1969).

89. C.A. Coulson, Nature, 221, 1106 (1969).

90. R.E. Rundle, Survey Progr. Chem., 1, 81 (1963).

91. For a detailed list of references, see ref. 91 and 92.

92. N. Rösch, V.H. Smith, Jr. and M.H. Whangbo, J. Amer. Chem. Soc., 96, 5384 (1974).

93. N. Rösch, V.H. Smith, Jr. and M.H. Whangbo, Inorg. Chem. (1976) to be published.

94. W.E. Dasent, Nonexisting Compounds (Dekker, New York, 1965) p. 153.

95. F.A. Cotton and G. Wilkinson, Advanced Inorganic Chemistry (Wiley-Interscience, New York, 1972) 3rd. ed.

96. M.M. Chen and R. Hoffmann, J. Amer. Chem. Soc., 98, 1647 (1976).

97. G.M. Schwenzer and H.F. Schaefer III, J. Amer. Chem. Soc., 97, 1393 (1975).

98. R.S. Mulliken, J. Chem. Phys., 23, 1833, 1841 (1955).

99. a) D.W. Turner, C. Baker, A.D. Baker and C.R. Brundle, Molecular Photoelectron Spectroscopy (Wiley, London, 1970);
b) A.W. Potts, H.J. Lempka, D.G. Streets and W.C. Price, Phil. Trans. Roy. Soc. London, A268, 59 (1970);
c) R.E. LaVilla, J. Chem. Phys., 57, 899 (1972);
d) U. Grelius, J. Electron. Spectr., 5, 985 (1974).

100. K. Siegbahn et al., ESCA Applied to Free Molecules (North-Holland, Amsterdam, 1969).

101. D.P. Santry and G.A. Segal, J. Chem. Phys., 47, 158 (1967).

102. a) G.L. Bendazzoli, P. Palmieri, B. Cadioli and U. Pincelli, Mol. Phys., 19, 865 (1970);
b) F.A. Gianturco, C. Guidotti, U. Lamanna and R. Moccia, Chem. Phys. Lett., 10, 269 (1971).
c) B. Roos, unpublished results, cited in b).

103. J.G. Fripiat, K.T. Chow, M. Boudart, J.B. Diamond and K.H. Johnson, J. Mol. Catal., 1, 59 (1975).

104. B.J. Ransil and J.J. Sinai, J. Chem. Phys., 46, 4050 (1967) and unpublished calculations by A.C. Wahl cited therein.

105. W. Kutzelnigg, V. Staemmler and M. Gelus, Chem. Phys. Lett., 13, 496 (1972).

106. P. Weinberger and K. Schwarz, Theor. Chim. Acta, 38, 231 (1975).

107. P. Weinberger and D.D. Konowalow, Int. J. Quantum Chem., S7, 353 (1973).

108. J. Weber and J.W.D. Connolly, Int. J. Quantum Chem., S9, 129 (1975).

109. J.B. Danese, J. Chem. Phys., 61, 3071 (1974).

110. J. Keller, Int. J. Quantum Chem., 9, 583 (1975).

111. D.E. Ellis and G.S. Painter, in ref. 11, p. 271, 277; Phys. Rev., B2, 2887 (1970).

112. E.J. Baerends, D.E. Ellis and P. Ros, Chem. Phys., 2, 41 (1973);
E.J. Baerends and P. Ros, Chem. Phys., 2 52 (1973); Chem. Phys. 8, 412 (1975).

113. C.B. Haselgrove, Math. Comp., 15, 323 (1961).

114. W. Heijser, A.T. van Kessel and E.J. Baerends, Chem. Phys. (1976), to be published.

115. H. Sambe and R.H. Felton, J. Chem. Phys., 62, 1122 (1975).

116. N. Rösch, W.G. Klemperer and K.H. Johnson, Chem. Phys. Lett., 23, 149 (1973).

117. D.A. Liberman, unpublished results.

118. F. Herman, A.R. Williams and K.H. Johnson, J. Chem. Phys., 61, 3508 (1974).

119. K.H. Johnson, F. Herman and R. Kjellander, in Electronic Structure of Polymers and Molecular Crystals, eds. J. André, J. Ladik and J. Delhalle (Plenum, New York, 1975) p. 601.

120. J. Norman, Jr., Mol. Phys., 31, 1191 (1976).

121. D.R. Salahub, R.P. Messmer and K.H. Johnson, Mol. Phys., 31, 529 (1976).

122. P.E. Cade, K.D. Sales and A.C. Wahl, J. Chem. Phys., 44, 1973 (1966).

123. A.C. Wahl, J. Chem. Phys., 41, 2600 (1964).

124. W.M. Huo, J. Chem. Phys., 43, 624 (1965).

125. G. Herzberg, Molecular Spectra and Molecular Structures I. Spectra of Diatomic Molecules (van Nostrand, Princeton, 1950).

126. M. Rosenblum, Chemistry of the Iron Group Metallocenes (Wiley-Interscience, New York, 1965).

127. D.A. Brown, in: Transition Metal Chemistry, Vol. 3, ed. R.L. Carlin (Dekker, New York, 1966).

128. C.J. Ballhausen and H.B. Gray, in: Chemistry of the Coordination Compounds, Vol. 1, ed. A.E. Martell (van Nostrand-Reinhold, New York, 1971).

129. a) E.M. Shustorovich and M.E. Dyatkina, J. Struct. Chem., 1, 109 (1960).
b) J.P. Dahl and C.J. Ballhausen, Kgl. Danske Videnskab Selskab Mat. Fys. Medd., 33, No. 5 (1961).
c) A.T. Armstrong, D.G. Carroll and S.P. McGlynn, J. Chem. Phys., 47, 1104 (1967);
d) J.H. Schachtschneider, R. Prins and P. Ros, Inorg. Chim. Acta, 1, 462 (1967);
e) R.F. Kirchner, G.H. Loew and U.T. Mueller-Westerhoff, Theor. Chim. Acta, 41, 1 (1976).

130. a) A.T. Armstrong, F. Smith, E. Elder and S.P. McGlynn, J. Chem. Phys., 46, 4321 (1967);
b) P.B. Stephenson and W.E. Winterrod, J. Chem. Phys., 52, 3308 (1970);
c) Y.S. Sohn, D.N. Hendrickson and H.B. Gray, J. Amer. Chem. Soc., 93, 3603 (1971).

131. R. Prins, Mol. Phys., 19, 603 (1970).

132. J.W. Rabalais, L.O. Werme, T. Bergmark, L. Karlsson, M. Hussain and K. Siegbahn, J. Chem. Phys., 57, 1185 (1972).

133. S. Evans, M.L.H. Green, B. Jewitt, A.F. Orchard and C.F. Pygall, JCS Faraday II, 68, 1847 (1972).

134. M.-M. Coutière, J. Demuynck and A. Veillard, Theor. Chim. Acta, 27, 281 (1972).

135. M.-M. Rohmer, A. Veillard and M. H. Wood, Chem. Phys. Lett., 29, 466 (1974).

136. P.S. Bagus, U.I. Wahlgren and J. Almlof, J. Chem. Phys., 64, 2324 (1976).

137. E.J. Baerends and P. Ros, Chem. Phys. Lett., 23, 391 (1973).

138. N. Rösch and K.H. Johnson, Chem. Phys. Lett., 24, 179 (1974).

139. R.K. Bohn and A. Haaland, J. Organometallic Chem., 5, 470 (1970); G.J. Palenik, Inorg. Chem., 9, 2424 (1970).

140. M.-M. Rohmer and A. Veillard, Chem. Phys. 11, 349 (1975).

141. M.J.S. Dewar, Bull. Soc. Chim. France, 18, C71 (1951); J. Chatt and L.A. Duncanson, J. Chem. Soc., 1953, 2939.

142. C.J. Ballhausen, Introduction to Ligand Field Theory (McGraw-Hill, New York, 1962).

143. J. Demuynck and A. Veillard, Theor. Chim. Acta, 28, 241 (1973).

144. R.B. Woodward and R. Hoffmann, The Conservation of Orbital Symmetry (Verlag Chemie, Weinheim, 1971).

145. e.g. N. Rösch and R. Hoffmann, Inorg. Chem., 13, 2654 (1974); M. Elian and R. Hoffmann, Inorg. Chem., 14, 365 (1975); J.W. Lauher and R. Hoffmann, J. Amer. Chem. Soc., 98, 1729 (1976); R. Hoffmann, J.M. Howell and A. Rossi, J. Amer. Chem. Soc., 98, 2484 (1976).

146. J.C. Slater and K.H. Johnson, Physics Today, October, 1974, p. 34.

147. S. Larsson, E.K.Viinikka, M.L. de Sequeira and J.W.D. Connolly, Int. J. Quantum Chem., S8, 145 (1974).

148. J.W.D. Connolly, H. Siegbahn, U. Gelius and C. Nordling, J. Chem. Phys., 58, 4265 (1973).

149. N. Rösch and V.H. Smith, Jr., work in progress.

150. S. Larsson and J.W.D. Connolly, J. Chem. Phys., 60, 1514 (1974); S. Larsson, Theor. Chim. Acta, 39, 173 (1975).

151. L. Noodleman, J. Chem. Phys., 64, 2343 (1976).

152. D. Dill and J.L. Dehmer, J. Chem. Phys., 61, 692 (1974); J.L. Dehmer and D. Dill, Phys. Rev. Lett., 35, 213 (1975).

153. J.W. Davenport, Phys. Rev. Lett., 36, 945 (1976).

154. R.P. Messmer, L.V. Interrante and K.H. Johnson, J. Amer. Chem. Soc., 96, 3847 (1974).

155. N. Rösch, R.P. Messmer and K.H. Johnson, J. Amer. Chem. Soc., 96, 3855 (1974).

156. K.H. Johnson and U. Wahlgren, Int. J. Quantum Chem., S6, 243 (1972).

157. A.P. Mortola, J.W. Moskowitz and N. Rösch, Int. J. Quantum Chem., S8, 161 (1974).

158. C.Y. Yang, K.H. Johnson, R.H. Holm and J.G. Norman, Jr., J. Amer. Chem. Soc., 97, 6596 (1975).

159. K. Schwarz and N. Rösch, J. Phys. C, 9, L433 (1976); N. Rösch and K. Schwarz, to be published.

160. T.E. Fischer, Physics, Today, 27, No. 5, 23 (1974).

161. Ed. E. Drauglis and R.I. Jaffee, The Physical Basis for Heterogeneous Catalysis, Proceedings of the Batelle Institute Materials Science Colloquium Gstaad 1974 (Plenum, New York, 1975).

162. D.A. Micha, Adv. Chem. Phys., 30, 7 (1975); W.H. Miller, ibid., p. 77.

163. R.B. Bernstein and R.D. Levine, Adv. At. Mol. Phys., 11, 216 (1975).

164. Various articles in Physics Today, 28, No. 4 (1975).

165. Interactions on Metal Surfaces, ed. R. Gomer, Topics in Applied Physics, Vol. 4 (Springer, Berlin, 1975).

166. Electronic Structure and Reactivity of Metal Surfaces, Proceedings of the NATO Advanced Study Institute at Namur 1975 (Plenum, New York, 1976).

167. T.N. Rhodin and D.S.Y. Tong, Physics Today, 28, No. 10 (1975).

168. A.M. Bradshaw, L.S. Cederbaum and W. Domcke, Structure and Bonding, Vol. 24 (Springer, Berlin, 1975).

169. a) J.E. Demuth, D.W. Jepsen and P.M. Marcus, Phys. Rev. Lett., 31, 540 (1973); Surf. Sci., 45, 433 (1975);
b) C.B. Duke, N.O. Lipari and G.E. Laramore, J. Vac. Sci. Technol., 11, 180 (1974); 12, 222 (1975);
c) J.E. Demuth and T.N. Rhodin, Surf. Sci., 45, 249 (1974).

170. D. Menzel, J. Vac. Sci. Technol., 12, 313 (1975).

171. a) J.R. Schrieffer, J. Vac. Sci. Technol., 9, 561 (1972); 13, 335 (1976).
b) A. Clark, The Chemisorptive Bond (Academic, New York, 1974).

172. D.M. Newns, Phys. Rev., B1, 3304 (1970).

173. a) D.M. Newns, Phys. Rev., 178, 1123 (1969);
b) G. Doyen and G. Ertl, Surf. Sci., 43, 197 (1974);
c) W. Brenig and K. Schönhammer, Z. Phys., 267, 201 (1974).

174. T.L. Einstein, Surf. Sci., 45, 713 (1974).

175. G.C. Bond, Discuss. Faraday Soc., 41, 200 (1966).

176. a) L.W. Anders, R.S. Hansen and L.S. Bartell, J. Chem. Phys., 59, 5277 (1973); 62, 1641 (1975);
b) D.J.M. Fassaert, H. Verbeek and A. van der Avoird, Surf. Sci., 29, 501 (1972);
c) A.B. Anderson and R. Hoffmann, J. Chem. Phys., 61, 4545 (1974);
d) A.B. Anderson, J. Chem. Phys., 64, 4046 (1976).

177. a) G. Blyholder, JCS Chem. Comm., 1973, 625;
b) G. Blyholder, J. Vac. Sci. Technol., 11, 865 (1974).

178. a) L.S. Cederbaum, W. Domcke, W. von Niessen and W. Brenig, Z. Phys., B21, 3811 (1975);
b) C.W. Bauschlicher, Jr., C.F. Bender, H.F. Schaefer III and P.S. Bagus, Chem. Phys., 15, 227 (1976).

179. D.E. Ellis, H. Adachi and F.W. Averill, to be published.

180. K.H. Johnson and R.P. Messmer, J. Vac. Sci. Technol., 11, 236 (1974).

181. a) T.B. Grimley and C. Pisani, J. Phys. C, 7, 2831 (1974);
b) T.B. Grimley, in ref. 166;
c) A. van der Avoird, S.P. Liebmann and D.J.M. Fassaert, Phys. Rev., B10, 1230 (1974);

181. d) E.A. Hyman, Phys. Rev., B11, 3739 (1975);
e) R.H. Paulson and T.N. Rhodin, Surf. Sci., 55, 61 (1976).

182. W. Kohn, Phys. Rev., B11,3756 (1975); N. Kar and P. P. Soven, Phys. Rev., B11, 3761 (1975).

183. R.H. Paulson and J.R. Schrieffer, Surf. Sci., 48, 329 (1975).

184. a) S.C. Ying, J.R. Smith and W. Kohn, Phys. Rev., B11, 1483 (1975);
b) L.M. Kahn and S.C. Ying, Solid State Commun., 16, 799 (1975).
c) N.D. Lang and A.R. Williams, Phys. Rev. Lett., 34, 531 (1975)
d) O. Gunnarsson and H. Hjelmberg, Phys. Scripta, 11, 97 (1975).

185. N. Rösch and D. Menzel, Chem. Phys., 13, 243 (1976).

186. a) J.W.D. Connolly, Phys. Rev., 159, 415 (1967);
b) J.F. Janak, A.R. Williams and V.L. Moruzzi, Phys. Rev., B11, 1522 (1975);
c) N.E. Christensen, Phys. Stat. Solidi, 31, 635 (1969).

187. S. Hüfner, G.K. Wertheim, N.V. Sunitz and M.M. Traum, Solid State Commun., 11, 329 (1972).

188. a) D.E. Eastman, J. Physique, C1, 293 (1971);
b) D.E. Eastman and J.K. Cashion, Phys. Rev. Lett., 24, 310 (1970);
c) A.D. MacLachlan, J. Liesegang, R.C.G. Leckey and J.G. Jenkin, Phys. Rev., B11, 2877 (1975).

189. R.P. Messmer, S.K. Knudson, K.H. Johnson, J.B. Diamond and C.Y. Yang, Phys. Rev., B13, 1396 (1976).

190. J. Callaway and C.S. Wang, Phys. Rev., B7, 1096 (1973).

191. J.L. Carter and J.H. Sinfelt, J. Catal., 10, 134 (1968).

192. R.O. Jones, P.J. Jennings and G.S. Painter, Surf. Sci., 53, 409 (1975).

193. R.C. Baetzold and R.E. Mack, J. Chem. Phys., 62, 1513 (1975).

194. a) R.C. Baetzold, J. Catal., 29, 129 (1973).
b) G. Blyholder, Surf. Sci., 42, 249 (1974).

195. R.P. Messmer and G.O. Watkins, Phys. Rev., B7, 2568 (1973).

196. R.P. Messmer and D.R. Salahub, Int. J. Quant. Chem., S10(1976).

197. D.E. Eastman and J.K. Cashion, Phys. Rev. Lett., 27, 1520 (1971).

198. a) H.H. Madden, J. Küppers and G. Ertl, J. Chem. Phys., 58, 3401 (1973); K. Christmann, J. Küppers and G. Ertl, J. Chem. Phys., 60, 4719 (1974);
b) F.A. Cotton, A.K. Fischer and G. Wilkinson, J. Amer. Chem. Soc., 81, 800 (1959).

199. J.C. Tracy, J. Chem. Phys., 56, 2736 (1972).

200. G. Blyholder, J. Phys. Chem., 68, 2772 (1964).

201. I.P. Batra and O. Robaux, J. Vac. Sci. Technol., 12, 242 (1975).

202. a) D.E. Eastman and J.E. Demuth, Jpn. J. Appl. Phys. Suppl. 2, 847 (1974);
b) T. Gustafsson, E.W. Plummer, D.E. Eastman and J.L. Freeouf, Solid State Commun., 17, 391 (1975);
c) J.C. Fuggle, M. Steinkilberg and D. Menzel, Chem. Phys., 11, 307 (1975);
d) P.M. Williams, P. Butcher, J. Wood and K. Jacobi, to be published;
e) P.R. Norton, R.L. Tapping and J.W. Goodale, Chem. Phys. Lett., 41, 247 (1976).

203. I.P. Batra and P.S. Bagus, Solid State Commun., 16, 1097 (1975).

204. I.P. Batra and C.R. Brundle, Surf. Sci., 57, 12 (1976).

205. a) A.M. Bradshaw, D. Menzel and M. Steinkilberg, JCS Faraday Disc., 58, 46 (1974);
b) S.J. Atkinson, C.R. Brundle and M.W. Roberts, ibid., 62;
c) S. Evans, E.L. Evans, D.E. Parrey, M.J. Fricker, M.J. Walters and J.M. Thomas, ibid., 97;
d) H. Conrad, G. Ertl, J. Küppers and E.E. Latta, ibid., 116.

206. H.A. Engelhardt, A.M. Bradshaw and D. Menzel, Surf. Sci., 40, 410 (1973); H.A. Engelhardt and D. Menzel, Surf. Sci., to be published.

207. W. Heiland, F. Iberl, E. Taglauer and D. Menzel, Surf. Sci., 53, 383 (1975).

208. J.C. Fuggle and D. Menzel, Surf. Sci., 53, 21 (1975)

209. a) I.P. Batra and O. Robaux, Surf. Sci., 49, 653 (1975);

209. b) S.J. Niemczyk, J. Vac. Sci. Technol., 12, 246 (1975).

210. H. Conrad, G. Ertl, J. Küppers and E.E. Latta, private communication.

211. T.N. Rhodin and D.L. Adams, in Adsorption of Gases on Solids, Vol. 6, Treatise on Solid State Chemistry, ed. N.B. Hannay (Plenum, New York, 1975).

212. H. Conrad, G. Ertl, J. Küppers and E. E. Latta, Solid State Commun., 17, 497 (1975).

213. R.P. Messmer, C.W. Tucker and K.H. Johnson, Surf. Sci., 42, 341 (1974).

ELECTRONIC STATES IN RANDOM SUBSTITUTIONAL ALLOYS: THE CPA AND BEYOND

B. L. GYORFFY AND G. M. STOCKS

University of Bristol

H. H. Wills Physics Laboratory, Royal Fort
Tyndall Avenue, Bristol BS8 1TL

I. INTRODUCTION

Most elements in the periodic table are metals. In a metallic environment the excess charge on an impurity is screened out locally with the result that the heats of mixing for most combinations of metals are small compared with those in systems with covalent or ionic binding[1]. Consequently, there are a countless number of metallic solid solutions[2]. As you might expect these display a vast variety of thermal, electrical, mechanical and magnetic properties. Hopefully, after these lectures, you will conclude that the study of metallic alloys is an interesting part of solid-state physics. Nevertheless, we would also like you to keep in mind that much of the thrust behind the work on alloys comes from the needs of industry. Progress in many important technologies is materials limited, finding the alloy with the right properties is frequently the key to a breakthrough.

A further consequence of the metallic environment ensuring that a local charge disturbance is screened out within a lattice spacing or so is that an atom sitting on a "wrong" site can not be the source of a large lattice strain energy. Hence, for two component metallic systems the ordering energies are rather small ($\sim .01$ eV) and they form disordered phases, even at stoichiometric concentrations, below their melting temperature ($k_B T_m \sim .05$ eV). It is these alloys which we shall be concerned with in these lectures.

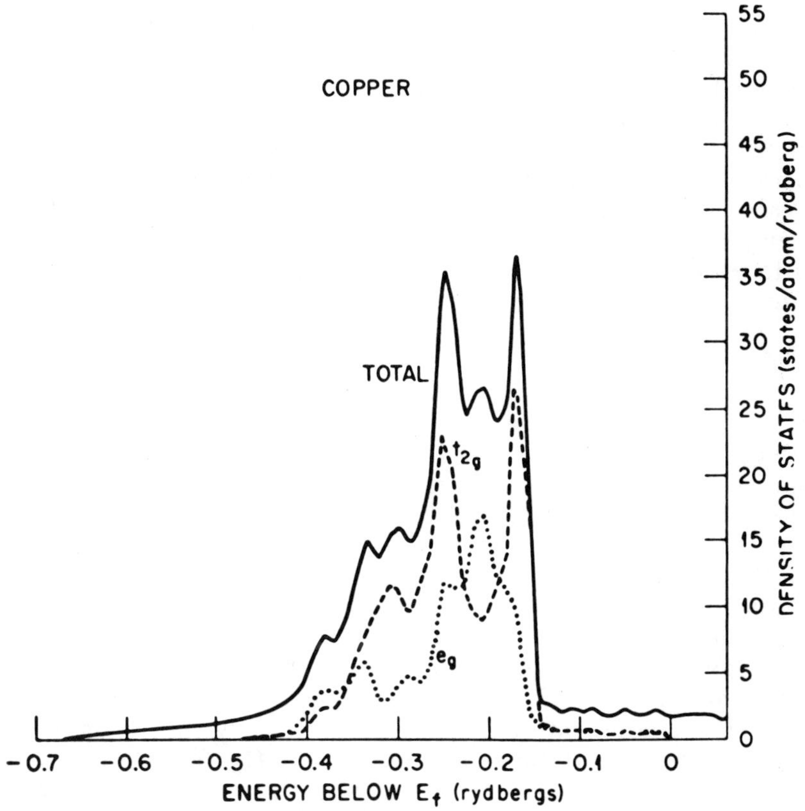

Fig. 1. Density of states in pure Cu as in Ref. 31.

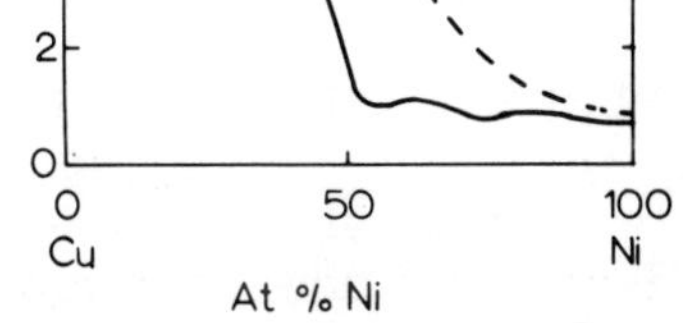

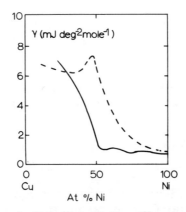

Fig. 2. The full line is the "rigid band" prediction for the electronic specific heat coefficient γ. The broken line follows the experimental points.

Having struck a practical note let us continue in this vain. Because of the very large number of systems involved even the most practical minded metallurgist would agree that we need a theory to guide us to the right alloy for a given application. Moreover, macroscopic theories such as thermodynamics and elasticity theory are not enough. As you probably know the range of stability of many alloy phases shows a striking correlation with the electron per atom ratio[3]. Thus we need to know about what the electrons are doing in microscopic detail.

A venerable technique which deals with the theory of one electron states in random alloys is the "rigid band approximation"[4]. Consider a pure Cu crystal and add some Ni to it. In the "rigid band" theory it is asserted that the energy bands of the pure materials remain unchanged on alloying and only the number of valence electrons filling those states changes. (Cu has 11 valence electrons and Ni has only 10). In Fig. 1 we show the density of states for pure Cu. The Fermi energy is also shown some 2 eV above the upper edge of the prominent d-band. Reducing the number of electrons means that the Fermi energy (ϵ_F) moves down. Thus we conclude that according to the rigid-band theory the density of states at the Fermi energy $n(\epsilon_F)$ does not change much initially as Ni is added. However, there will be a concentration C_0 for which ϵ_F has moved down to the upper edge of the Cu d-band, beyond that concentration $n(\epsilon_F)$ will start rising rapidly. Recall now that the electronic component of the specific heat is given by $c_{el} = \gamma T$ where $\gamma = (1/3)2\pi k^2 n(\epsilon_F)$. Thus, using the rigid-band idea we can predict the variation of γ with the Ni concentration c. In Fig. 2 we show the result of an actual calculation of γ using the density of states curve in Fig. 1. The broken line represents experimental points. For a long time one might have considered this as a successful application of the rigid-band theory to Cu rich Ni-Cu alloys[5]. Even in the latest edition of Kittel's book the spin susceptibility of Ni-Cu is understood in these terms. However, nothing can be further from the truth.

Roughly speaking in a photoemission experiment one measures the density of states below the Fermi energy. In Fig. 3 we show the density of states so obtained for Cu and a few Cu-Ni alloys. Clearly ϵ_F does not move with respect to the upper edge of the Cu d-band as the Ni concentration is increased. The effect of Ni impurities seems to be a subsidiary peak between the Cu d-band and ϵ_F. The size of this

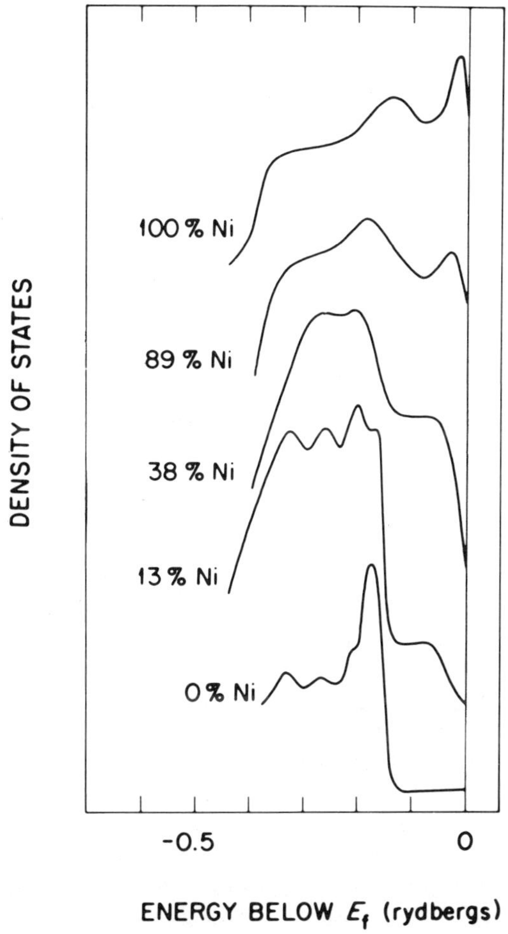

Fig. 3. Optical densities of states as obtained from photoemission experiments[6] for pure Cu and various Cu-Ni alloys.

impurity peak increases as the Ni concentration increases and at some concentration ($c \sim c_0 = .38$) it begins to overlap with the Fermi energy ϵ_F. Obviously, at this concentration $n(\epsilon_F)$ begins to rise rapidly in agreement with the findings of specific heat measurement. Thus the rigid-band idea completely misconstrued the effect of alloying on the density of states at the Fermi energy in this system. The agreement with experiments was a fluke.

Another simple theory of the electrons in random alloys is the virtual crystal approximation of Nordheim and Muto[7]. The crystal potential for a particular configuration of a random alloy may be written as $V(\underline{r}) = \sum_i v_i(\underline{r} - \underline{R}_i)$ where the potential function $v_i(\underline{r} - \underline{R}_i)$ on every $\underline{R}_i$ is either $v_A(\underline{r} - \underline{R}_i)$ or $v_B(\underline{r} - \underline{R}_i)$ in a random fashion. The total number of A sites is cN and the total number of B sites is $(1 - c)N$ where N is the total number of sites ($N \sim \infty$ for an infinite crystal). This potential does not have the periodicity of the lattice. Consequently, the solutions of the corresponding Schrödingers equation do not satisfy the Bloch theorem and therefore the usual methods of band theory cannot be applied to solve for the energy eigen values and eigen functions. However, one can recover the Bloch theorem by considering the configurationally averaged potential $\overline{V}(\underline{r}) = \sum_i \overline{v}(\underline{r} - \underline{R}_i)$ where $\overline{v}(\underline{r} - \underline{R}_i) = c\, v_A(\underline{r} - \underline{R}_i) + (1 - c)v_B(\underline{r} - \underline{R}_i)$. Clearly, $\overline{V}(\underline{r})$ has the full crystal symmetry and the corresponding Schrödingers equation can be solved by the standard methods developed for pure systems (APW, KKR etc.).

This is an attractive approach since it is obviously correct for $c = 0$ and $c = 1$. However, for the densities of states for alloys of Cu and Ni it leads to predictions similar to those of the rigid-band model. In Fig. 4 we show densities of states for a number of Cu-Ni alloys as calculated using the virtual crystal approximation. Obviously a plot of γ against c using these densities of states would lead again to a picture similar to Fig. 2. However, once again, the agreement with experiments is fortuitous since for Cu rich alloys we do not have the impurity band seen in the experiment. Thus, for these alloys, the virtual crystal approximation is not a useful procedure.

It is a simple exercise in perturbation theory to show that the virtual crystal approximation is correct to the order[8] $\epsilon_F^{-2}(v_A - v_B)^2$. That is to say for alloys where the potentials characteristic of the constituents are very close together on

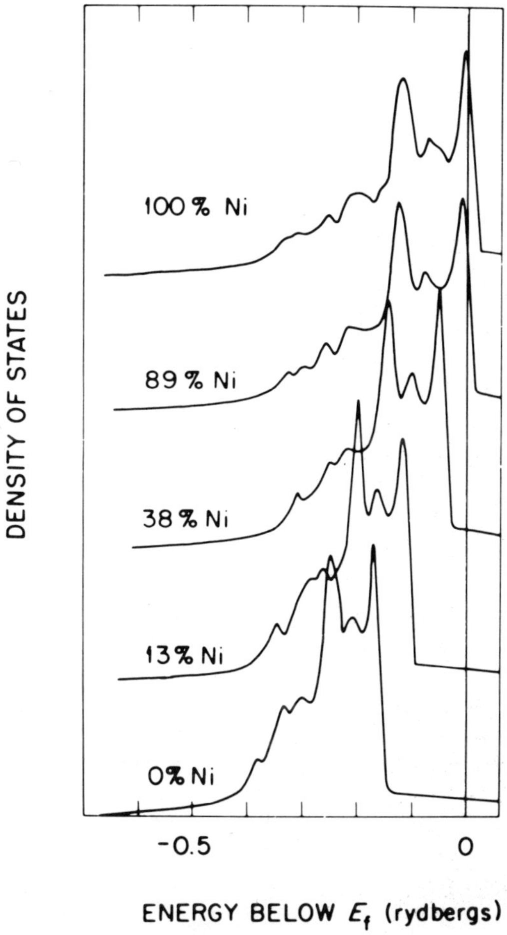

Fig. 4. The density of states for pure copper and a number of Cu-Ni alloys calculated in the virtual crystal approximation.

the scale of the Fermi energy ϵ_F. Evidently, this is not the case in the Cu-Ni alloys.

Perturbation theory is also useful when the potentials $v_A(\underline{r} - \underline{R}_i)$ and $v_B(\underline{r} - \underline{R}_i)$ are themselves small irrespective of their differences. In this case one can start with free electrons and, for an arbitrary configuration [9], formally calculate any physical observable in finite order perturbation theory using the crystal potential as a small parameter ($\epsilon_F^{-1}V$). The averaging over all possible configurations can then be performed by elementary means. As is well known for simple metals it is always possible to describe the electron-ion interaction by a relatively weak pseudo potential [10]. Therefore, for alloys of these metals there is no need for a special theory of randomness. The final result of a typical calculation is at most second order in the crystal potential and the average $<|V|^2>$ is easily formed even in the presence of short range order.

In alloys involving transition metals or rare-earths the situation is much more complicated. In this case neither the crystal potential nor the potential difference between sites occupied by different constituents are small in any sense of the word. Although one still can use perturbation theory and in each order the averaging can be carried out by elementary means - at least in principle - this process now has to be continued to all orders. Meaningful results follow only when the term by term averaged infinite order perturbation theory is resumed into an analytic form. Thus we encounter the usual technical difficulties of field theory. Therefore, it should not come as a surprise that, until recently, the theory of electronic states in such alloys has been the preserve of the more mathematically inclined theoretical physicist [11].

The basic issue here is "how to find the averaged behavior of a system whose Hamiltonian contains random parameters". This problem turns up with the same force in connection with other elementary excitations in solids not only for electrons. Thus, we have the problem of phonons in crystals where the masses and force constants vary from site to site in a random fashion, [12] or the problem of spin waves in systems with spin defects and random exchange integrals [13]. The same questions also arise when we consider Frenkel exitons in mixed crystals [14]. Initially, all these problems were treated in isolation and frequently identical methods were developed to deal with them. Nowadays it is customary to talk about a theory of disordered systems [15].

Once the problem is defined it becomes clear that we cannot, in general, find an exact solution except for the most artifical systems like an alloy on a Cayley tree[16]. Consequently, most of the early work in the field was concerned with trying to find an approximation scheme which does justice to the main qualitative features of the problem and at the same time remains tractible for fairly realistic Hamiltonians. It is a measure of the difficulties involved that while the first paper on the subject appeared in the early 30's [17] the first such approximation scheme was introduced only in 1968. The Coherent Potential Approximation (CPA) of Soven [18], Taylor [19] and Onodera [20] marked the beginning of a new era where the theory finally could make contact with experiments. In particular it opened up the possibility of realistic calculations of electronic states in random substitutional, metallic alloys. The outlines of such a theory is the main subject of these lectures.

Although it would be educational we shall not have the time to discuss the history before the advent of the CPA. Fortunately, this subject has been well treated in recent review articles by Elliott, Leath and Krumhansl [21] and also by Ehrenreich and Schwartz [22]. However, before turning to our main task we shall now describe the CPA idea in the context of a simple tight binding model Hamiltonian.

II. THE BASIC IDEA OF THE COHERENT POTENTIAL APPROXIMATION

1. The model Hamiltonian

For a system with a strongly scattering crystal potential the usual way of obtaining a simple description is to use the tight binding method for solving the Schrödingers equation

$$(-\nabla^2 + \sum_i v(\underline{r} - \underline{R}_i))\, \psi(\underline{r}) = \epsilon\psi(\underline{r}) \qquad \text{II-1}$$

where $v(\underline{r} - R_i)$ is the atomic potential seen by an electron in the isolated atom, and for the moment, it is the same on every site. So, we are talking about a pure metal. The method consists of assuming that

$$\psi(\underline{r}) = \sum_{j,\mu'} a_{j,\mu'} \varphi_{\mu'}(\underline{r} - \underline{R}_j) \qquad \text{II-2}$$

where

$$(-\nabla^2 + v(\underline{r}))\, \varphi_\mu(\underline{r}) = \epsilon_\mu \varphi_\mu(\underline{r}) \qquad \text{II-3}$$

and determining $a_{i,\mu}$ from the condition that Eq. II. 2 should be a solution of Eq. II. 1. Substituting Eq. II. 2 into Eq. II. 1, multiplying both sides by $\varphi_{\mu}(\underline{r} - \underline{R}_i)$ and integrating over $\underline{r}$ gives

$$\sum_{j,\mu'} (\epsilon_{\mu}\, \delta_{\mu,\mu'}\, \delta_{i,j} - J_{\mu,\mu'}(i,j))a_{j,\mu'} = \epsilon\, a_{i,\mu} \qquad \text{II-4}$$

where the overlap integral

$$J_{\mu,\mu'}(i,j) = \int dr^3\; \varphi^*_{\mu}(\underline{r} - \underline{R}_i)\; v(\underline{r} - \underline{R}_i)\; \varphi_{\mu'}(\underline{r} - \underline{R}_j)$$

and the three centre integrals as well as the orthogonality terms ($\int dr^3\, \varphi^*_{\mu'}(\underline{r} - \underline{R}_i)\, \varphi_{\mu'}(\underline{r} - \underline{R}_j)$) have been neglected. If the $\varphi_{\mu}(\underline{r})$'s form a more or less complete set this method can be made very accurate (LCAO). However, here we are interested in deriving a simple model. Therefore, we take a single atom wave function only (for simplicity we take an s-function, but we have in mind the d-bands of transition metals) and assume that the overlap integral J(i, j) is only non zero for sites i and j which are nearest neighbours. We then have the simple eigenvalue problem

$$\sum_j [(\epsilon - \epsilon_0)\,\delta_{ij} + J(i,j)]\; a_j = 0 \qquad \text{II-5}$$

Taking the lattice Fourier transform of this equation yields the eigenvalues immediately. For a simple cubic lattice $\epsilon_q = \epsilon_0 + 1/3\, J(\cos q_x a + \cos q_y a + \cos q_z a)$ where a is the lattice constant. Note that the band width is 2 J and the band is centred on the atomic level ϵ_0.

Consider now an alloy composed of two types of atoms denoted by A and B. Clearly on an A site $\epsilon_0 = \epsilon_A$ and on a B site $\epsilon_0 = \epsilon_B$. If the pure metals of the constituents have roughly the same band width we may assume that $J_{AA} = J_{BB} = J_{AB} = J$ where the notation should be self explanatory.

For a given configuration the energy spectrum is given by the eigenvalue problem

$$\sum_j [(\epsilon - \epsilon_i)\,\delta_{i,j} + J(i,j)]\; a_j = 0 \qquad \text{II-6}$$

where ϵ_i is either ϵ_A or ϵ_B as required by the configuration considered. Obviously the eigenvalues ϵ_n will depend on the configuration $\epsilon_n = \epsilon_n(\epsilon_1, \epsilon_2, \epsilon_3, \ldots, \epsilon_N)$ where N is the number of sites in the lattice.

It is a fundamental principle of statistical physics that a physical measurement can not differentiate between configurations. Therefore, after a certain physical observable has been calculated as a function of the configurations we must average it over the ensemble of all configurations. The energy of our system of electrons is an observable. In the one electron approximation $E = \int_0^{\epsilon_F} d\epsilon\, \epsilon\, n(\epsilon)$ where $n(\epsilon)$ is the density of states defined as

$$n(\epsilon) = \sum_n \delta\,(\epsilon - \epsilon_n(\epsilon_1, \epsilon_2, \epsilon_3, \ldots \epsilon_N))$$

Thus $n(\epsilon)$, is an 'observable'. In fact it is the simplest 'observable' and therefore it is the ensemble averaged density of states we shall first want to calculate in a disordered system problem. This is given by

$$\overline{n}(\epsilon) = \sum_{\epsilon_1, \epsilon_2, \epsilon_3 \cdots \epsilon_N} P(\epsilon_1, \epsilon_2, \epsilon_3, \ldots \epsilon_N)\, n(\epsilon; \epsilon_1, \epsilon_2, \epsilon_3, \ldots \epsilon_N)$$

where $P(\epsilon_1, \epsilon_2, \epsilon_3, \ldots \epsilon_N)$ is the joint probability that the energy on site 1 is ϵ_1, on site 2 is ϵ_2 etc. and the summation is over the two values each energy can take ϵ_A and ϵ_B. While the correlations between occupations of different sites are interesting and frequently physically important [23] we shall assume, for now, that the sites are uncorrelated. That is to say $P(\epsilon_1, \epsilon_2, \epsilon_3, \ldots \epsilon_N) = \prod_{i=0} P(\epsilon_i)$ and $P(\epsilon_A) = c$, $P(\epsilon_B) = 1-c$ where c is the concentration of A sites.

Thus the problem we shall try to solve is defined by the eigenvalue equation Eq. II-6 and the prescription for finding the averaged density of states $\overline{n}(\epsilon)$. Clearly, the problem of finding the averaged spectrum of a random Hamiltonian is a very general one and the discussion that will follow has wide application beyond the theory of alloys.

2. The Greens function, perturbation theory and the self energy

An efficient way of tackling the problem defined in the previous section is to consider the Greens function $G(i,j;\epsilon)$ of the Schrödingers equation given in Eq. II-6. This is found by solving the equation

$$\sum_k [\,(\epsilon + i\eta - \epsilon_i)\delta_{i,k} + J(i,k)]\; G(k,j;\epsilon) = \delta_{i,j} \qquad \text{II-7}$$

where η is a positive infinitesimal quantity. The reason for working with the Greens function is that from it we can obtain the density of states directly by noting that

$$n(\epsilon) = -\frac{1}{N\pi} \sum_{i=0}^{N} \mathrm{Im}\, G(i, i; \epsilon) \qquad \text{II-8}$$

(To prove this discard mathematical inhibition and evaluate the trace of the matrix function $\delta(\epsilon - H)$ where H is the Hamiltonian matrix $\epsilon_i \delta_{ij} - J(i,j)$. Using the eigenvectors of H it is easy to show that $\mathrm{tr}\, \delta(\epsilon - H) = \sum_n \langle n| \delta(\epsilon - H)| n\rangle = \sum_n \delta(\epsilon - \epsilon_n) = \bar{n}(\epsilon)$. Note now that the matrix $G(i,j;\epsilon)$ is equal to $[\epsilon + i\eta - H]^{-1}_{ij}$ and therefore $\delta(\epsilon - H) = -\frac{1}{\pi} \mathrm{Im}\, G(\epsilon)$, hence Eq. II-8 follows). Thus to find $n(\epsilon)$ we have to calculate the configurationally averaged Greens function.

It is useful at this stage to separate out a part of the problem which has the full lattice symmetry. Consider for instance the virtual crystal described by the following Greens function equation

$$\sum_k [(\epsilon + i\eta - \bar{v}) \delta_{i,k} + J(i,k)]\, G_0(k,j;\epsilon) = \delta_{i,j} \qquad \text{II-9}$$

where $\bar{v} = c\epsilon_A + (1-c)\epsilon_B$. Since $G(i,j;\epsilon)$ depends only on $\underline{R}_i - \underline{R}_j$ Eq. II-9 is solved by

$$G(i,j;\epsilon) = \sum_{\underline{q}} G_0(\underline{q};\epsilon) e^{i\underline{q}\cdot(\underline{R}_i - \underline{R}_j)} \text{ and } G_0(\underline{q},\epsilon) = \frac{1}{\epsilon - \epsilon_{\underline{q}} + i\eta} \qquad \text{II-10}$$

where

$$\epsilon_{\underline{q}} = \bar{v} - J(\underline{q})$$

The full Greens function $G(i,j;\epsilon)$ can also be written in the crystal momentum representation

$$G(i,j;\epsilon) = \sum_{\underline{q},\underline{q}'} G(\underline{q},\underline{q}';\epsilon) e^{i\underline{q}\cdot\underline{R}_i - i\underline{q}'\cdot\underline{R}_j} \qquad \text{II-11}$$

however since $G(i,j;\epsilon)$ depends on $\underline{R}_i$ and $\underline{R}_j$ separately for a given configuration it will not be diagonal. It is now the matter of simple algebra to write an integral equation for $G(\underline{q},\underline{q}';\epsilon)$ in terms of $G_0(\underline{q},\epsilon)$. Using Eq. II-9, II-10 and II-11 we obtain

$$G(\underline{q},\underline{q}',\epsilon) = G_0(\underline{q};\epsilon)\delta_{\underline{q},\underline{q}'} + \sum_{\underline{q}'',i} G_0(\underline{q};\epsilon) v_i(\underline{q},\underline{q}'') G(\underline{q}'',\underline{q}';\epsilon) \qquad \text{II-12}$$

where $v_i(\underline{q},\underline{q}') = e^{i(\underline{q}-\underline{q}')\cdot\underline{R}_i} v_i$; with $v_i = \epsilon_i - \bar{v}$. The physical picture represented by Eq. II-12 is that an electron of the virtual crystal described by $G_0(\underline{q},\epsilon)$ propagates from site to site and is scattered at each site by the deviation from the averaged "potential" $v_i = \epsilon_i - \bar{v}$. Clearly if these deviations are very small it is a good approximation to take $\overline{G}(\underline{q},\underline{q}';\epsilon) = G_0(\underline{q};\epsilon)\delta_{\underline{q},\underline{q}'}$. This gives $G(\underline{q},\underline{q}';\epsilon) = G_0(\underline{q};\epsilon)\delta_{\underline{q},\underline{q}'}$ and $\overline{n}(\epsilon) = -(1/\pi)\mathrm{Im}\sum_{\underline{q}} G_0(\underline{q};\epsilon)$. This is the virtual crystal approximation discussed in the introduction.

If the scattering potentials v_i are not small we must use infinite order perturbation theory. Thus we must resume the infinite series

$$\overline{G}(\underline{q},\underline{q}';\epsilon) = G_0(\underline{q};\epsilon)\delta_{\underline{q},\underline{q}'} + \langle\sum_i G_0(\underline{q};\epsilon)v_i(\underline{q},\underline{q}')G_0(\underline{q}';\epsilon)\rangle$$

$$\langle\sum_{\underline{q}''}\sum_{i,j} G_0(\underline{q};\epsilon)v^i(\underline{q},\underline{q}'')G_0(\underline{q}'';\epsilon)v^j(\underline{q}'',\underline{q}')G_0(\underline{q}';\epsilon)\rangle + \ldots \qquad \text{II-14}$$

which was obtained from Eq. II-12 by iteration and averaging the result.

The averaged Greens function will again have the translational symmetry of the lattice. Therefore $\overline{G}(\underline{q},\underline{q}';\epsilon) = \overline{G}(\underline{q};\epsilon)\delta_{\underline{q},\underline{q}'}$. It is also generally the case that the series in Eq. II-14 resums into the form

$$\overline{G}(\underline{q};\epsilon) = \frac{1}{\epsilon - \epsilon^0_{\underline{q}} - \Sigma(\underline{q},\epsilon)} \qquad \text{II-15}$$

where the quantity $\Sigma(\underline{q},\epsilon)$ is called the self energy and it is defined by the requirement that the right hand side of Eq. II-15 should be the same as the right hand side of Eq. II-15.

At this point several comments are in order. From Eq. II-10 we see that the energy eigenvalues for the virtual crystal are given by the poles of the Greens function $G_0(\underline{q};\epsilon)$. This is a special instance of a very general fact that the excitation energies of a system are given by the poles of an appropriate Greens function. The poles of $\overline{G}(\underline{q};\epsilon)$ are given by $\epsilon - \epsilon^0_{\underline{q}} - \Sigma(\underline{q},\epsilon) = 0$. Since $\Sigma(\underline{q},\epsilon)$ will be complex in general the poles Z_q will also be complex. Writing $Z_{\underline{q}}$ in the form $\epsilon_{\underline{q}} + i\,\Gamma_q$ what we can say is that provided $\epsilon_{\underline{q}}^{-1}\,\Gamma_{\underline{q}} \ll 1$ we still have a well defined excitation with the wave number $\underline{q}$. However, it will have a life time $\tau_{\underline{q}} = \hbar/\Gamma_{\underline{q}}$. This life time represents a smearing of the levels of an ordered system and is a direct consequence of disorder ; that is to say, the fact that in a disordered system $\underline{q}$ is not a good quantum number. In a metallic alloy this lifetime gives rise to a residual resistivity at low temperatures. In Fig. 5 we show the residual resistivity of $Pt_c\,Pd_{1-c}$. Clearly the lifetime is the shortest ($\rho_0 = (\frac{m}{ne^2})\tau^{-1}$) at mid-concentration where we expect the

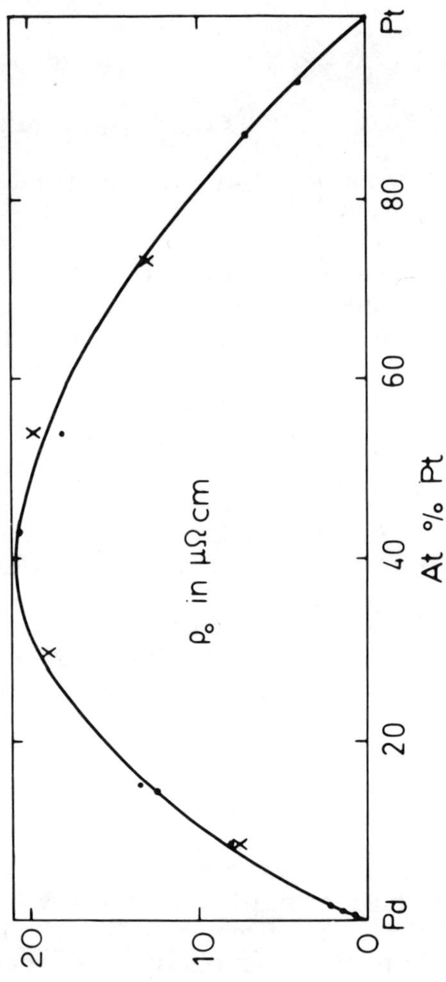

Fig. 5. The residual resistivity of Pt_c Pd_{1-c} alloys as measured by Woods et al.[69]

largest amount of disorder. It is one of important failings of the virtual crystal approximation that it allows $\underline{q}$ to be a good quantum number. The Z_q are therefore real and hence the life-times are infinite. One might think that the crystal potential in Pt and Pd are rather similar. Nevertheless, as can be seen from Fig. 5 there is still quite an increase in the resistivity as the disorder increases indicating a reducation of the life-time of the states due to disorder.

Another way of looking at the smearing discussed above is to consider the spectral function defined as $A(\underline{q},\epsilon) = -\frac{1}{\pi} \operatorname{Im} \bar{G}(\underline{q};\epsilon)$. For an ordered system $A(\underline{q},\epsilon)$ as a function of ϵ is a set of delta functions at $\epsilon = \epsilon_{\underline{q}}$. (Clearly this is also the case in the virtual crystal approximation). However, in a disordered system as the disorder increases these delta functions broaden into peaks with a width $\Gamma_{\underline{q}}$.

Let us now get back to our main preoccupation of calculating $\bar{n}(\epsilon)$. In the present formulation what we have to do is to find $\Sigma(\underline{q},\epsilon)$ by resumming the series in Eq. II-14 and calculate $\bar{n}(\epsilon)$ from

$$\bar{n}(\epsilon) = -\frac{1}{\pi} \sum_{\underline{q}} \operatorname{Im} \frac{1}{\epsilon - \epsilon^0_{\underline{q}} - \Sigma(\underline{q},\epsilon)} \qquad \text{II-16}$$

As we mentioned in the introduction this turns out to be a very difficult problem even if we are satisfied with something quite a bit less than an exact solution. In fact the first, all around sensible approximation was proposed only relatively recently. This is the Coherent Potential Approximation. Although it can be obtained by judiciously selecting the most important terms in the perturbation series given in Eq. II-14 and resumming the result to all orders in v_i [24] here we shall derive it in another way.

3. The Coherent Potential Idea

Let us rewrite Eq. II-15 as

$$\bar{G}(\underline{q};\epsilon) = G_0(\underline{q};\epsilon) + G_0(\underline{q};\epsilon)\Sigma(\underline{q},\epsilon)G(\underline{q};\epsilon) \qquad \text{II-17}$$

On comparing this expression with Eq. II-12 we see that the self energy $\Sigma(\underline{q},\epsilon)$ can be interpreted as complex energy dependent effective crystal potential which gives rise to the averaged Greens function directly. It is analogous to $V_{\underline{q},\underline{q}'} = \sum_i v_i e^{i(\underline{q}-\underline{q}')\cdot\underline{R}_i}$ in Eq. II-12. We can regard $\Sigma(\underline{q},\epsilon)$

as a potential diagonal in $\underline{q}$ and hence in $\underline{r}$-space it has the translational symmetry of the lattice. From a careful analysis of the perturbation series in Eq. II-14 it follows that $\Sigma(\underline{q},\epsilon)$ is independent of $\underline{q}$. This is not surprising since $\Sigma(\underline{q},\epsilon)$ is a consequence of the perturbation $V_{\underline{q},\underline{q}'}$ which if made translationally invariant by setting all v_i's equal is also diagonal and $\underline{q}$ independent. In fact $\Sigma(\epsilon)$ may be regarded as the lattice Fourier transform of the real space potential $\sum_i \Sigma(\epsilon)\delta_{ij}$. Thus, finding the exact averaged Greens function for our random lattice is equivalent to finding the Greens function for an effective ordered lattice with an energy dependent and complex site "potential" $\Sigma(\epsilon)$ (site energy). That is to say in real space we must write

$$\sum_k [(\epsilon - \Sigma(\epsilon))\delta_{i,k} + J(i,k)]\, G(k,j;\epsilon) = \delta_{ij} \qquad \text{II-18}$$

While exact this is of course not very useful if the effective potential has to be calculated via Eq. II-14 and II-12. However, this is precisely the picture which suggested the coherent potential approximation (CPA) to Soven. What he did is the following : he assumed a coherent potential $\Sigma_c(G)$ on every site and instead of trying to find it by summing up a perturbation series he sought to determine it by imposing some physically reasonable condition on it. What turned out to be a useful condition is the requirement that an A impurity in the coherent potential lattice, on a given site with the probability c, and a B impurity, on the same site with probability 1 – c, should not scatter the electrons of the coherent potential lattice on the average.

At the first sight this condition appears all too complicated, but when put into mathematical form it becomes quite transparent. Consider an A impurity in the coherent potential lattice. This means that we must perturb the coherent potential Greens function $G_c(\underline{q};\epsilon)$ by the potential $V^{A,i}_{\underline{q},\underline{q}'} = (v_A - \Sigma_c)e^{i(\underline{q}-\underline{q}')\cdot\underline{R}_i}$. The perturbed Greens function $G^A(\underline{q},\underline{q}';\epsilon)$ will then satisfy the equation

$$G^A(\underline{q},\underline{q}';\epsilon) = G_c(\underline{q};\epsilon)\delta_{\underline{q},\underline{q}'} + \sum_{\underline{q}''} G_c(\underline{q},\epsilon)\bar{V}^{A,i}_{\underline{q},\underline{q}''}\, G^{A,i}(\underline{q}'',\underline{q}';\epsilon) \qquad \text{II-19}$$

A way of solving this equation is to define the t-matrix[25] (Transition matrix) $\bar{t}^{A,i}_{\underline{q},\underline{q}'}(\epsilon)$ by the relation

$$G^A(\underline{q},\underline{q}';\epsilon) = G_c(\underline{q};\epsilon) + G_c(\underline{q};\epsilon)\bar{t}^{A,i}_{\underline{q},\underline{q}'}(\epsilon)G_c(\underline{q}';\epsilon) \qquad \text{II-20}$$

and attempt to solve

$$\bar{t}^{A,i}_{\underline{q},\underline{q}'}(\epsilon) = \bar{V}^{A,i}_{\underline{q},\underline{q}'} + \sum_{\underline{q}''} \bar{V}^{A,i}_{\underline{q},\underline{q}''} G_c(\underline{q}'', \epsilon)\bar{t}^{A,i}_{\underline{q}'',\underline{q}'}(\epsilon) \qquad \text{II-21}$$

which follows from the definition in Eq. II-20 and Eq. II-19. It is trivial to show that the solution of Eq. II-21 is

$$\bar{t}^{A,i}_{\underline{q},\underline{q}} = \left(\frac{v_A - \Sigma_c(\epsilon)}{1 - (v_A - \Sigma_c(\epsilon))G_c(i,i,\epsilon)}\right) e^{i(\underline{q}-\underline{q}').\underline{R}_i} \qquad \text{II-22}$$

where $G_c(i,i,\epsilon) = \sum_{\underline{q}} G_c(\underline{q};\epsilon)$

Thus the scattering at the impurity site is described by $\bar{t}^{A,i}_{\underline{q},\underline{q}'}(\epsilon)$ in the sense that $\bar{t}^{A,i}_{\underline{q},\underline{q}'}(\epsilon)$ is the probability amplitude that an electron in the $\underline{q}$ state of the coherent potential lattice scatters to the state $\underline{q}'$ due to the fact that at the site i we have replaced the coherent potential $\Sigma_c(\epsilon)$ by v^A. Obviously, the scattering by a B impurity at $\underline{R}_i$ is described by $\bar{t}^{B,i}_{\underline{q},\underline{q}'}(\epsilon)$ given by Eq. II-22 with A everywhere replaced by B. The CPA condition that there should be no scattering on the average from the impurity site is achieved if we choose the coherent potential $\Sigma_c(\epsilon)$ such that

$$c\,\bar{t}^{A,i}_{\underline{q},\underline{q}'}(\epsilon) + (1-c)\,\bar{t}^{B,i}_{\underline{q},\underline{q}'}(\epsilon) = 0 \qquad \text{II-23}$$

After some rearrangement of terms using Eq. II-22 this condition may be written in the following computationally more suitable form[26]

$$\Sigma_c(\epsilon) = (v_A - \Sigma_c(\epsilon))\, G_c(i,i;\epsilon)\,(v_B - \Sigma_c(\epsilon)) \qquad \text{II-24}$$

Together with the statement that $\bar{G}(\underline{q};\epsilon) = G_c(\underline{q},\epsilon)$ Eq. II-24 is the CPA for our model. Note that it determines the self energy $\Sigma_c(\epsilon)$ in terms of the centre of the A and B bands ϵ_A and ϵ_B respectively and the band shape function $J(\underline{q})$ which is the lattice Fourier transform of the non random overlap integral $J(i,j)$. When these ingredients are given Eq. II-24 can be solved rather easily by numerical iteration.

Since its introduction by Soven[18], Taylor[19], and Onodera[20] and Toyazawa independently a great deal of work has gone into assessing the range and quality of its validity. The general conclusion which has emerged is that the CPA gives a very good overall guide to the averaged eigenvalue spectra of random Hamiltonians[21,22]. We shall now offer some evidence in support of this claim :

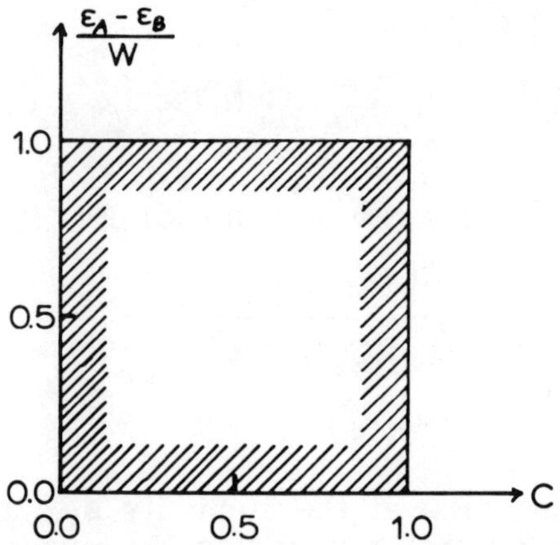

Fig. 6. Range of validity of the CPA in parameter space for the simple tight-binding model discussed in the text. In the shaded regions the theory becomes exact as the boundary approached

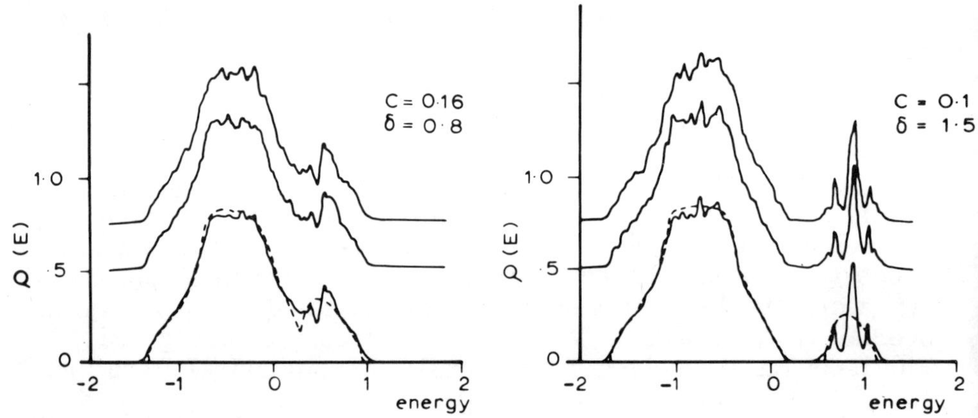

Fig. 7. The full lines are the results of exact calculations of the density of states using the simple model Hamiltonian described in the text for a 7980 site simple cubic lattice. The broken curve is the CPA result for an infinite lattice described by the same parameters. For details see Ref. 30.

a) In Fig. 6 we show the range of validity of CPA for the present model in parameter space[27]. What this picture says is that the theory is exact to order c or $1-c$ if expanded about $c = 0$ or $c = 1$ respectively. It is also exact in the weak scattering $W^{-1}(\epsilon_A - \epsilon_B) \ll 1$ and the strong scattering $W^{-1}(\epsilon_A - \epsilon_B) \gg 1$ limit to order $W^{-1}(\epsilon_A - \epsilon_B)$ and $W(\epsilon_A - \epsilon_B)^{-1}$ respectively. A theory which interpolates between these widely different limits is bound to be more or less right for intermediate values of the parameters.

b) As was shown by Siggia and Schwartz[28], for the present model, there is a small parameter in terms of which the CPA is the lowest order approximation. This small parameter is $1/Z$ where Z is the number of the nearest neighbours. Thus the CPA is a kind of mean field theory. As such it can be expected to work rather well away from critical regions surrounding "phase transitions". A kind of phase transition in this problem is the Anderson localization. Indeed the CPA has been shown not to be able to describe localization[29].

c) There have been many model calculations comparing the predictions of CPA with results of exact calculations. From the point of view of electronic states in real alloys perhaps the most telling of these is that by Krenker, Schwartz and Bloom[30]. They used the simple tight binding model discussed above for a simple cubic lattice and compared the CPA with the results of an exact calculation for a large but finite lattice (7980 sites). The comparison is shown in Fig. 7. As is evident the CPA gives a good overall agreement. Note, however, that the structure in the impurity band in frame b is due to clustering of the impurity atoms. In the CPA the self consistency is established on a single site whose environment is always the averaged environment. It is not allowed to fluctuate. Thus, it can not describe the effect of an improbable cluster forming.

d) The most tangible achievement of CPA in connection with random alloys is the explanation it provided for the photo-emission results in Cu rich Cu-Ni alloys. The calculation was done by Stocks, Williams, and Faulkner[31]. Ni and Cu has roughly the same band width ($W_{Ni} \sim 3.5$ eV, $W_{Cu} \sim 3.0$ eV). This is particularly true if one performs a band theory calculation for Ni on a Cu lattice. Therefore, the randomness of the overlap integrals can be neglected and with some modifications the model discussed in this section should be applicable. What made a fairly realistic calculation possible is the reali-

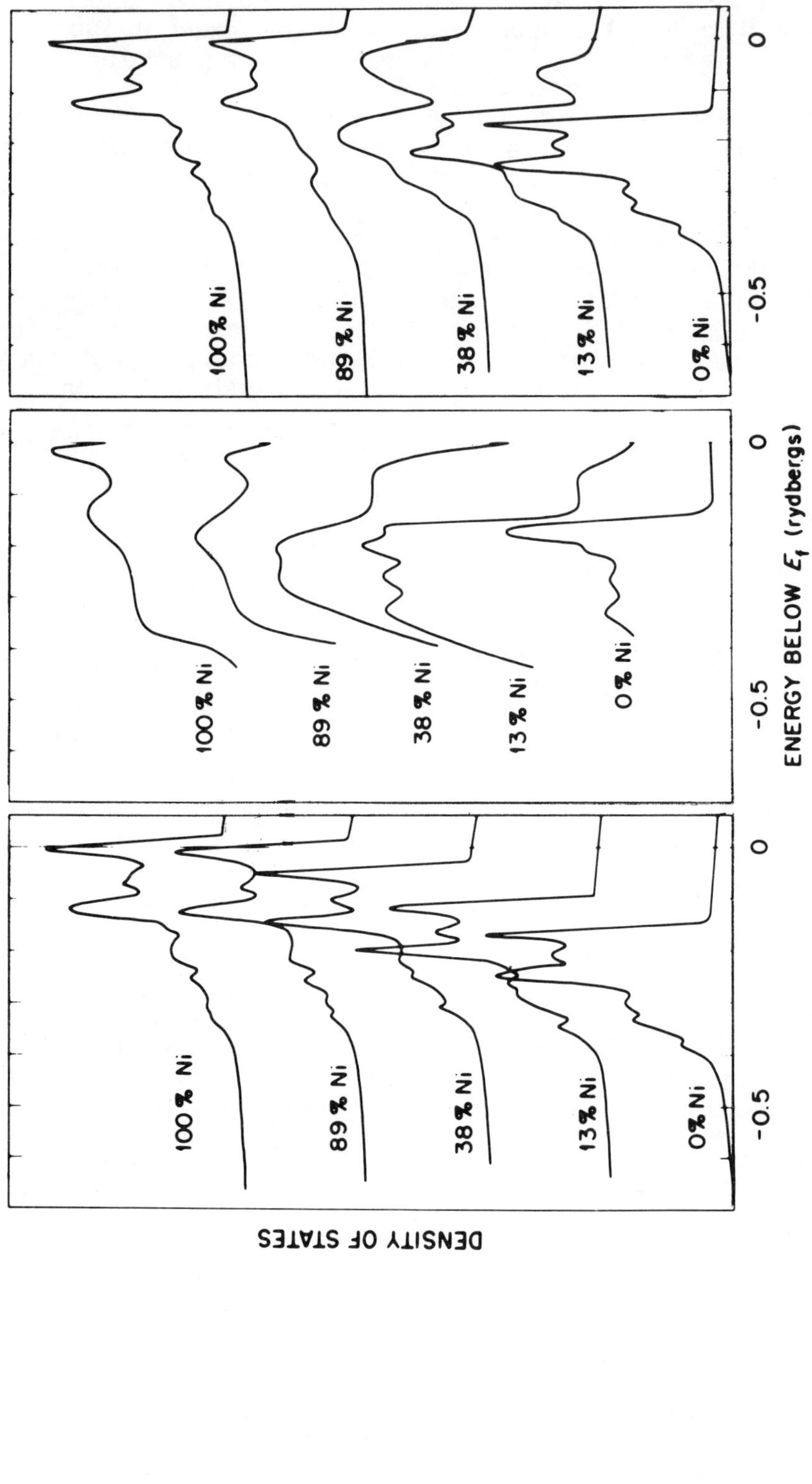

Fig. 8. Comparison of the density of states as calculated by Stocks, Williams, and Faulkner in the CPA and the photoemission density of states for Cu rich Cu-Ni alloys.

zation by Kirkpatrick Velický and Ehrenreich[32] that for similar band shapes and $\underline{q}$ independent self-energies $G_c(i,i;\epsilon)$ may be generated from the pure metal density of states by the Kramers-Kronig relation

$$G_c(i,i;\epsilon) = F(\epsilon - \Sigma(\epsilon)), \quad F(\epsilon) = \frac{1}{\pi}\int_{-\infty}^{\infty} d\epsilon' \frac{\mathrm{Im}\, G(i,i;\epsilon')}{\epsilon - \epsilon'} = \int_{-\infty}^{\infty} d\epsilon' \frac{n(\epsilon')}{\epsilon - \epsilon'} \qquad \text{II-26}$$

where the fact that $\bar{n}(\epsilon) = -\frac{1}{\pi}\,\mathrm{Im}\; G(i,i;\epsilon)$ has been used. In this way one can use $n(\epsilon)$ obtained from a relatistic band structure calculation and one does not have to rely on a single overlap integral to reproduce the band shape (alternatively one could use more overlap integrals and provided they are not allowed to vary with concentration or be random, one can determine them by requiring that they give rise to bands which fit the bands of the pure metals[33]). To take some account of the fact that in these metals the d-band is composed of five bands Stocks et al[31] introduced two self energies one for the bands with e_g symmetry and another for the three t_{2g} bands. The s-band was not treated in the CPA and was added to the final result for the density of states (they are rather similar for pure Ni and pure Cu). Of course, with two self energies one obtains two coupled equations instead of Eq. II-14. These were decoupled by using the fact that $W^{-1}(\epsilon_0^{Cu} - \epsilon_0^{Ni}) \sim .5$ is small. A comparison between their calculation and the photoemission "density of states" is shown in Fig. 8. Note that not only do they get a Ni impurity peak at the right place between the Fermi energy and the upper edge of the Cu d-band but, in agreement with the experiment, the position of this peak remains remarkably constant as the Ni concentration increases.

There are three different attitudes one can adopt towards the mounting evidence in support of the CPA : one can say that the main problem of the electrons in a random potential has been solved and go off to work on neutron stars ; one can attempt to go beyond CPA and try to find approximations which can treat localization and clustering ; and finally one can argue that the CPA has been proven to be a very sensible idea for treating disorder on a lattice and therefore it is worthwhile to implement it for more realistic descriptions of the crystal potential than the one provided by a simple tight-binding model Hamiltonian.

In these lectures we shall adopt the last point of view. The work we shall report on presumes that more can be learned about electrons in random alloys at this stage by doing a better CPA rather than trying to do better than CPA. Our aim is to implement the CPA program for as realistic a description of the electron ion system as is currently used in the band theory of pure metals. In broad terms the justification for such an approach is two fold ; firstly we feel that only by doing realistic calculations will we reap the full benefit of the CPA from the point of view of understanding the behavior of alloys and secondly only by eliminating the main uncertainties which have to do with the model Hamiltonian and not with our treatment of randomness will we be able to establish the most significant weaknesses of the CPA.

To be more specific we shall now briefly summarise the main limitations of the model calculations discussed so far.

In the first place a better CPA must treat random overlap integrals since there are only a few alloys like Ni-Cu or AgPd[34] whose constituents have roughly equal band widths. There has been some progress in this direction and there are some preliminary useful calculations in connection with the analogous problem of random force constants in lattice dynamics[35]. However, in the tight binding framework there is great deal of ambiguity about what to take for the J_{AB} overlap integral and therefore off diagonal randomness can not be treated on equal footing with diagonal randomness.

Secondly, we must treat random hybridization coefficients if we want to handle the s-p and d-band consistently. The adhoc procedure of Stroud and Ehrenreich[36] does not appear to have general validity.

Thirdly we should attempt to abandon the tight binding framework altogether for the first principles methods of band theory. This is a most important point. A real tight binding calculation with overlap integrals obtained from atomic potentials does not give realistic energy bands for pure metals unless very many (10-20) overlap integrals are included and therefore cannot be trusted for alloys. In most current tight binding calculations the parameters are fitted to first principle bands and work well only as interpolation schemes. This option is not open in an alloy at mid-concentration where volume changes and charge transfer make the energy bands of the constituent pure metals largely irrelevant. Therefore in an attempt to make a tight binding calculation more realistic, except for isolated cases, one is likely to end up having too many fitting parameters and therefore endless difficulties in interpreting the results. One way out is to do a full LCAO calculation for the first principles crystal potentials of band theory[37]. This removes the above objection without altering too drastically the general shape of the theory. However, the large basis sets necessary to make such a calculation realistic would make the CPA program too difficult to carry out. In the following lecture we shall suggest that the KKR band theory based on first principles crystal potentials can be generalized to disordered systems and provides a tractable way out of the impass described above.

Finally, to make contact with most experiments we shall need to be able to calculate matrix elements. Even if the overlap integrals have been fitted to first principles bands one can conclude next to nothing from these parameters as to the wave functions. Atomic wave functions are again unreliable. Thus, in this approach one is forced to treat the matrix elements as further adjustible parameters.

To remove these limitations of applying the CPA to real alloys we have generalized the KKR method of band theory to make it applicable to random lattices and developed a calculational scheme which we shall call the KKR-CPA. Before describing this theory however we shall briefly review the band theory program and the KKR for pure metals.

III. BAND THEORY FOR PURE SYSTEMS : the KKR

1) The band theory program

The modern band theory programme consists of three steps : the construction of the crystal potential, the solution of the corresponding Schrödinger equation and the calculation of physical observables. The second step is done numerically and virtually exactly. Typical methods are the KKR, APW LCAO[38]. While some ten years ago there was considerable controversy about the relative merits of these methods this has now subsided as the connection between the methods are reasonably well understood. When carried to convergence they are more or less equivalent[38] .

The rationale behind solving the Schrödinger equation exactly is two-fold : a) this allows one to concentrate on the real, unsolved problem of constructing the crystal potential b) it allows one to interpret quantitatively the complex data provided by the measurements of the de Haas-van Alfen oscillations, soft X-ray spectra, Compton profile etc. Clearly, simple models cannot reproduce the complicated details measured in modern sophisticated experiments and if you are not going to interpret them why measure them ? Moreover, if the Schrödinger equation is solved only approximately and an observable is calculated only approximately then by comparing the results with experiments one is not likely to learn much about the crystal potential.

The crystal potential did not always occupy such a central position in the theory. The point of view permeating the book of Mott and Jones[3] or the famous article of Wigner and Seitz on metallic cohesion[39], is that the one electron picture is only a rough guide and the real problem is how to put the electron-electron interaction back into the calculations. This is the kind of thinking which leads to the Hubbard Hamiltonian description of the electron correlations in transition metals.[40] The relatively new attitude expressed in the first three paragraphs of this section is largely due to the reformulation of the interacting inhomogenious electron liquid problem by Hohenberg and Kohn[41] and Kohn and Sham[42]. They showed that, at least as far as the ground state is concerned, the many electron problem can be reduced to an effective one electron problem which has to be solved self-consistently like a Hartree-Fock calculation. This formulation involves a precise definition of the crystal potential for which the self-consistent solution of the Schrödinger equation solves the many body problem exactly. More significantly they have introduced the local density approximation for this crystal potential[42]. This is exact in the high density and in the slowly varying density limit. In a metal it is reasonable to expect the electron density to be high near the nuclear sites and slowly varying in the interstitial region. The local density approximation appears to be a good interpolation between these limits. The important thing about this approximation is that it treats exchange and correlation on equal footing and it is tractable.

In these lectures we shall not have the time even to sketch the outlines of this major development. We mention it only to indicate that the attitude described at the beginning of this section now has a firm intellectual foundation. There is a crystal potential which is relevant to the full many electron problem and therefore it makes sense to want to discover what it is. Attempting to relate different physical properties to different crystal potentials has been fruitful and remains a promising way of making progress in metal physics.

This 'thumb nail' sketch of the history of our subject is seriously distorted in one important respect. Slater and his collaborators[43] seem to have held the opinion which we described as modern ever since the early fifties[44]. Their sustained efforts to develop efficient ways of solving the Schrödinger equation were always motivated by the desire to solve the many electron problem in the Hartree-Fock-Slater sense. A number of very revealing total energy calculations in the literature are a tribute to these efforts[45].

Having said all this we do not want to leave you with the impression that the question of how to construct the crystal potential is settled. There are many competing methods of construction currently in use : Mattheiss prescription, $x - \alpha$, $x - \alpha - \beta$, the method of Hedin and Lundquist to mention but a few. However, we do not intend to dwell on their relative merits in these lectures. We merely wanted to point out that in modern band theory the construction of the crystal potential is one of the central issues. Others are the calculation of the interesting observables, such as the cohesive energy, the dielectric function[45], the electron-phonon matrix elements[46] etc. On the other hand improving our methods for solving the Schrödinger equation is no longer a main preoccupation of band theorists.

The theory we shall be presenting in these lectures should be seen in the light of the above remarks. Until now there was not a sufficiently accurate method of solving the Schrödinger equation for a random alloy, thus the band theory program as outlined above could not be contemplated for such systems. By developing the KKR-CPA we are hoping to create a situation where the attention can be focussed on the problem of constructing an alloy crystal potential. Except for the "renormalized atom" proposal of Hodges, Watson and Ehrenreich[46] this is so far a virgin territory. Yet, only by developing a workable scheme , which can be tested against many different kinds of experiments, for the calculation of the alloy potential can we understand such important effects as charge transfer and by implication electronegativity differences in the solid state. Furthermore, if we can obtain an alloy potential reliably and we can solve the Schrödinger equation reasonably accurately we should be in the position to study the energetics of alloy formation, and even magnetism within the ground rules of the Kohn and Sham theory and its generalizations[47]. It is important to have these aims in mind if one is not to recoil in horror when faced with the computational effort required to carry the program forward.

2) The crystal potential in the muffin-tin form

For the purposes of illustration we shall now consider the Mattheiss prescription[48,49]. According to this method of constructing the crystal potential one places the Hartree-Fock-Slater atomic charge densities (these are given in the Herman-Skillman tables for most of the elements whose alloys we are concerned with) on every site. After adding up all the charges that fall within a unit cell (overlaps can come from as far as from the 5-th or 6-th nearest neighbour shell) the Poisson equation is solved to determine the electrostatic potential in the unit cell. One then adds to this potential the nuclear Coulomb potential and an exchange contribution of the form $-6[3\rho(r)/8\pi]^{1/3}$ and spherically averages the result about each non equivalent site. Around each site this is considered as the crystal potential out to the muffin tin radius r_{MT} which is usually taken to be half of the nearest neighbour distance. The touching spheres so constructed leave an interstitial region where the potential is set equal to a constant which is the average of the potential between r_{MT} and the Wigner-Seitz radius r_{WS}.

Roughly speaking this construction can be looked on as overlapping atomic potentials and flattening out the interstitial region at the value equal to the average in this region, namely the muffin tin zero, V_{MTZ}. If the atoms are sufficiently far apart that the potentials seen by the outer electrons do not overlap then the muffin tin zero is clearly the same as the atomic zero of the potential (absolute zero) below which are the atomic levels. As the atoms are brought closer together V_{MTZ} will sink rapidly. Typically, for a transition metal at the equilibrium lattice spacing V_{MTZ} is below the atomic d-level from which the d-band is formed. This construction is illustrated in Fig. 9.

We shall solve the Schrödingers equation using V_{MTZ} as our energy zero. Since a valence band which originates from a tightly bound atomic state is roughly centred about the energy of that state most of the valence band will be above V_{MTZ}. That is to say we shall be solving the Schrödinger equation at positive energies. Hence it is natural to make use of the language of scattering theory. We shall now describe those elements of this theory[50] which are relevant to our calculation.

3) Description of a single scatterer

Consider a single muffin tin potential shown in Fig. 10. Since the potential is spherically symmetric the solution of Schrödinger equation

$$(-\nabla^2 + v(\underline{r}))\,\Psi(\underline{r}) = \epsilon\,\Psi(\underline{r}) \qquad \text{III-1}$$

can be written in the form

$$\Psi(\underline{r};\epsilon) = \sum_L a_L(\epsilon)\,R_\ell(r;\epsilon)Y_L(\hat{r}) \qquad \text{III-2}$$

where $Y_L(\hat{r})$ is a spherical harmonic with polar and azimuthal quantum numbers $(\ell, m) \equiv L$[50] and $R_\ell(r;\epsilon)$ is the solution of the radial Schrödingers equation. As usual $\hat{r} = \underline{r}/r$.

$$\left[-\frac{1}{r}\frac{d^2}{dr}r + \frac{\ell(\ell+1)}{r^2} + v(r)\right]R_\ell(r;\epsilon) = \epsilon R_\ell(r;\epsilon) \qquad \text{III-3}$$

For $r > r_{MT}$ Eq. III-3 is the radial Schrödinger equation for free electrons, whose two linearly independent solutions are the spherical Bessel and Neumann functions $j_\ell(\sqrt{\epsilon}\,r)$ and $n_\ell(\sqrt{\epsilon}r)$ respectively. Thus for $r > r_{MT}$ $R_\ell(r;\epsilon)$ is a linear combination of j_ℓ and n_ℓ. Therefore, without loss of generality we may define $R_\ell(r;\epsilon)$ as those solutions of Eq. III-3 which satisfies the boundary condition that

$$R_\ell(r;\epsilon) = \cos\delta_\ell\, j_\ell(\sqrt{\epsilon}\,r) - \sin\delta_\ell\, n_\ell(\sqrt{\epsilon}\,r) \qquad \text{III-4}$$

Inside the muffin-tin sphere $R_\ell(r;\epsilon)$ is that solution of Eq. III-3 which is regular at the origin and at $r = r_{MT}$ its logarithmic derivative $\gamma_\ell(\epsilon) = R_\ell^{-1}\,(dR_\ell/dr)|_{r=r_{MT}}$ must be equal to the logarithmic derivative of $R_\ell(r;\epsilon)$ on the outside given in Eq. III-4. From this condition it follows that

$$\delta_\ell(\epsilon) = \cot^{-1}\left[\frac{\sqrt{\epsilon}\,n_\ell'(\sqrt{\epsilon}\,r_{MT}) - \gamma_\ell(\epsilon)n_\ell(\sqrt{\epsilon}\,r_{MT})}{\sqrt{\epsilon}\,j_\ell'(\sqrt{\epsilon}\,r_{MT}) - \gamma_\ell(\epsilon)\,j_\ell(\sqrt{\epsilon}\,r_{MT})}\right] \qquad \text{III-5}$$

where $n_\ell'(x) = d\,n_\ell(x)/dx$. Note that asymptotically

$$\lim_{r\to\infty} R_\ell(r;\epsilon) = \frac{1}{\sqrt{\epsilon}\,r}\sin\left(\sqrt{\epsilon}r - \frac{\ell\pi}{2} + \delta_\ell(\epsilon)\right) \qquad \text{III-6}$$

differs from the free electron solution only by a shift of phase

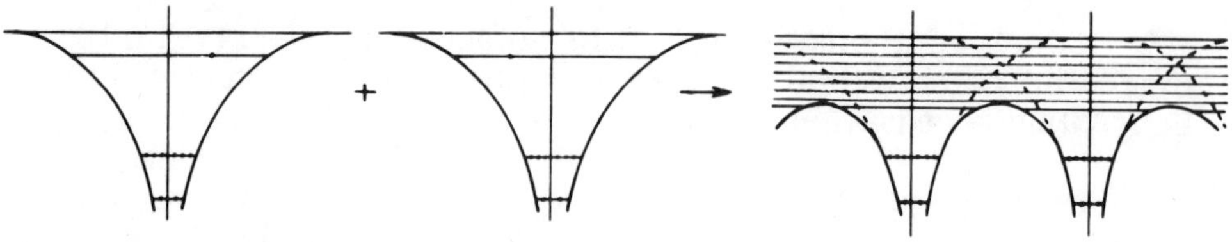

Fig. 9. A schematic illustration of the construction of the muffin tin potential for Na. The solid dots represents the bound core electrons. The open circle is the "conduction" electron. The shaded area is the conduction band in the metal.

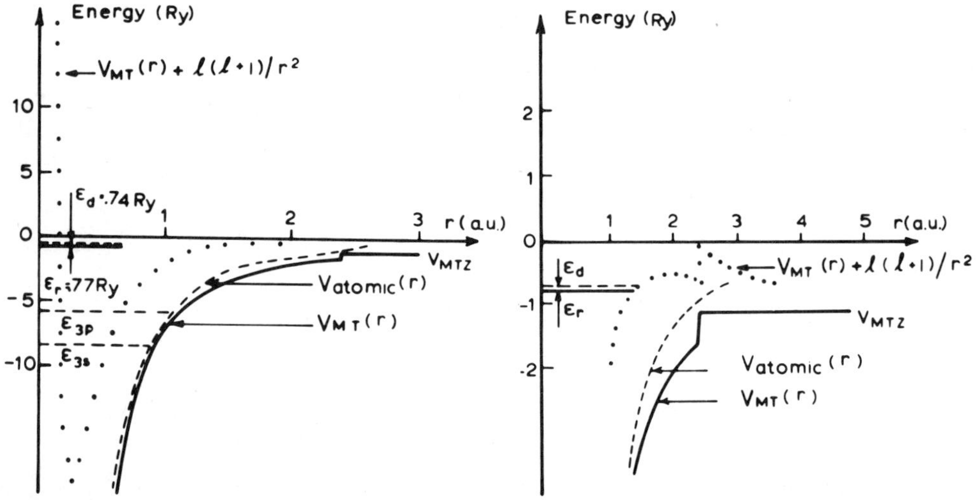

Fig. 10a. The muffin-tin potential $v_{MT}(r)$, the atomic potential $v_{atomic}(r)$ and the sum of $v_{MT}(r)$ and the centrifugal term $\ell(\ell + 1)/r^2$ for a Cu crystal. Fig. 10b is the same as 10a but on an enlarged scale to show what happens at r_{MT}.

$\delta_\ell(\epsilon)$. Hence $\delta_\ell(\epsilon)$ is usually referred to as the phase shift which, as indicated, is a function of the energy.

The connection between the phase shifts and scattering theory can be seen if we recall that in scattering theory one seeks the solution of Eq. III-1 such that $\Psi_{\underline{k}}(r)$ is an eigen-function with energy $\epsilon = k^2$ and asymptotically it can be written as the superposition of an incident plane wave and an out-going spherical wave e.g.[50]

$$\Psi_{\underline{k}}(\underline{r}) = e^{i\underline{k}\cdot\underline{r}} + f(k^2,\theta)\frac{e^{ikr}}{kr} \qquad \text{III-7}$$

where θ is the angle between the vectors $\underline{k}$ and $\underline{r}$. Setting $\epsilon = k^2$ in Eq. III-2 and requiring that $\Psi(\underline{r};k^2)$ has the asymptotic form given in Eq. III-7 we find $a_L(k^2) = 4\pi i\, e^{i\delta_\ell} Y_L^*(\hat{k})$. Furthermore

$$\Psi_{\underline{k}}(\underline{r}) = 4\pi \sum_L i^\ell e^{i\delta_\ell(k^2)} R_\ell(r;k^2)\, Y_L(\hat{r})\, Y_L^*(\hat{k}) \qquad \text{III-8}$$

and the scattering amplitude $f(k^2;\theta)$ is given by

$$f(\epsilon;\theta) = \sum_\ell (2\ell+1)\, f_\ell(\epsilon)\, P_\ell(\cos\theta) \qquad \text{III-9}$$

where

$$f_\ell(\epsilon) = \sin\delta_\ell(\epsilon)\, e^{i\delta_\ell(\epsilon)} \qquad \text{III-10}$$

Using the definition that the total scattering cross section is given by $\sigma = 2\pi \int_{-\pi}^{\pi} d\theta\, |f|^2$ one can also show that

$$\sigma = \frac{1}{\sqrt{\epsilon}} \sum_\ell (2\ell+1)\sin^2\delta_\ell(\epsilon) \qquad \text{III-11}$$

Thus a phase shift near $\pi/2$ means very strong scattering while δ_ℓ near 0 or π means weak scattering.

As is evident from Eqs. III-7, III-9 and III-10 the phase shifts completely determine the scattering properties of the potential function. To put it another way, the effect of the potential within the muffin tin sphere on the wave function outside of this region is completely determined by the δ_ℓ's. Thus the phase shifts are a very efficient description of the potential function for problems where we are interested in the wave functions only outside the scattering region. This is particularly so in the case of muffin tin potentials one encounters in most metals. It turns out that except for the f-band metals

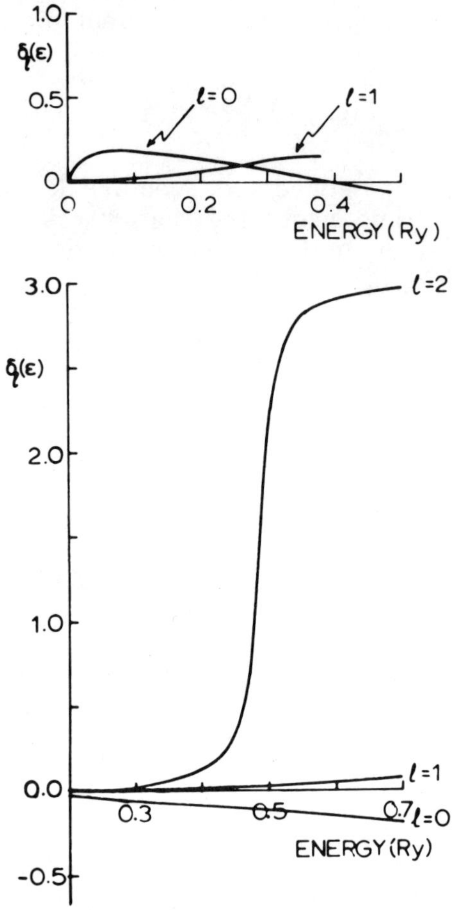

Fig. 11. The phase shifts $\delta_\ell(\epsilon)$ for the muffin-tin potential wells of Na and Cu constructed for the observed lattice spacing.

$\delta_\ell(\epsilon)$ is practically zero for $\ell > 2$ in the energy range of the valence band. As will be seen later the energy bands of a pure metal depend on the muffin tin potential only through the phase shifts. Thus, three simple functions of energy and the crystal structure determines the band structure (but not the wave functions) completely. Clearly, this is a much more concise description than that afforded by an even halfway realistic tight binding Hamiltonian with many overlap integrals.

The phase shifts for a Na and a Cu muffin-tin potentials are shown in Fig. 11. Note that the sodium muffin-tin potential, like a pseudo-potential, is a weak scatterer inspite of the fact that it is sufficiently strong to bind all the core electrons (20). On the other hand the Cu potential is a very strong scatterer at energies near 49 Ry above the muffin tin zero since $\delta_2(\epsilon)$ passes through $\pi/2$.

A phase shift rising from 0 to π within a small energy interval Γ about ϵ_r where it is equal to $\pi/2$ is called a scattering resonance. It is customary to refer to ϵ_r and Γ as the position and the width of the resonance and parametrize the corresponding scattering amplitude by the Breit-Wigner form

$$f(\epsilon) = \frac{\Gamma}{(\epsilon - \epsilon_r) + i\Gamma} \qquad \text{III-12}$$

Alternatively one can write $\tan \delta(\epsilon) = \Gamma(\epsilon - \epsilon_r)^{-1}$. Clearly, the contribution of the resonance to the scattering cross section is a Lorentzian peak $\sigma \sim \Gamma[(\epsilon - \epsilon_r)^2 + \Gamma^2]^{-1}$ which describes the rise and fall of the scattering strength of the potential as ϵ passes through ϵ_r.

An elegant result in scattering theory is that the time spent by an electron at the scattering centre, the Wigner delay time τ_w, is given by $\tau_w = 2\hbar \frac{\partial}{\partial \epsilon} \delta_\ell(\epsilon)$. Thus an electron with an energy near the resonance will spend a long time at the scattering centre. As can be seen in Cu the d-band straddles the resonant energy where $\delta_2^{cu}(\epsilon_r) = \frac{\pi}{2}$. Thus, an electron with energy in the d-band will spend a long time at each site as it travels through the lattice. Moreover, while at a site its wave function will have predominantly d-character about that site. Clearly this is the same physical picture as provided by the tight binding theory where one assumes that a d-electron in a narrow d-band, like that of Cu, is localized in an atomic d-state most of the time and hops to another site only occasionally. The deceptively different language is due to the fact that in the tight binding picture one deals with bound states at negative energies which decay exponentially in space, while in scattering theory we work at positive energies where the scattered wave travels off to infinity with only its phase shifted by the interaction at a site.

To make this connection even more explicit we note that the $\ell = 2$ resonance in the Cu potential arises because the potential $v(r) + \ell(\ell + 1)/r^2$ in Eq. III-3 can almost form a bound state behind the centrifugal barrier shown in Fig. 10. Because we are at positive energies this state is metastable since an electron can always tunnel out. In fact the tunnelling probability gives the width of the resonance. The long Wigner delay time is merely a reflection of the fact that the electron can be trapped into this metastable bound state. Let us now remove the neighbouring atoms around a scattering centre. From the way the muffin-tin potential has been constructed it is clear that V_{MTZ} will rise to the atomic zero of the energy scale but the potential will remain relatively unchanged (we are removing the tails of the atomic potentials on the neighbouring sites). Consequently, at the same absolute energy where previously we had a resonance we will now have a real bound state whose wave function decays exponentially in space. This is the atomic bound state whose wave function enters into the tight-binding theory. Thus the difference in language is due to a different choice for the energy zero.

From this somewhat long winded discussion it should be clear that the muffin-tin potential for all transition metals should have an $\ell = 2$ resonance. Indeed this is the most characteristic feature of the transition metal crystal potentials. Furthermore, all the characteristic transition metal properties like high resistivity, large cohesive energy, high superconducting transition temperature, metallic magnetism etc. can be fairly directly attributed to the presence of this d-resonance.

Having discovered the scattering resonance in the muffin tin potential of transition metals it is now clear why we cannot use perturbation theory to describe their band structure. In fact there is a theorem in scattering theory which says that the Born series does not converge at the resonance and therefore no finite order perturbation theory has any meaning. Thus, for alloys involving transition metals one is forced to consider some non perturbative method of treating disorder as a matter of principle.

Let us now turn to the discussion of the Greens function and the t-matrix in the language of scattering theory.

Almost by definition the Greens function of Eq. III-1 is

$$G(\underline{r}, \underline{r}'; \epsilon) = \sum_n \frac{\Psi_n^*(\underline{r})\, \Psi_n(\underline{r}')}{\epsilon - \epsilon_n + i\eta} \qquad \text{III-13}$$

where the $\Psi_n(r)$'s form an orthonormal complete set and are eigenfunctions of the Hamiltonian $H = -\nabla^2 + v(\underline{r})$. It is a fairly complicated matter but it can be shown that the functions $\Psi_{\underline{k}}(\underline{r})$ in Eq. III-8 for all $\underline{k}$ form a complete set provided the potential $v(\underline{r})$ has no bound states. However, they must be treated with care since their norm, like that of plane waves, is infinite e.g. $\langle\Psi_{\underline{k}}|\Psi_{\underline{k}'}\rangle = (2\pi)^3\delta(\underline{k} - \underline{k}')$. Nevertheless, in the interest of getting on allow us to assert that they can be used to evaluate Eq. III-13 for positive energies. This gives us

$$G(\underline{r}, \underline{r}'; \epsilon) = \int \frac{d^3k}{(2\pi)^3} \frac{\Psi_{\underline{k}}^*(\underline{r})\, \Psi_{\underline{k}}(\underline{r}')}{\epsilon - k^2 + i\eta} \qquad \text{III-14}$$

and

$$\text{Im}\, G(\underline{r}, \underline{r}'; \epsilon) = -\sqrt{\epsilon} \sum_L R_\ell(r; \epsilon) R_\ell(r', \epsilon) Y_L^*(\hat{r}) Y_L(\hat{r}') \qquad \text{III-15}$$

where we have made use of Eq. III-8 and that $\epsilon = k^2$. If there was no potential at the origin, that is to say for free electrons Eq. III-15 would read

$$\text{Im}\, G(\underline{r}, \underline{r}'; \epsilon) = -\sqrt{\epsilon} \sum_L j_\ell(\sqrt{\epsilon}\, r) j_\ell(\sqrt{\epsilon}\, r') Y_L^*(\hat{r}) Y_L(\hat{r}') \qquad \text{III-16}$$

and hence the change in the density of states due to the introduction of the potential is

$$\Delta n(\epsilon) = n(\epsilon) - n_0(\epsilon) = \frac{\sqrt{\epsilon}}{\pi} \sum_\ell (2\ell + 1) \int_0^\infty dr\, r^2 (R_\ell^2(r;\epsilon) - j_\ell^2(\sqrt{\epsilon}\, r))$$

$$= \frac{1}{\pi} \sum_\ell (2\ell + 1) \frac{\partial}{\partial \epsilon} \delta_\ell(\epsilon) \qquad \text{III-17}$$

where the details of the calculation are given in the Ref. 50.

This is the famous Friedel sum and it shows how a phase shift varying rapidly with energy can give rise to a large

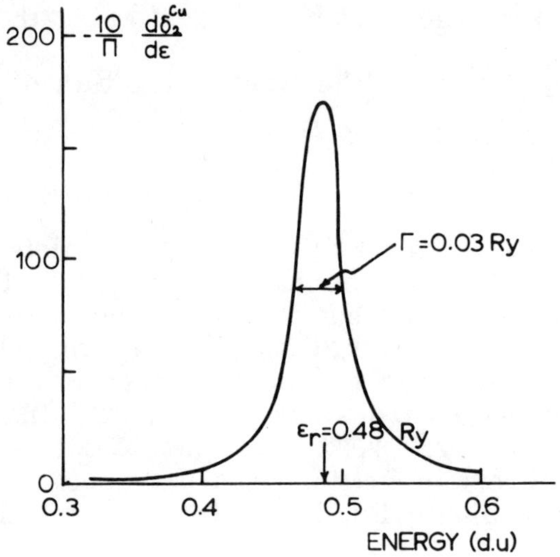

Fig. 12. The Friedel sum contribution to the density of states for the Cu muffin-tin potential discussed in the text.

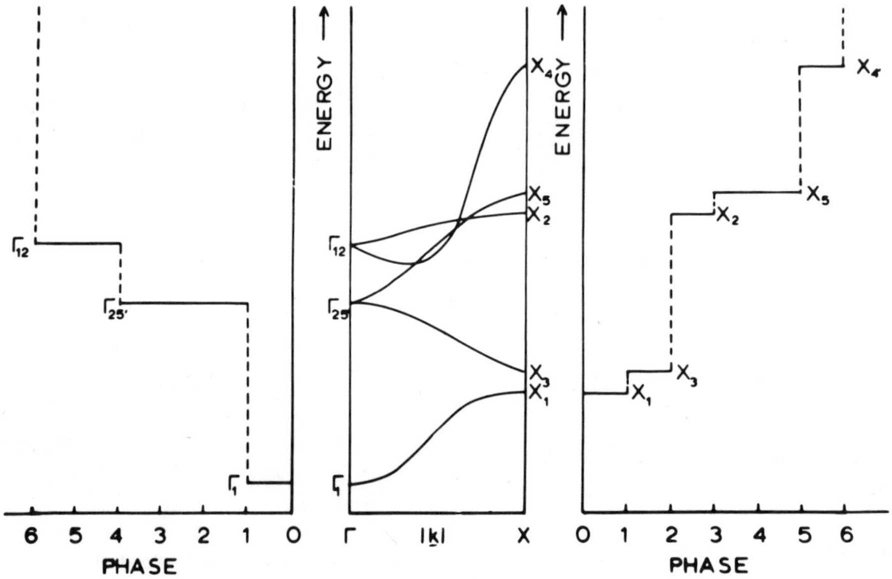

Fig. 13. The Energy bands of pure Cu in the Δ direction and the phase of the KKR determinant at the Γ and X points.

density of states near the resonance energy. Using the phase shift for Cu we show $\Delta n(\epsilon)$ in Fig. 12. Of course the density of states curve in Fig. 12 is not to be confused with the d-band density of states in Cu. Here, Δn is due to a single Cu muffin-tin potential in a flat potentials background. Its width is the width of the d-resonance Γ and not the widt of the d-band. However, when scattering from more potentials are included, (the multiple scattering correction), this resonance peak will broaden out into the d-band. All the same, it remains true that a sharp resonance will give rise to a narrow d-band and a broad resonance a broad one. The point of this short discussion was to give you an inkling of how resonance scattering is related to large densities of states. Obviously, in the scattering theory parlance, it is this connection which replaces the small-overlap-integral, flat-band, high density of states argument of tight binding theory.

As an alternative to constructing the Greens function from the eigenfunctions of Eq. III-1 as in Eq. III-13 we could start with the integral equation

$$G(\underline{r},\underline{r}';\epsilon) = G_0(\underline{r},\underline{r}';\epsilon) + \int d^3r_1 G_0(\underline{r},\underline{r};\epsilon)v(\underline{r}_1)G(\underline{r}_1,\underline{r}';\epsilon) \quad \text{III-18}'$$

which follows from Eq. III-1 and Eq. III-13 if one notes that the free particle $G_0(\underline{r},\underline{r}';\epsilon)$ is defined by the equation

$$(+\nabla^2 + \epsilon + i\eta)G_0(\underline{r}-\underline{r}';\epsilon) = \delta(\underline{r}-\underline{r}') \quad \text{III-19}$$

A way of solving Eq. III-18 is to define the t-matrix by the relation

$$G(\underline{r},\underline{r}';\epsilon) = G_0(\underline{r}-\underline{r}';\epsilon) + \int d^3r_1 \int d^3r_2 G_0(\underline{r}-\underline{r}_1)t(\underline{r}_1,\underline{r}_2;\epsilon)$$

$$G_0(\underline{r}_2-\underline{r}';\epsilon) \quad \text{III-20}$$

and attempt to solve the integral equation

$$t(\underline{r},\underline{r}',\epsilon) = v(\underline{r})\delta(\underline{r}-\underline{r}') + \int d^3r_1 v(\underline{r})G_0(\underline{r}-\underline{r}_1;\epsilon)t(\underline{r}_1,\underline{r}';\epsilon) \quad \text{III-21}$$

which follows from Eqs. III-18 and III-20.

In its full generality Eq. III-21 is quite difficult to solve, however, the "on the energy shell" angular momentum components of $t(\underline{r},\underline{r}';\epsilon)$ turn out to be simple quantities. To define these consider the matrix elements of the t-matrix in the plane wave representation.

$$t(\underline{k},\underline{k}';\epsilon) = \int d^3r_1 \int d^3r_2\, e^{-i\underline{k}\cdot\underline{r}_1}\, t(\underline{r}_1,\underline{r}_2;\epsilon)\, e^{i\underline{k}'\cdot\underline{r}_2} \qquad \text{III-22}$$

In general the above matrix elements are defined for all $\underline{k}$ and $\underline{k}'$, that is to say for arbitrary incoming and outgoing plane wave energies k^2 and k'^2 respectively. If $k^2 = k'^2 = \epsilon$ we are describing elastic scattering and $t(\underline{k},\underline{k}';\epsilon)$ is the probability amplitude that a plane wave $|\underline{k}\rangle$ scatters into the state $|\underline{k}'\rangle$. This situation is often referred to as being "on the energy shell". Using the expansion $e^{i\underline{k}\cdot\underline{r}} = 4\pi \sum_L i^\ell j_\ell(kr) Y_L^*(\hat{k}) Y_L(\hat{r})$ it is easy to see that

$$t(\underline{k},\underline{k}';\epsilon) = \sum_{L,L'} i^{-\ell+\ell'} (4\pi)^2 Y_L(\hat{k})\, t_{L,L'}(\epsilon)\, Y_{L'}(\hat{k}') \qquad \text{III-23}$$

where the angular momentum components of the "on the energy shell" t-matrix $t_{L,L'}$ are defined as

$$t_{L,L'}(\epsilon) = \int d^3r_1 \int d^3r_2\, j_\ell(\sqrt{\epsilon}\, r_1) Y_L(\hat{r}_1)\, t(\underline{r}_1,\underline{r}_2;\epsilon)\, Y_{L'}^*(\hat{r}_2)\, j_{\ell'}(\sqrt{\epsilon}\, r_2) \qquad \text{III-23}$$

We note in passing that for a general potential we should be talking about a matrix $t_{L,L'}(\epsilon)$. However, for a spherically symmetric potential $v(r)$ the t-matrix is diagonal and it is sufficient to define $t_L(\epsilon) = t_{LL}(\epsilon)$. Given what we said about the meaning of $t(\underline{k},\underline{k}';\epsilon)$ "on the energy shell" it should not come as a surprise that

$$t_L(\epsilon) = -\frac{1}{\sqrt{\epsilon}} f_\ell(\epsilon) = -\frac{1}{\sqrt{\epsilon}} \sin\delta_\ell\, e^{i\delta_\ell} \qquad \text{III-24}$$

This is shown in Ref. 50. Following a somewhat inprecise practice we shall refer to both $t_L(\epsilon)$ and $f_\ell(\epsilon)$ as scattering amplitudes.

Using Eq. III-16 and III-24 we can thus write :

$$\mathrm{Im}\, G(\underline{r},\underline{r}';\epsilon) = \sum_L \Delta_L(\underline{r};\epsilon)\, \Delta_L(\underline{r}';\epsilon)\, \mathrm{Im}\, t_L(\epsilon) \qquad \text{III-25}$$

where

$$\Delta_L(\underline{r},\epsilon) = -\sqrt{\epsilon}\, \frac{R_\ell(r;\epsilon)}{\sin\delta_\ell(\epsilon)}\, Y_L(\hat{r}) \qquad \text{III-26}$$

It might come as a let down to end the presentation of a sequence of sophisticated results with this rather artificial expression. However, bear with us. In the next section we shall show how Eq. III-25 generalizes very nicely when we consider the imaginary part of the Greens function at a site in the presence of scattering centres at all other sides. Then the expression analogous to Eq. III-25 will be a non trivial result.

In summary the density of states for a single scatterer is given in terms of the phase shifts in Eq. III-17. Physical observables which depend on the wave function as well as the eigenvalues can frequently be calculated from Im $G(\underline{r},\underline{r}';\epsilon)$. For such purposes Eq. III-25 can be used. Clearly, Im $G(\underline{r},\underline{r}';\epsilon)$ depends on the wave functions $\Delta_L(\underline{r},\epsilon)$ as well as on the phase shifts.

As a final remark we note that in scattering theory we have a continuum of states even for one scatterer. This is evident from Fig. 12 or from the definition of the functions $\Psi_{\underline{k}}(\underline{r})$. This continuity is due to the fact that we are working at positive energies. In the tight-binding formulation one is solving the Schrödinger equation at negative energies and therefore, in the case of the one atom problem, the spectrum is discrete. It is then a very subtle mathematical process by which this discrete spectrum turns into a continuum when the possibility of hopping from site to site all the way to infinity is introduced. No finite cluster will give a continuum. This is an important technical difference which must be borne in mind when approximations in the two different pictures are compared.

4) Multiple Scattering Theory

Let us now consider the Schrödinger equation for the full crystal lattice with the appropriate muffin-tin potential on every site. As was the case in the previous section this can be done in one of three ways. One can solve the Schrödinger equation for the wave functions. Equivalently, one can attempt solving Eq. III-18 with $v(\underline{r})$ replaced by the full crystal potential $V(\underline{r}) = \sum_i v(\underline{r} - \underline{R}_i)$. Finally, one can calculate the Greens function from the t-matrix for the full crystal potential defined as

$$G(\underline{r},\underline{r}';\epsilon) = G_0(\underline{r}-\underline{r}';\epsilon) + \int d^3r_1 \int d^3r_2 G_0(\underline{r}-\underline{r}_1;\epsilon) \cdot T(\underline{r}_1,\underline{r}_2;\epsilon)\, G_0(\underline{r}_2-\underline{r}';\epsilon) \qquad \text{III-27}$$

In this case the problem is to find $T(\underline{r},\underline{r}';\epsilon)$ by solving

$$T(\underline{r},\underline{r}';\epsilon) = V(\underline{r})\delta(\underline{r}-\underline{r}') + \int d^3r_1 V(\underline{r})G_0(\underline{r}-\underline{r}_1;\epsilon)T(\underline{r}_1,\underline{r}';\epsilon) \qquad \text{III-28}$$

An efficient way of proceeding is to follow this last method.

To solve Eq. III-28 we introduce the scattering path operator[51] by the relation

$$\mathcal{T}^{i,j}(\epsilon) = v_i\, \delta_{ij} + \sum_k v_i\, G_0(\epsilon)\mathcal{T}^{k,j}(\epsilon) \qquad \text{III-29}$$

where we used the operator notation to simplify the algebra. Obviously, $v(\underline{r})\,\delta(\underline{r} - \underline{r}')$ is the matrix element $\langle \underline{r}|v|\underline{r}'\rangle$ and $G_0(\underline{r} - \underline{r}';\epsilon) \equiv \langle \underline{r}|G_0(\epsilon)|\underline{r}'\rangle$ etc. Furthermore by taking similar matrix elements of Eq. III-28 we obtain an integral equation like Eq. III-28.

Summing over i and j in Eq. III-29 one obtains the operator version of Eq. III-28 if one identifies $\sum_{i,j} \mathcal{T}^{i,j}$ with T. Thus solving Eq. III-29 is equivalent to solving Eq. III-28.

By taking the term with k = i on the right hand side of Eq. III-29 over to the left and dividing both sides of the equation by the operator $1 - v_i G_0$ we can rewrite Eq. III-29 as

$$\mathcal{T}^{i,j} = t_i\, \delta_{i,j} + \sum_{k\neq i} t_i\, G_0 \mathcal{T}^{k,j} \qquad \text{III-30}$$

where the t-matrix operator t_i is defined as $(1 - v_i G_0)^{-1} v_i$. Its matrix elements $\langle \underline{r}|t_i(\epsilon)|\underline{r}'\rangle$ are the t-matrix defined in Eq. III-20 with the exception that now the muffin tin potential $v(\underline{r} - \underline{R}_i)$ is centred at $\underline{R}_i$ and not at the origin.

In our version of the multiple scattering theory[50] Eq. III-30 is the fundamental equation. It is a way of constructing the solution to the many scatterer problem namely $T(\underline{r}, \underline{r}'; \epsilon) \equiv \langle \underline{r}|T(\epsilon)|\underline{r}'\rangle$ from the knowledge of the solution to the individual scattering problems $\langle \underline{r}|t_i(\epsilon)|\underline{r}'\rangle$. Its meaning becomes apparent if we note that quite generally the scattering solution to a Schrödinger equation can be written as $\Psi_{\underline{K}} = \phi_{\underline{K}} + \Phi_{\underline{K}}$ where $\phi_{\underline{K}}$ is the incident wave and $\Phi_{\underline{K}}$ is the scattered wave as in III-7 and it can be shown that $|\Phi_{\underline{K}}\rangle = t\, G_0\, |\phi_{\underline{K}}\rangle$. Thus, the t-matrix operator generates the scattered wave from the incident wave. More specifically t_i generates the scattered wave due to the potential $v(\underline{r} - \underline{R}_i)$ and T generates the scattered wave due to the whole crystal potential. A little further thought along these lines reveals that the scattering path operator $\mathcal{T}^{i,j}$ when it operates on the wave incident at the site $\underline{R}_j$ it gives the scattered waves emanating from the site at $\underline{R}_i$ and includes the effects of all the scattering inbetween. Evidently $\sum_{ij} \mathcal{T}^{i,j}$ takes the incident waves arriving at all the sites in the lattice turns them into scattered waves emanating from all other sites and adds up all the scattered waves. This is precisely what T does, hence the equality. In these terms Eq. III-30 is merely a self consistency requirement on $\mathfrak{J}^{i,j}$: a wave incident at $\underline{R}_j$ becomes a scattered wave at $\underline{R}_k$ then propagates to $\underline{R}_i$ according to the free particle propagator G_0, to become an incident wave there and there to be turned into a scattered wave from $\underline{R}_i$ by t_i, the sum of all such processes must be equal to $\mathfrak{J}^{i,j}$ again for $i \neq j$. For $i = j$ there is also a direct scattering term : t_i giving the scattered wave due to the direct scattering at $\underline{R}_i$ and $\sum_{k\neq i} t_i G_0 \mathfrak{J}^{k,i}$ giving the contribution to the scattered wave which arises because the scattered particle has returned to the site $\underline{R}_i$, after multiple scattering at the other sites, there to be scattered according to t_i.

The importance of this formulation of the problem to the crystal potential which consists of non-overlapping muffin-tin wells is that for non overlapping potentials the "on the energy shell" part of Eq. III-30 decouples from the rest and this gives rise to a particularly powerful tool for calculating the energy bands.

To see what is meant by this statement let us define, in analogy with Eq. III-23 the "on the energy shell" components of the matrix elements $\langle \underline{r} | \mathfrak{J}^{i,j}(\epsilon) | \underline{r}' \rangle \equiv \mathfrak{J}^{i,j}(\underline{r},\underline{r}';\epsilon)$ of the scattering path operator $\mathfrak{J}^{i,j}_{L,L'}(\epsilon)$ as

$$\mathfrak{J}^{i,j}_{L,L'}(\epsilon) = \int d^3r_i \int d^3r_j \, Y_L(\hat{r}_i) j_\ell(\sqrt{\epsilon}\, r_i) \mathfrak{J}^{i,j}(\underline{r}_i, \underline{r}_j'; \epsilon) j_{\ell'}(\sqrt{\epsilon}\, r_j') Y^*_{L'}(\hat{r}_j') \qquad \text{III-31}$$

where L is the angular momentum about the site i and L' is that about the site $\underline{R}_j$, $\underline{r}_i = \underline{r} - \underline{R}_i$ and $\underline{r}_j' = \underline{r}' - \underline{R}_j$. The use of these spatial variables is legitimate since it follows from the integral equation version of Eq. III-30 that $J^{i,j}(\underline{r},\underline{r}';\epsilon)$ depends on $\underline{r}$ and $\underline{r}'$ only through $\underline{r} - \underline{R}_i$ and $\underline{r}' - \underline{R}_j$.

We now want to find an expression for $\mathfrak{J}^{i,j}_{L,L'}(\epsilon)$ in terms $t_L(\epsilon)$ from Eq. III-30. To do this take the matrix elements of this equation between the states $|\underline{r}\rangle$ and $|\underline{r}'\rangle$ inserting the identity $1 = \int d^3r \, |\underline{r}\rangle\langle \underline{r}|$ between operator products and then take the "on the energy shell" components of both sides of the equation by the operation defined in Eq. III-31. Recognising that for $\underline{r}$ in the muffin-tin sphere surrounding $\underline{R}_i$ and $\underline{r}'$ in the muffin-tin sphere surrounding $\underline{R}_j$ with $i \neq j$

$$G_0(\underline{r} - \underline{r}';\epsilon) = \sum_{L,L'} Y_L(\hat{r}_i) j_\ell(\sqrt{\epsilon}\, r_i) G_{L,L'}(\underline{R}_i - \underline{R}_j;\epsilon) \, j_{\ell'}(\sqrt{\epsilon}\, r_j) Y_{L'}(\hat{r}_j) \qquad \text{III-32}$$

where

$$G_{L,L'}(\underline{R}_i - \underline{R}_j;\epsilon) = 4\pi\sqrt{\epsilon} \sum_{L''} C^{L''}_{L,L'} (i)^{\ell''} h^{+}_{\ell}(\sqrt{\epsilon}\,|\underline{R}_i - \underline{R}_j|) \, Y_{L''}(\widehat{\underline{R}_i - \underline{R}_j}) \qquad \text{III-34}$$

with h^{+}_{ℓ} a Hankel function and $C^{L''}_{L,L'}$ a Gaunt number defined by $C^{L''}_{L,L'} = \int d\Omega Y_L(\Omega) Y_{L''}(\Omega) Y^{*}_{L'}(\Omega)$ it is entirely straightforward to show that

$$\mathcal{T}^{i,j}_{L,L'}(\epsilon) = t_L(\epsilon)\delta_{i,j} + \sum_{k\neq i,L''} t_L(\epsilon) G_{L,L''}(\underline{R}_i - \underline{R}_k;\epsilon) \mathcal{T}^{k,j}_{L'',L'}(\epsilon) \qquad \text{III-35}$$

This is, then, our fundamental multiple scattering equations "on the energy shell". It gives $\mathcal{T}^{i,j}_{L,L'}$ in terms $t_L(\epsilon)$, that is to say in terms of the phase shifts, and the structure constants $G_{L,L'}(\underline{R}_i - \underline{R}_j;\epsilon)$ which do not depend on the potential function and are determined completely by the spatial arrangement of the scattering sites. Eq. III-34 is valid for any arrangement of potentials even if at each site we have a different scatterer (in this case we would have to replace $t_L(\epsilon)$ by $t_{i,L}(\epsilon)$). Thus, it is a good starting point to discuss pure metals, liquids as well as random alloys. The only limitation on its validity is the requirement that the potential wells at the different sites may not overlap. If they are allowed to overlap we need an expression for $G_0(\underline{r}, \underline{r}';\epsilon)$ which is valid for $\underline{r} = \underline{r}'$ where Eq. III-32 cannot be used. The unfortunate consequence of overlapping potentials is that the "on the energy shell" part of the problem becomes coupled to the off the energy shell part and except for some simplified models the problem becomes intractable. The physical reason for the "off the energy shell" component coming in for overlapping potentials is easy to see. For nonoverlapping potentials one elastic scattering process is completely over before the next one begins. So, you stay on the energy shell, and the only information about the potential which matters are the phase shifts. For overlapping scatterers before a scattering process is over, e. g. one is still within the scattering region, the next one begins because the tail of the next potential hangs into the scattering region of the first. Under these circumstances the inelastic virtual processes begin to play a role. Fortunately, we will not need the theory under such adverse circumstances.

Before closing this section we would like to introduce a notation which will simplify many manipulations later on in these lectures. Consider the space spanned by the abstract vectors $|i, L\rangle$ and regard $\mathcal{T}^{i,j}_{L,L'}(\epsilon)$ as matrix elements of the operator $\mathcal{T}(\epsilon)$ in that space. Namely write

$$\mathcal{T}^{i,j}_{L,L'}(\epsilon) \equiv \langle i, L|\mathcal{T}(\epsilon)|j, L\rangle$$

$$t_{i,L}(\epsilon) \equiv \langle i, L|t(\epsilon)|i, L\rangle$$

$$\begin{aligned} G_{L,L'}(\underline{R}_i - \underline{R}_j;\epsilon) &\equiv \langle i, L|G_0(\epsilon)|j, L\rangle \quad i \neq j \\ &\equiv 0 \qquad\qquad\qquad\qquad i \neq j \end{aligned} \qquad \text{III-36}$$

and regard

$$\mathcal{T}(\epsilon) = t(\epsilon) + t(\epsilon)G_0(\epsilon)\mathcal{T}(\epsilon) \qquad \text{III-37}$$

as the operator equivalent of Eq. III-35 in the site-angular momentum space spanned by $|i, L\rangle$

It is interesting to note that

$$[t^{-1}(\epsilon) - G_0(\epsilon)]\mathcal{T}(\epsilon) = I \qquad \text{III-36}$$

and therefore

$$\sum_{k,L''} (t^{-1}(\epsilon)\delta_{L,L'}\delta_{i,j} - G_{L,L''}(\underline{R}_i - \underline{R}_k;\epsilon)\mathcal{T}^{k,j}_{L'',L'}(\epsilon) = \delta_{i,j}\delta_{L,L'} \qquad \text{III-37}$$

On comparing this expression with Eq. II-7 we see that in this language the multiple scattering theory on the energy shell has precisely the same form as the tight binding theory. $\mathcal{T}^{i,j}_{L,L'}(\epsilon)$ plays the role of the Greens function $G(i,j;\epsilon)$ and $(\epsilon - \epsilon_i)\delta_{i,j}$ is replaced by $t^{-1}_{i,L}\delta_{i,j}$. For resonant scatterers this connection is even closer since then $t^{-1}_L(\epsilon) = \Gamma^{-1}(\epsilon - \epsilon_r) - i\sqrt{\epsilon}$. However, this connection is mainly formal. The physical content is very different. For instance the "overlap integral" $G_{L,L'}(R_i - R_j;\epsilon)$ is infinite ranged and energy dependent but is independent of the potential. This fact is of enormous significance from the point of view of the CPA since, as we shall see, it eliminates the necessity of dealing with off-diagonal randomness without making any approximations. (The approximation is made when the crystal potential is constructed but not in solving the Schrödinger equation). However, we are getting ahead of ourselves. It is more useful at this stage if we now turn to the discussion of pure metals.

5) The KKR band theory for pure metals

If the t-matrix $t_L(\epsilon)$ is the same on every site $\mathcal{T}^{i,j}_{L,L'}(\epsilon)$ depends only on $\underline{R}_{i,j}$ and, for an infinite lattice, Eq. III-35 can be easily solved by the method of lattice Fourier transforms. Defining $\mathcal{T}_{L,L'}(\underline{q},\epsilon)$ and $G_{L,L'}(\underline{q},\epsilon)$ by the relations

$$\mathcal{T}_{L,L'}(\underline{q},\epsilon) = \frac{1}{N} \sum_{i} e^{-i\underline{q}\cdot(\underline{R}_i - \underline{R}_j)} \mathcal{T}^{i,j}_{L,L'}(\epsilon) \qquad \text{III-38}$$

and

$$G_{L,L'}(\underline{q},\epsilon) = \frac{1}{N} \sum_{i} e^{-i\underline{q}\cdot(\underline{R}_i - \underline{R}_j)} G_{LL'}(\underline{R}_i - \underline{R}_j;\epsilon) \qquad \text{III-39}$$

we find

$$\mathcal{T}_{LL'}(\underline{q},\epsilon) = \left[(t^{-1}(\epsilon) - G_0(\underline{q},\epsilon))^{-1} \right]_{LL'} \qquad \text{III-40}$$

where t^{-1} is a diagonal matrix with diagonal elements $t^{-1}_L(\epsilon)$ and $G_0(\underline{q},\epsilon)$ stands for the matrix $G_{LL'}(\underline{q},\epsilon)$.

As we have mentioned before T and therefore $\mathcal{T}$ generates the scattered waves from the incident waves. Thus $\mathcal{T}$ is like a response function. Where it diverges there will be a scattered wave even in the absence of an incident wave. Evidently these will be the stationary states of the systems, that is to say the energy bands. It follows from the rules of matrix inversion

$$\mathcal{T}_{L,L'}(\underline{q},\epsilon) = \frac{1}{||t^{-1} - G(\underline{q},\epsilon)||} \operatorname{cof}\left(t^{-1}_L(\epsilon)\delta_{L,L'} - G_{L,L'}(\underline{q},\epsilon)\right) \qquad \text{III-41}$$

that $\mathcal{T}_{L,L'}(\underline{q},\epsilon)$ will diverge where the determinant

$$||t^{-1}_L(\epsilon)\delta_{L,L'} - G_{L,L'}(\underline{q},\epsilon)|| = 0 \qquad \text{III-42}$$

This is well known KKR condition[52]. It can yield the energy bands in two ways. For a given $\underline{q}$ one can search in ϵ and the energies at which Eq. III-42 is satisfied will be the eigen energies $\epsilon_{q,n}$ at that $\underline{q}$. Depending how far up in energy one is willing to go one will find all the bands on the way. This mode of operation is usually referred to as the "constant $\underline{q}$ search". Alternatively, one can do a constant energy search . Operating the KKR in this mode consists of fixing the energy and searching for all the zeros of the KKR determinant in the Brillouin zone. This is done direction by direction. Starting at te zone centre the KKR determinant is evaluated on a given mesh out to the zone boundary and in this way all its zeros, in that direction, are determined. When this is done for a sufficiently large number of directions (561 in $1/48^{-th}$ of the Brillouin zone) the $\underline{q}$ coordinates of the zeros map out a constant energy surface. The volume in q-space below this surface is the integrated density of states.

In fact the integrated density of states may be written as

$$N(\epsilon) = N_0(\epsilon) - \frac{1}{\pi\Omega_{BZ}} \text{Im} \int_{BZ} d^3q \, \ln ||t_L^{-1}\delta_{LL'} - G_{LL'}(\underline{q},\epsilon)|| \quad \text{III-43}$$

which is the integral of the phase of the K.K.R. determinant over the Brillouin zone (BZ) whose volume is Ω_{BZ} plus the free electron electron contribution N_0. It can be regarded as the generalization of the Friedel sum for an infinite lattice. We give the proof of Eq. III-43 in Appendix I. However, that the Brillouin zone integral of the phase of the determinant $||t_L^{-1}\delta_{LL'} - G_{LL'}(\underline{q},\epsilon)||$ is the integrated density of states can be seen without such derivation. Take it from us that for cubic systems the KKR matrix $t_L^{-1}\delta_{LL'} - G_{LL'}(\underline{q},\epsilon)$ is real. Thus, its determinant when regarded as a complex number will have a phase which is an integer multiple of π. If this phase is defined as $\varphi(\underline{q},\epsilon)$ $\text{Im} \ln ||t_L^{-1}\delta_{LL'} - G_{LL'}(\underline{q},\epsilon)||$ it will change discontinuously by π every time the determinant passes through 0 with either $\underline{q}$ or ϵ changing. Thus by determining the phase difference between the bottom of the band and an arbitrary ϵ for a given $\underline{q}$ we have found a number which is π times the number of states below that energy. Hence, Eq. III-43 is manifestly the integrated density of states. To illustrate this 'counting' in Fig. 13 we have plotted the energy bands of pure Cu along the Δ direction and the corresponding phase as a function energy for the zone centre (Γ-point) and the zone boundary (X-point). The singlet Γ_1 gives rise to a change of phase of π whilst the triplet $\Gamma_{25'}$ and doublet Γ_{12} states give rise to charges in the phase of 3π and 2π respectively, thus accounting correctly for the degeneracy of the levels and in the process yielding at all energies the phase with respect to the bottom of the band.

Observe that the possibility of a constant energy search differentiates sharply between the scattering theory approach to solving the Schrödinger equation and the Rayleigh-Ritz type of methods for solving eigenvalue problems. As is evident from the foregoing discussion using the KKR in the constant energy search mode allows us to determine the eigen states at a given energy without knowing all the states below. This makes this method particularly efficient for accurate determination of the Fermi energy, Fermi surface, and the density of states at the Fermi energy. It will also facilitate the solution of the random alloy problem energy by energy.

Of course there is no mystery connected with the above remarks. The Greens function is determined by all the eigenvalues and eigen-functions of the Hamiltonian. All these must be known to calculate the Greens function at a given energy if the construction in Eq. III-13 is used. However in scattering theory we can calculate the Greens function at a given energy directly and the eigen solutions at other energies are not needed. The mathematical theorem which ensures the equivalence of these two approaches is the Mittag-Leffler theorem[54] which relates the values of a meromorphic function to the positions of its poles.

It will amplify these remarks and will be useful later if we note that for $\underline{r}$ and $\underline{r}'$ within the same muffin-tin sphere at $\underline{R}_i$

$$\mathrm{Im}\, G(\underline{r},\underline{r}',\epsilon) = \sum_{L,L'} \Delta_L(\underline{r};\epsilon)\,\Delta_L(\underline{r}';\epsilon)\,\mathrm{Im}\,\mathcal{T}^{i,i}_{L,L'}(\epsilon) \qquad \text{III-44}$$

where $\Delta_L(\underline{r};\epsilon)$ is defined in Eq. III-26. This expression is clearly the analogue of Eq. III-26 in the case of many scatterers and is shown to be valid for arbitrary number and arrangement of non-overlapping scattering centres in Ref. 51. For a finite group of scatterers or for a random system where the potential wells on different sites are different $\mathcal{T}^{i,i}_{L,L'}(\epsilon)$ is given by the solution of Eq. III-35. In the case of an ordered system of scatterers

$$\mathcal{T}^{i,i}_{L,L'}(\epsilon) = \frac{1}{\Omega_{BZ}} \int d^3q\, \mathcal{T}_{LL'}(\underline{q},\epsilon) \qquad \text{III-45}$$

Clearly, in going from Eq. III-25 to Eq. III-44 we have merely replaced the single site t-matrix on the energy shell $t_L(\epsilon)$ with the multiple scattering t-matrix at the site in question $\mathcal{T}^{i,i}_{L,L}(\epsilon)$ which is also on the energy shell. As we mentioned before $\mathcal{T}^{i,i}_{L,L}(\epsilon)$ gives the scattered wave from the site at $\underline{R}_i$ and includes the contribution from all processes in which the electron scattered from the potential well at $\underline{R}_i$ scatters repeatedly at other sites and then returns to the site at $\underline{R}_i$ to be scattered there again. It is appealing to the intuition that the modification of the Greens function within the muffin tin-well at $\underline{R}_i$ due to the introduction of the other scattering centres only depends on the scattering properties, namely the phase shifts, at those new sites.

The Greens function contains much more information than the energy bands. One can calculate many useful observables from the Greens function within a muffin-tin well. Thus the significance of Eq. III-44 is that it allows us to calculate such wave function dependent quantities with $R_\ell(r;\epsilon)$ the solution of the radial Schrödinger equation (Eq. III-3) for a single potential well as the only extra information needed beyond the phase shifts and the structure constants. Unfortunately, the Brillouin zone integration in Eq. III-45 is rather difficult to carry out and its evaluation by numerical means has become possible only recently[55]. It is, still, a reasonably tractable calculation for selected energies only. Some of the interesting observables which can be calculated in this way will be discussed in the next section in connection with random alloys.

IV. RANDOM ALLOYS AND THE KKR-CPA

1. Multiple scattering on the energy shell for random alloys

Consider a typical alloy configuration. If we were to construct the crystal potential all through the lattice we would find that not all the muffin-tin potentials on the A sites are the same. Due to the fact that the atomic charge distributions at the neighbouring sites will have tails which overhang into the muffin-tin sphere at the site in question the muffin-tin potential there will depend on the local environment. For most configuration this will vary from A site to A site. Of course, the same is true on the B sites. Consequently, at least in principle, we shall have to describe the "on the energy shell" scattering on A and B sites by sequences of scattering amplitudes $t^{\nu}_{A,L}(\epsilon)$ and $t^{\nu}_{B,L}(\epsilon)$ where the superscript ν labels all possible configurations within the first few nearest neighbour shells (as many shells as used in the construction of the potential). In a later lecture we shall return to discuss this complication and will show how it can be used to describe short range order within the context of the CPA. However, for now, we shall assume that all the A scattering amplitudes are the same and all the B scattering amplitudes are the same. Namely we set $t^{\nu}_{A,L} = t_{A,L}$ and $t^{\nu}_{B,L} = t_{B,L}$ for all ν. That this is a reasonable approximation can be seen from Fig. 10. Comparing the atomic potential and the muffin-tin potential it is evident that the effect of having neighbours is relatively small well within the muffin-tin radius. Furthermore, it is precisely this part of the potential which determines the position of the resonance. Since this is the most dominant aspect of our potential, as a first approximation, it is reasonable to neglect the slight variation in the potential due to the variation in the environment.

Let us then consider an alloy crystal potential which for each configuration is $v_A(\underline{r} - \underline{R}_i)$ on an A site and $v_B(\underline{r} - \underline{R}_j)$ on a B site. These potential functions may be constructed for the average environment and therefore depend on the concentration, the lattice spacing and the atomic charge distribution of the constituent atoms but are independent of the short range order. Under these circumstances the fundamental equations of multiple scattering on the energy shell are

$$\mathcal{T}^{i,j}_{L,L'}(\epsilon) = t_{i,L}(\epsilon)\,\delta_{i,j}\,\delta_{L,L'} + \sum_{L''}\sum_{k\neq i} t_{i,L}(\epsilon) G_{L,L'}(\underline{R}_i - \underline{R}_k;\epsilon)\mathcal{T}^{k,j}_{L'',L'}(\epsilon) \qquad \text{IV-1}$$

where $t_{i,L} = t_{A,L}$ or $t_{B,L}$ as prescribed by the configuration in question.

Thus, the problem we want to solve consists of three parts : solution of Eq. IV-1, the calculation of an interesting observable for a given configuration and averaging of the observable over the ensemble of all configurations consistent with a given concentration and any further information which is determined by the measuring process.

One of the simplest "observables" we shall want to calculate is the density of states. As is shown in Appendix I the real space analogue of Eq. III-43 is [56]

$$N(\epsilon) = N_0(\epsilon) - \frac{1}{\pi N}\,\mathrm{Im}\,\ell n\,\left|\left| t^{-1}_{i,L}(\epsilon) - G_{L,L'}(\underline{R}_i - \underline{R}_j;\epsilon)\right|\right| \qquad \text{IV-2}$$

where the process of taking the determinant has to be carried out with respect to the site index as well as the angular momentum index L. Taking the derivative of Eq. IV-2 and averaging the result gives

$$\bar{n}(\epsilon) = n_0(\epsilon) - \frac{1}{\pi N}\Big[\sum_{i,L}\Big(\frac{\partial}{\partial\epsilon} t^{-1}_{A,L}\Big)\langle\mathcal{T}^{ii}_{LL}\rangle_A + \sum_{i,L}\Big(\frac{\partial}{\partial\epsilon} t^{-1}_{B,L}\Big)\langle\mathcal{T}^{ii}_{LL}\rangle_B$$

$$+ \sum_{i,j}\sum_{L,L'}\Big(\frac{\partial}{\partial\epsilon} G_{L,L'}(\underline{R}_i - \underline{R}_j;\epsilon)\Big)\langle\mathcal{T}^{i,j}_{L,L'}\rangle \qquad \text{IV-3}$$

where we have used the matrix identity $\ell n\,||\mathcal{T}^{-1}|| = \mathrm{tr}\,\ell n\,\mathcal{T}^{-1}$ and the relation $(\mathcal{T}^{-1})_{iL,jL'} = t^{-1}_{iL}\delta_{i,j}\,\delta_{L,L'} - G_{LL'}(\underline{R}_i - \underline{R}_j;\epsilon)$ which follows from Eq. III-37. By $\langle\mathcal{T}^{ii}_{LL'}\rangle_A$ and $\langle\mathcal{T}^{ii}_{LL'}\rangle_B$ we mean the average of $\mathcal{J}^{ii}_{LL'}$ over all configurations which have an A and a B atom respectively at the site i. $<\mathcal{J}^{ij}>$ means an average without any restriction except that implied by the concentration. Thus our task is to find $<J^{ii}>_A$, $<J^{ii}>_B$ and $<J^{ij}_{LL'}>$.

2. The averaged t-matrix approximation (ATA)

Having formulated the alloy problem entirely in terms of scattering amplitudes the obvious first approximation to make is to replace the scattering amplitude $t_{i,L}(\epsilon)$ on every site by the averaged scattering amplitude $\bar{t}_{i,L} = c\, t_{A,L}(\epsilon) + (1-c)t_{B,L}(\epsilon)$. As is customary we shall refer to this procedure as the averaged t-matrix approximation or ATA,[57] In this approximation Eq. IV-1 reduces to

$$\bar{\mathcal{J}}^{i,j}_{L,L'}(\epsilon) = \bar{t}_L(\epsilon)\,\delta_{i,j}\,\delta_{L,L'} + \sum_{k\neq i}\sum_{L''}\bar{t}_L(\epsilon)G_{LL''}(\underline{R}_i - \underline{R}_k;\epsilon)\bar{\mathcal{J}}^{k,j}_{L'',L'}(\epsilon) \qquad \text{IV-4}$$

which is the multiple scattering equation for an ordered set of effective scatterers each of which is described by the on the energy shell t-matrix $\bar{t}_L(\epsilon)$. In a sense, which we shall describe presently, this effective ordered lattice will have to describe the averaged behaviour of the original random lattice.

This sounds like the virtual crystal idea discussed in the introduction. However, the ATA turns out to be a much more sophisticated approximation. A way of seeing this is to consider the Argand plot for the averaged scattering amplitude. $\bar{f}_\ell = -\sqrt{\epsilon}\,\bar{t}_L = c\, f^A_\ell + (1-c)f^B_\ell$. This is a plot of Im $\bar{f}_\ell$ vs Rel $\bar{f}_\ell$ for various values of the energy. As we shall show presently $\bar{f}_\ell$ describes an inelastic scatterer. In order to facilitate our discussion of this fact we shall now briefly digress to introduce some basic notions regarding such scatterers.

An elastic scatterer is always described by real phase shifts. From the definition of the scattering amplitude

$$f_\ell(\epsilon) = \sin\delta_\ell\, e^{i\delta_\ell} = \frac{1}{2i}(e^{i\delta_\ell} - 1) \qquad \text{IV-5}$$

it follows that

$$(\text{Rel } f_\ell)^2 + (\text{Im } f_\ell - \frac{1}{2})^2 = \frac{1}{4} \qquad \text{IV-6}$$

Thus, for a resonant phase shift down in Fig. 14a the Argand plot is a circle with a radius 1/2 centered at Rel $f_\ell = 0$,

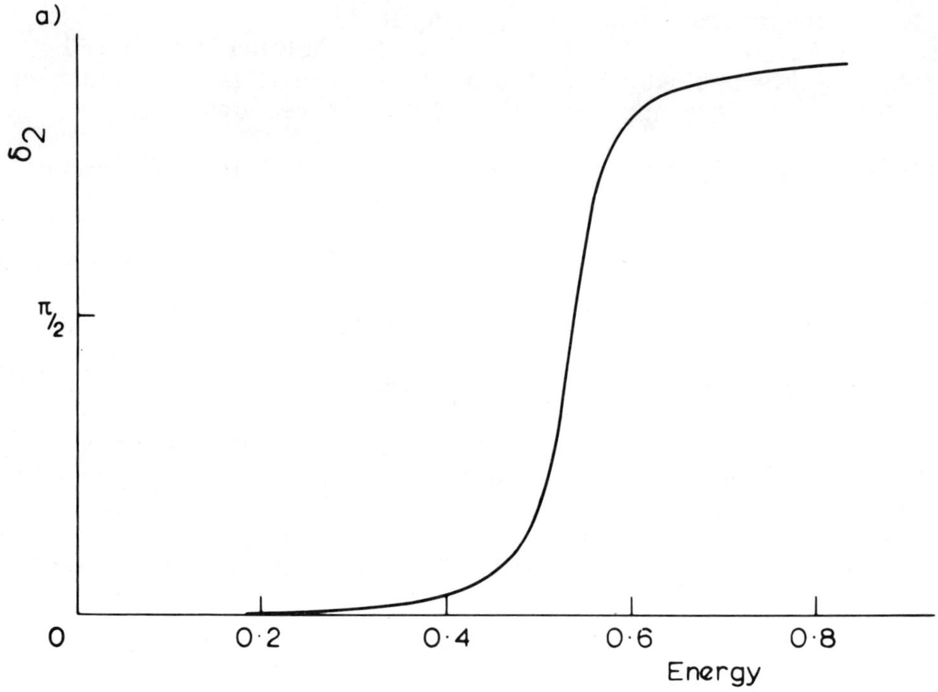

Fig. 14a. A generic resonant phase-shift.

b)

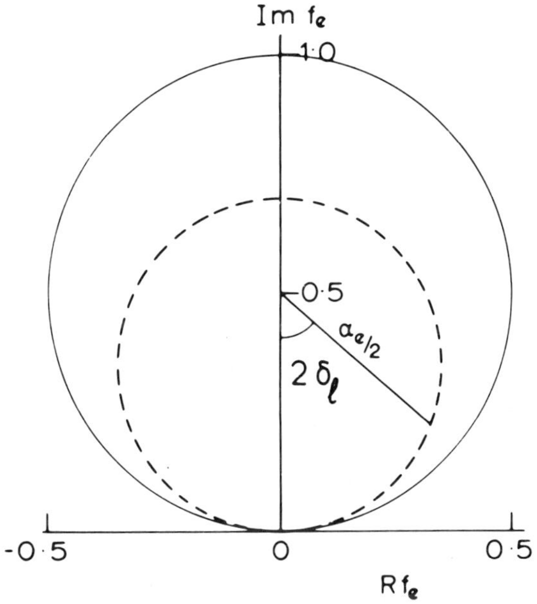

Fig. 14b. The full line is the unitarity circle which is the Argand plot of the elastic scattering amplitude $\sin\delta_\ell(\epsilon) \times e^{i\delta_\ell(\epsilon)}$ for $\delta_\ell(\epsilon)$ given in 14a. The dashed circle is the Argand plot for an inelastic scattering amplitude $(1/2i)(\alpha_\ell(\epsilon)e^{i2\delta_\ell} - 1)$ with $\alpha < 1$.

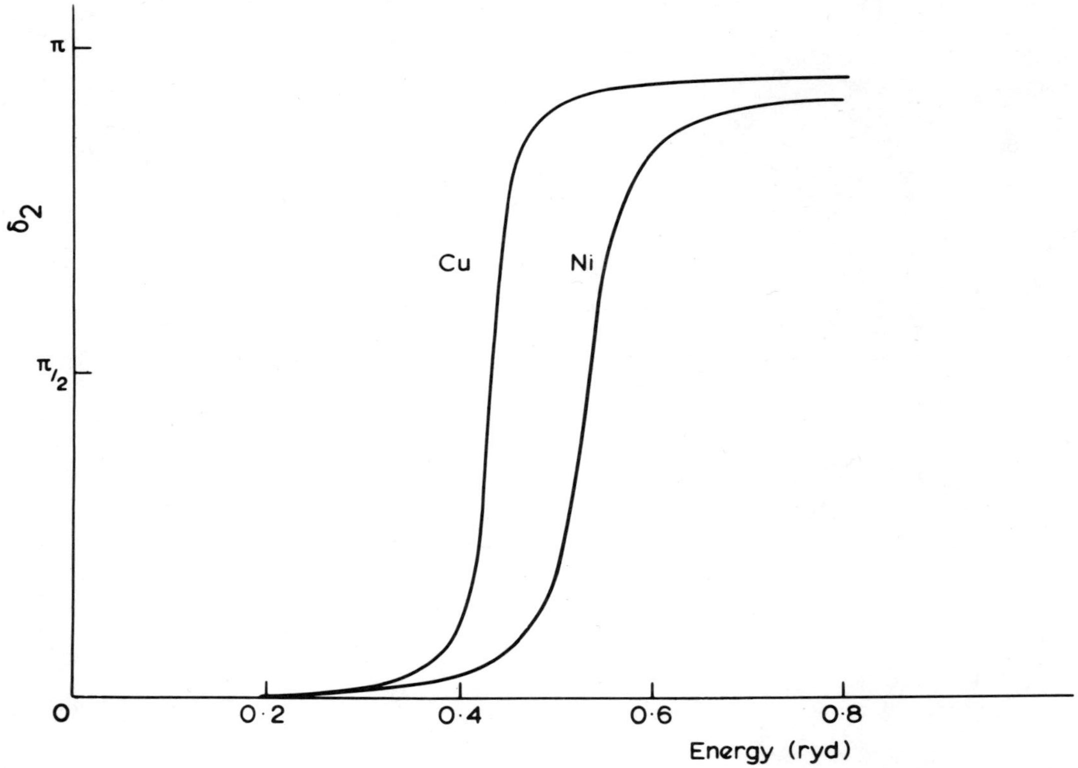

Fig. 15. The $\ell = 2$ phase shifts used in our calculation to describe the scattering at Cu and Ni sites respectively. They are the same as in pure Cu and Ni except here they are plotted with respect to an alloy V_{MTZ} whose choice is explained in the text.

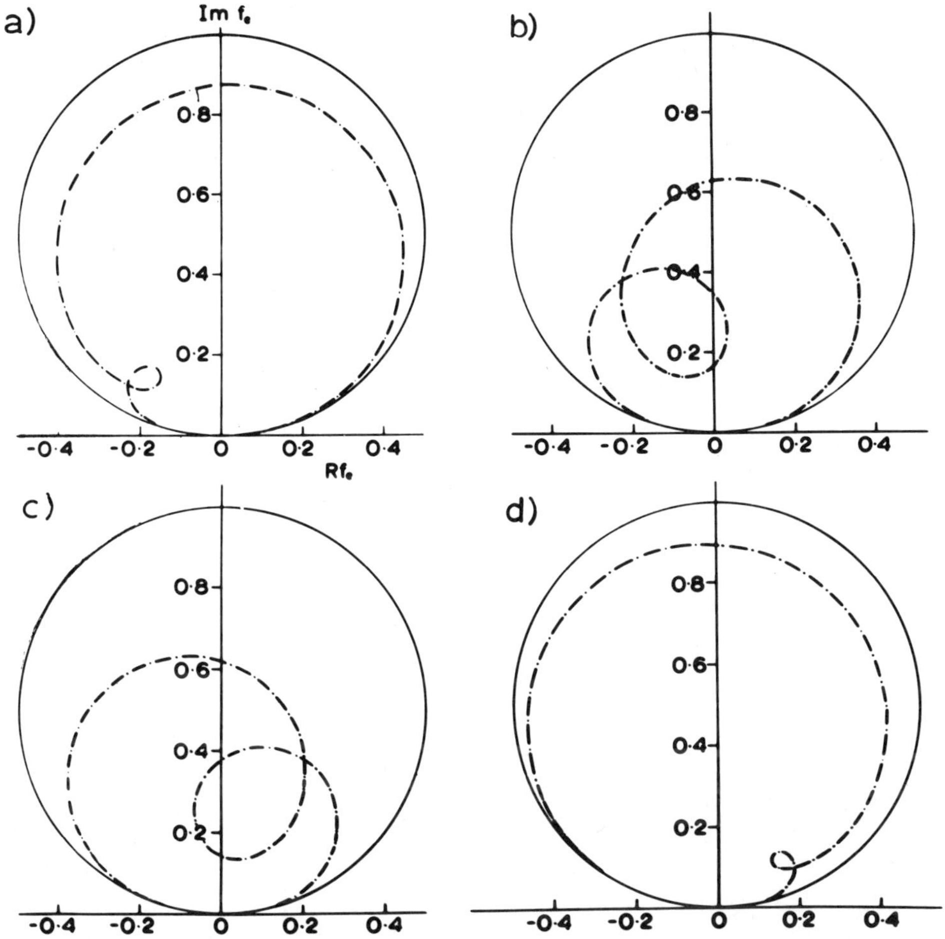

Fig. 15a, b, c, d. The Argand plots are the $\ell = 2$ averaged scattering amplitudes $f_2 = cf_2^{Cu} + (1-c)f_2^{Ni}$ for $Cu_{.9}Ni_{.1}$, $Cu_{.6}Ni_{.4}$, $Cu_{.4}Ni_{.6}$, $Cu_{.1}Ni_{.9}$.

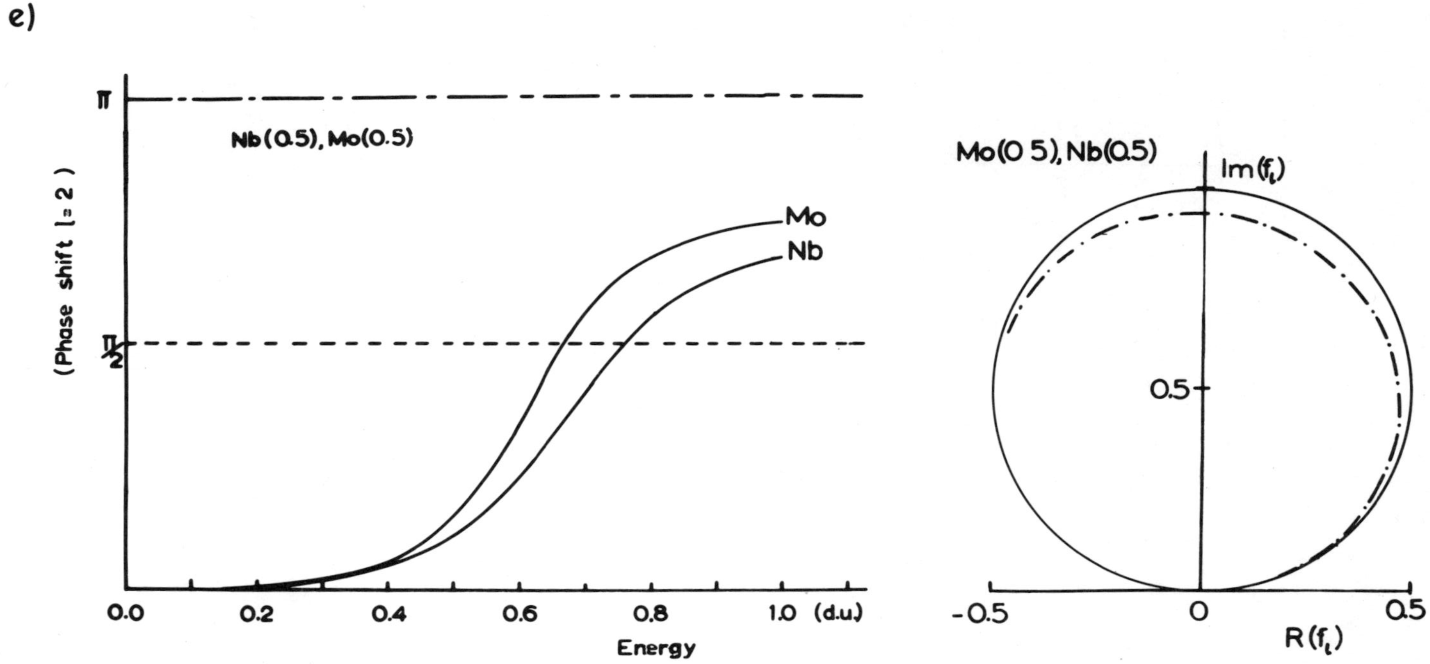

Fig. 15e. The l = 2 phase shift and the corresponding averaged scattering amplitude Argand plot for $Nb_{.5}Mo_{.5}$. The corresponding muffin-tin potentials were constructed according to the Mattheiss prescription using an average environment at 50 - 50 concentration and the appropriate lattice spacing.

f)

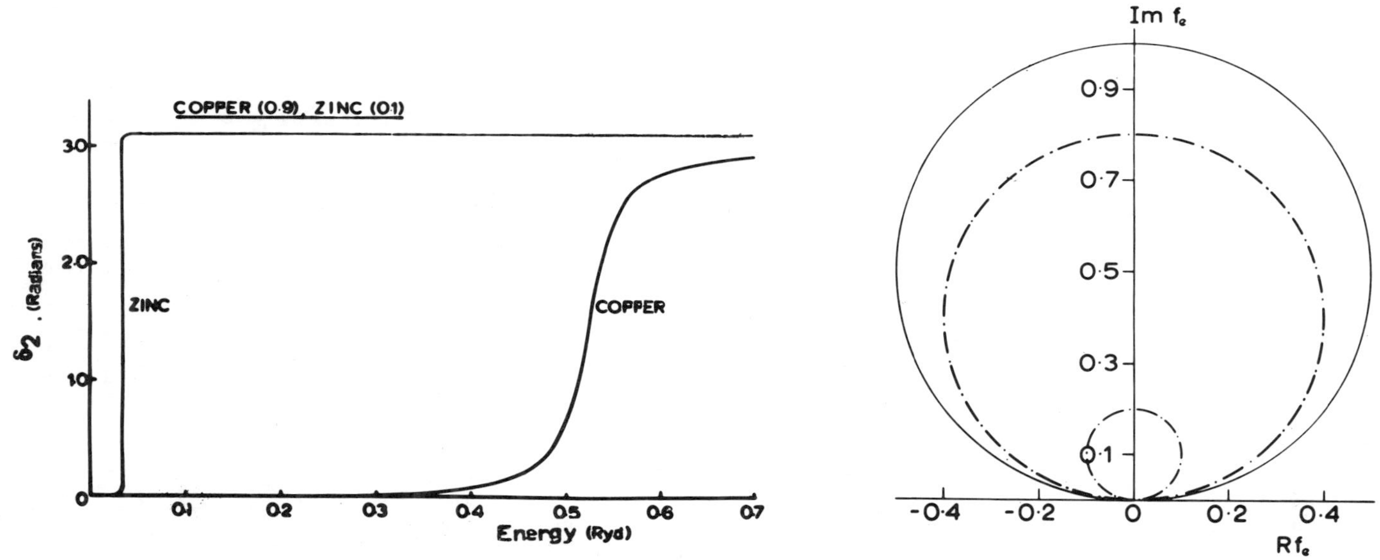

Fig. 15f. The $l = 2$ phase shift and the corresponding averaged scattering amplitude Argand plot for $Cu_{.9}Zn_{.1}$. The corresponding muffin-tin potentials were again constructed according to the Mattheiss prescription using an average environment at 50 - 50 concentration and the appropriate lattice spacing.

$\mathrm{Im}\, f_\ell = 1/2$ as shown in Fig. 14b. It is customary to refer to this circle as the unitary circle. Note that f_ℓ as defined above automatically satisfies the optical theorem $\mathrm{Im}\, f_\ell = |f_\ell|^2$

On the other hand an inelastic scatterer is described by a complex phase shift, or equivalently it may be represented as

$$f_\ell = \frac{1}{2i}\,(\alpha_\ell(\epsilon)e^{i\delta_\ell(\epsilon)} - 1) \qquad \text{IV-7}$$

where $\delta_\ell(\epsilon)$ is a real phase shift which describes the scattering power of the underlying potential and $\alpha_\ell(\epsilon)$ is a real function of energy which, for an absorptive scatterer, is less than one. Recall that the optical theorem is the consequence of the conservation of the probability of finding the particle somewhere after the scattering process. In an inelastic process this conservation law is violated. It is easy to show that $\mathrm{Im}\, f_\ell = |f_\ell|^2 - 1/4\,(1 - \alpha^2)$. Hence the probability that the particle was absorbed in the scattering process is $1/4(1 - \alpha^2)$. The Argand plot for a typical inelastic scattering amplitude is shown in Fig. 14b as marked.

Let us now return to the averaged scattering amplitude. In Fig. 15 we show the $\ell = 2$ phase shifts for pure Ni and pure Cu. In Fig. 15b, c, d we show the Argand plots of $\bar{f}_2(\epsilon)$ for $Cu_{87}Ni_{13}$, $Cu_{50}Ni_{50}$, $Cu_{19}Ni_{81}$. The points corresponding to the bottom of the pure Cu d-band and that for the top of the Ni d-band is marked by radial slashes.

The first point to note about these Argand plots is the fact that in the relevant energy range our effective scatterers are inelastic. This means that even though we have an ordered array of them the state with a given $\underline{q}$ will have a finite life time. This is most unlike the case in the virtual crystal approximation. Moreover, the inelasticity is relatively small for $c \sim 0$ and $c \sim 1$ and large for $c = .5$. Clearly, this should be interpreted as long life time at the two ends of the concentration range and short life time for mid concentration, as one would expect in a sensible theory.

The physical reason for the inelasticity is the randomness in the problem. Loosely speaking a particle wave arrives to a site and scatters there according to $t_{A,L}$ with a probability c and according to $t_{B,L}$ with a probability $1 - c$. The emerging wave is a random superposition of the two kinds of scattered waves. As a result of destructive interference between these

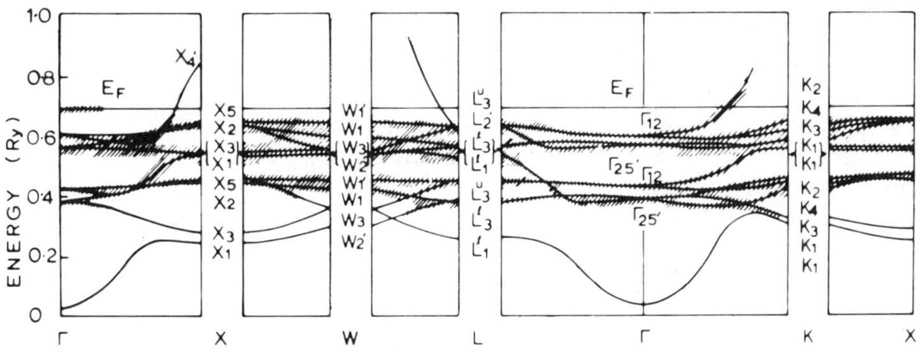

Fig. 16. Complex energy bands for $Cu_{.7}$ $Ni_{.3}$ (Ref. 22). The shading of the bands corresponds to twice the imaginary part of the complex energies. (The shading around the Fermi energy represents the average damping on the Fermi surface). The energy zero is taken to be – 0.8341 Ry, which is the muffin-tin zero for pure Cu.

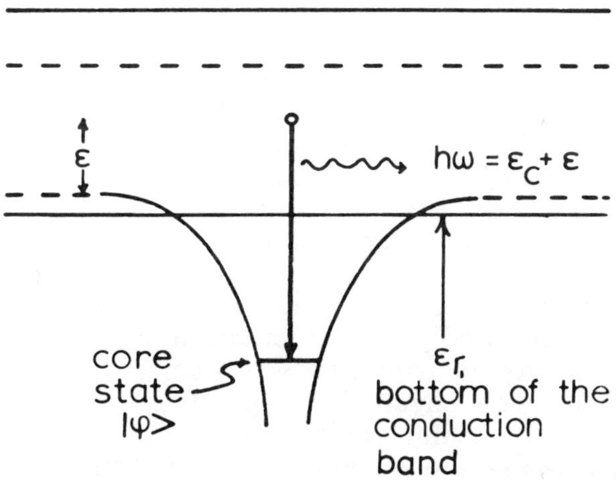

Fig. 17. A schematic representation of the soft X-ray emission process.

components the beam is attenuated.

We recover the more conventional description of randomness if we search for the zeros of the determinant $\|t_L^{-1}\delta_{L,L'} - G_{L,L'}(\underline{q}, \epsilon)\|$. Unlike in the case of pure metals the matrix $t_L \delta_{L,L'} - G_{L,L'}(\underline{q}, \epsilon)$ is not real even for cubic systems and for real $\underline{q}$ the eigenvalues $\epsilon_{\underline{q},n}$ are complex. The complex energy bands calculated in this way by Schwartz, Bansil and Ehrenreich for $Cu_{.7}Ni_{.3}$ are shown in Fig. 16.

The Argand plots in Fig. 15 b, c, d, deserve some further comments. In the parlance of inelastic scattering theory two loops in the Argand plot implies two resonances. The radius which can be associated with a loop measures the strength with which the scattered particle is coupled into that resonance. A loop is shifted from the centre of the unitarity circle because of background scattering due to other resonances. Clearly, all three Argand plots in Fig. 15 have two resonances : the one at lower energy is associated with the Cu resonance and the higher energy loop is due to the Ni resonance. For c = .87 the Cu resonance is more eleastic and the coupling into this resonance is stronger than the coupling into the more severely damped Ni resonance. In this case we would expect the band structure to consist of fairly well defined Cu bands and smeared out Ni impurity bands. At the Ni rich end of the concentration range Fig. 15d predicts the reverse situation. At c = .5 both set of bands should be there but they will be severly damped. Of course, if the broadening is large enough no separate bands can be distinguished.

Evidently, all this makes a good deal of sense and suggests that the ATA is a reasonable first pass theory. Though we shall present evidence that it is inadequate in many respects, our more sophisticated calculations will tend to support the qualitative picture we have discussed above.

At the risk of belabouring a point we would like to note that we have done hardly any calculation yet. We have merely averaged two scattering amplitudes. The fact that we have been able to build up such physically appealing qualitative picture with so little effort is due to the great power and formal economy of the scattering theory language. For exercise interpret Fig. 15 e, f.

For those of you familiar with the ATA in the context of tight-binding model Hamiltonians the suggestion that it is not so bad should come as a surprise. Unfortunately, we have time here only to assert that ATA has quite different status in scattering theory, where it was first proposed, [57] than in tight-binding theory. The reason for this difference has to do with the different choice of energy zero in the two approaches. In scattering theory the new states put in by the introduction of the scattering centres overlap and hybridize with an ever present free electron continuum. In general there are no gaps in the spectrum at any stage of the calculation. On the other hand, in the tight-binding approach one is building up the band out of discrete atomic states with genuine gaps in between them. Apparently this feature of the model can introduce unphysical gaps in the ATA in the final result making the whole calculation more suspect than is warranted.

To make further headway with the theory we must calculate the averaged density of states. Unfortunately, this is not straightforward and at the moment the prescription for the density of states in the scattering theory ATA is somewhat controversial. (In the tight binding theory there is no corresponding ambiguity). The attractive thing to do would be to use $\bar{t}_L^{-1}$ in place of t_L^{-1} and evaluate Eq. III-43. This turns out to be an easy calculation, however it is wrong. The trouble is that the energy derivative of $N(\epsilon)$ in Eq. III-43 as well as in Eq. IV-2 is not the same as the trace of the Greens function, for an energy dependent potential. Unhappily the averaged t-matrix $\bar{t}_L(\epsilon)$ corresponds to an energy dependent complex potential. Consequently, it is not the same approximation to evaluate Eq. III-43 using $\bar{t}_L(\epsilon)$ or to find an expression for the trace of the Greens function directly in terms of $\bar{t}_L(\epsilon)$ and evaluate it. Indeed, the first of these procedures does not give a sensible density of states. We believe that a reasonable thing to do is to evaluate Eq. IV-3 (this is the trace of the averaged Greens function) by taking for $<\mathcal{T}^{i,j}_{L,L'}>$ the solution of Eq. IV-4 and for $<\mathcal{T}^{i,i}_{L,L'}>_A$, $<\mathcal{T}^{i,i}_{L,L'}>_B$ use the solution for the problem of A, B impurity respectively in the averaged t-matrix lattice. Namely take $<\mathcal{T}^{ii}_{L,L'}>_A = \bar{\mathcal{T}}^{A,ii}_{L,L'}$ where $\bar{\mathcal{T}}^{A,ii}_{L,L'}$ is the solution of Eq. IV-5 with $t_{j,L} = \bar{t}_L + (t_{A,L} - \bar{t}_L)\delta_{i,j}$ and for $<\mathcal{T}^{i,i}_{L,L}> = \bar{\mathcal{T}}^{B,ii}_{L,L}$ where $\bar{\mathcal{T}}^{B,ii}_{L,L}$ is defined similarly to $\bar{\mathcal{T}}^{A,ii}_{L,L'}$. It is only the matter of some simple matrix algebra to show that

$$\bar{\mathcal{T}}^{A,ii}_{L,L'} = \left[\frac{1}{1 - (t_A^{-1} - \bar{t}^{-1})\bar{\mathcal{T}}^{ii}} \bar{\mathcal{T}}^{ii} \right]_{L,L'} \qquad \text{IV-8}$$

$$\bar{\mathcal{T}}^{B,ii}_{L,L'} = \left[\frac{1}{1 - (t_B^{-1} - \bar{t}^{-1})\bar{\mathcal{T}}^{ii}} \bar{\mathcal{T}}^{ii} \right]_{L,L'}$$

where the matrix $\bar{\mathcal{T}}^{ii}$ is given by

$$\bar{\mathcal{T}}^{ii}_{LL'}(\epsilon) = \frac{1}{\Omega_{BZ}} \int d^3q \left[\frac{1}{\bar{t}^{-1} - G(\underline{q}, \epsilon)} \right]_{L,L'} \qquad \text{IV-9}$$

Unfortunately, the Brillouin zone integral in Eq. IV-9 is more difficult to evaluate than the one in Eq. III-43 and no calculation using the above scheme has yet been reported.

There are two less evident but more tractable ways of evaluating the density of states in the ATA. We shall discuss these in the next section.

3. The KKR-CPA

By now it should not come as a surprise that the way we propose to do CPA for non overlapping muffin-tins is to search for an ordered lattice of effective scatterers which describe the average behaviour of the random lattice. We bypass the problem of finding a coherent potential by assuming that it has the non overlapping form and seek to find an effective scattering amplitude $t_{c,L}(\epsilon)$ which describes the scattering at each site in the coherent potential lattice. By concentrating on reproducing the "on the energy shell" scattering behaviour of the random lattice we shall render the CPA program tractable. An attractive feature of this approach as opposed to some other ways of simplifying the problem is that it leaves the predictive power of the theory virtually unaffected with respect to a large class of useful observables.

The way to proceed is now fairly straightforward[58,59,60] Consider an A impurity at $\underline{R}_0$ in the coherent potential lattice. The scattering from that site is described by $\mathcal{T}^{A,ii}_{L,L'}$ which is the solution of Eq. IV-1 with $t_{j,L} = t_{c,L} + (t_{A,L} - t_{c,L})\delta_{i,j}$ Similarly the scattering from a B impurity at $\underline{R}_0$ is described by $\mathcal{T}^{B,00}_{L,L'}(\epsilon)$. As in Eqs. IV-8 and IV-9 we find that

$$\mathcal{T}^{A,00}_{L,L'}(\epsilon) = \left[\frac{1}{1 - (t_A^{-1} - t_c^{-1})\mathcal{T}^{c,ii}} \mathcal{T}^{c,ii} \right]_{L,L'} \qquad \text{IV-10}$$

$$\mathcal{T}^{B,00}_{L,L'}(\epsilon) = \left[\frac{1}{1 - (t_B^{-1} - t_c^{-1})\mathcal{T}^{c,ii}} \mathcal{T}^{c,ii} \right]_{L,L'}$$

where

$$\mathcal{T}^{c,ii}_{L,L'}(\epsilon) = \frac{1}{\Omega_{BZ}} \int d^3q \left[\frac{1}{t_c^{-1} - G(\underline{q}, \epsilon)} \right]_{L,L'} \qquad \text{IV-11}$$

The CPA condition that the averaged scattering from the site at $\underline{R}_0$ should be the same as in the pure coherent potential lattice now reads as

$$c\,\mathcal{T}^{A,00}_{L,L'}(\epsilon) + (1-c)\,\mathcal{T}^{B,00}_{L,L'}(\epsilon) = J^{c,00}_{L,L'}(\epsilon) \qquad \text{IV-12}$$

Using Eq. IV-10 it is a simple matter to rearrange this condition into the following more useful form[61].

$$t^{-1}_{c,L} = c\,t^{-1}_{A,L} + (1-c)t^{-1}_{B,L} + (t^{-1}_{c,L} - t^{-1}_{A,L})\,\mathcal{T}^{c,00}_{L,L}(t^{-1}_{c,L} - t^{-1}_{B,L}) \qquad \text{IV-13}$$

This is then the fundamental equation for the effective scattering amplitude in the KKR-CPA. It determines $t_{c,L}(\epsilon)$ in terms of the A site and B site scattering amplitudes $t_{A,L}(\epsilon)$, $t_{B,L}(\epsilon)$ and the crystal structure which fixes the structure constants $G_{L,L'}(\underline{q}, \epsilon)$.

It is worth remembering that $t_{A,L}$ and $t_{B,L}$ need not be the same scattering amplitudes which determine the band structures of the pure metals A and B respectively. Normally, they would be constructed at each concentration according to one of the prescriptions mentioned in Sec. IV. 1. Thus, the KKR-CPA is a first principles theory of the electronic states in random alloys in the same sense as one talks about first principles band theory for pure systems. As we mentioned in connection with our critique of the tight-binding model Hamiltonian approach in Sec. III the construction of such a theory was one of our aims.

Obviously, there is no limitation to the applicability of the theory arising from the size and energy dependence of the scattering amplitudes. They can correspond to pure metals band structures with unequal band widths. Moreover there can be any number of such bands overlapping and hybridizing in an arbitrary fashion. To put it another way we are treating randomness in site energies, overlap integrals (arbitrary number of them) and hybridization coefficients on an equal footing. This removes another set of objections raised in connection with the tight-binding model Hamiltonian.

As does the KKR, our method makes full use of the fact that for most metals only the first few phase shifts matter. Since $t_{c,L}(\epsilon)$ will have the cubic symmetry of the crystal lattice, for $\ell \leq 2$ we will have at the most four coupled equations to solve. Frequently, the alloying behaviour is dominated by the difference between the $\ell = 2$ phase shifts on the A and B sites. Then, we shall have to solve only for an e_g and a t_{2g} scattering amplitudes. Clearly $t_{c,L}$ as $\bar{t}_L$ in the previous section, will correspond to an inelastic scatterer. For physical reasons this will have to be an absorber and hence the effective scattering amplitudes $f_{c,L} = -\sqrt{\epsilon}\, t_{c,L}$ will have to be within the unitarity circle. Thus, at each energy Eq. IV-13 needs to be solved only for a limited set of complex numbers whose values are restricted, a priori, to a very limited part of the complex plane. As you shall see all these features play an important role in making the KKR-CPA a tractable proposition.

Model calculations and general theorems suggest that Eq. IV-13 can always be solved by iteration. One assumes some value for $t_{c,L}$ and calculates $\mathcal{T}^{c,00}_{L,L'}$ numerically using Eq. IV-11. A new value of $t_{c,L}$ is obtained by evaluating the right hand side of Eq. IV-13 for the assumed value of $t_{c,L}$. This process is continued until the new value is the same as the assumed value at the beginning of the last step. The difficult part of the calculation is the evaluation of the Brillouin zone integral in Eq. IV-11 at each iteration. While we believe that this method of solution can be carried out by stretching to its limit our present day calculational capabilities in these lectures we shall present an alternative procedure. Although it is less clear but it appears to work for the systems we have tried it on and it reduces the computational effort to manageable size.

Of course, finding $t_{c,L}$ is not enough. We must calculate observables. Consider, at first, the density of states. As in the case of ATA we cannot use Eq. III-43 with t_L^{-1} replaced by $t_{c,L}^{-1}$ on account of the fact that $t_{c,L}$ corresponds to an energy dependent coherent potential. Thus, we must turn to Eq. IV-3 which gives rigorously the trace of the averaged Greens function. Following the line of argument at the end of Sec. IV.1 we may make the substitutions

$$\langle \mathcal{T}^{ii}_{LL'} \rangle_A = \mathcal{T}^{A,ii}_{L,L'} \; ; \; \langle \mathcal{T}^{ii}_{LL'} \rangle_B = \mathcal{T}^{B,ii}_{L,L'} \; ; \; \langle \mathcal{T}^{ij}_{LL'} \rangle = \mathcal{T}^{c,ij}_{L,L'} \qquad \text{IV-14}$$

Substituting these expressions into Eq. IV-3 we can write our CPA prediction for the averaged density of states as

$$\bar{n}(\epsilon) = n_0(\epsilon) - \frac{1}{\pi N} \operatorname{Im} \Big\{ \sum_{i,j} \sum_{L,L'} \Big[\Big(\frac{\partial}{\partial\epsilon} t^{-1}_{c,L}\Big) \delta_{L,L'} \delta_{i,j} -$$

$$- \Big(\frac{\partial}{\partial\epsilon} G_{L,L'}(\underline{R}_i - \underline{R}_j ; \epsilon)\Big)\Big] \mathcal{T}^{c,ij}_{L,L'} + \sum_{i,L} c\Big(\frac{\partial}{\partial\epsilon} t^{-1}_{A,L}\Big) \mathcal{T}^{A,ii}_{L,L}$$

$$+ \sum_{i,L} (1-c)\Big(\frac{\partial}{\partial\epsilon} t^{-1}_{B,L}\Big) \mathcal{T}^{B,ii}_{L,L} - \sum_{i,L} \Big(\frac{\partial}{\partial\epsilon} t^{-1}_{c,L}\Big) \mathcal{T}^{c,ii}_{L,L} \Big\} . \qquad \text{IV-15}$$

The first term in this expression is immediately recognized as the energy derivative of the real space KKR-CPA determinant $|| t^{-1}_{c,L} \delta_{L,L'} - G_{L,L'}(\underline{q}, \epsilon) ||$. The rest of the terms may be rearranged as follows. Note that, using matrix notation in the angular momentum but not in the site index, Eq. IV-10 may be recast as

$$(\mathcal{T}^{A,ii})^{-1} = (\mathcal{T}^{c,ii})^{-1} + t_A^{-1} - t_c^{-1} ; \quad (\mathcal{T}^{B,ii})^{-1} = (\mathcal{T}^{c,ii})^{-1} + t_B^{-1} - t_c^{-1} \qquad \text{IV-16}$$

and therefore

$$c\Big(\frac{\partial}{\partial\epsilon} t_A^{-1}\Big)\mathcal{T}^{A,ii} + (1-c)\Big(\frac{\partial}{\partial\epsilon} t_B^{-1}\Big)\mathcal{T}^{B,ii} = c \frac{\partial}{\partial\epsilon} \ln (\mathcal{T}^{A,ii})^{-1}$$

$$+ (1-c) \frac{\partial}{\partial\epsilon} \ln (\mathcal{T}^{B,ii})^{-1} = \Big[-\Big(\frac{\partial}{\partial\epsilon} t_c^{-1}\Big) + \frac{\partial}{\partial\epsilon} (\mathcal{T}^{c,ii})^{-1} \Big] \times \qquad \text{IV-17}$$

$$\times \big[c\,\mathcal{T}^{A,ii} + (1-c)\mathcal{T}^{B,ii} \big] = \Big[-\Big(\frac{\partial}{\partial\epsilon} t_c^{-1}\Big) + \frac{\partial}{\partial\epsilon} (\mathcal{T}^{c,ii})^{-1} \Big] \mathcal{T}^{c,ii}.$$

Hence

$$n(\epsilon) = n_0(\epsilon) - \operatorname{Im} \frac{\partial}{\partial\epsilon} \frac{1}{\pi\Omega_{BZ}} \int_{BZ} d^3q \, \ln || t^{-1}_{c,L} \delta_{L,L'} - G_{L,L'}(\underline{q}, \epsilon) ||$$

$$+ \operatorname{Im} \frac{1}{\pi} \operatorname{tr} \Big[c \frac{\partial}{\partial\epsilon} \ln (\mathcal{T}^{A,00})^{-1} + (1-c) \frac{\partial}{\partial\epsilon} \ln (\mathcal{T}^{B,00})^{-1}$$

$$- \frac{\partial}{\partial\epsilon} \ln (\mathcal{T}^{c,00})^{-1} \Big] . \qquad \text{IV-18}$$

Since $\operatorname{tr} \ln (\mathcal{T}^{A,00})^{-1} = \ln ||(\mathcal{T}^{A,00})||^{-1}$ and the same relations hold with A replaced by B, we may write the averaged integrated density of states whose derivative is $\bar{n}(\epsilon)$ as

$$\bar{N}(\epsilon) = N_0(\epsilon) - \operatorname{Im} \frac{1}{\pi\Omega_{BZ}} \int_{BZ} d^3q \, \ln || t^{-1}_{c,L} \delta_{L,L'} - G_{L,L'}(\underline{q}, \epsilon) ||$$

$$+ \operatorname{Im} \frac{c}{\pi} \ln || 1 + (t_A^{-1} - t_c^{-1}) \mathcal{T}^{c,00} ||$$

$$+ \operatorname{Im} \frac{1-c}{\pi} \ln || 1 + (t_B^{-1} - t_c^{-1}) \mathcal{T}^{c,00} || \qquad \text{IV-19}$$

We now use the CPA equations of Eq. IV-13 in the form

$$(t_A^{-1} - t_c^{-1})\mathcal{T}^{c,00} = \frac{t_c^{-1} - \langle t^{-1}\rangle}{t_B^{-1} - t_c^{-1}} \; ; \; (t_B^{-1} - t_c^{-1}) = \frac{t_c^{-1} - \langle t^{-1}\rangle}{t_A^{-1} - t_c^{-1}} \qquad \text{IV-20}$$

and our final result is

$$\bar{N}(\epsilon) = N_0(\epsilon) - \mathrm{Im}\,\frac{1}{\pi\Omega_{BZ}} \int_{BZ} d^3q \; \ell n \,\| t_{c,L}^{-1}\,\delta_{L,L'} - G_{L,L'}(\underline{q},\epsilon)\|$$

$$+ \mathrm{Im}\,\frac{c}{\pi} \sum_L \left(\frac{t_{B,L}^{-1} - \langle t_L^{-1}\rangle}{t_{B,L}^{-1} - t_{c,L}^{-1}}\right) + \mathrm{Im}\,\frac{1-c}{\pi} \sum_L \left(\frac{t_{A,L}^{-1} - \langle t_L^{-1}\rangle}{t_{A,L}^{-1} - t_{c,L}^{-1}}\right) \qquad \text{IV-21}$$

Note that the first terms is just the integrated density of states for an ordered set of effective scatterers. The correction to this term arises because the effective scattering amplitudes correspond to an energy dependent potential.

As mentioned in Sec. III. 5 the phase of the KKR determinant for a pure system consists of monotonically increasing steps as shown in Fig. 13. For our effective scatterers these steps will be rounded off but, except for a limited energy range, the phase remains a monotonically increasing function. Such functions are relatively easy to integrate over the Brillouin zone. By contrast Im $\mathcal{T}^c(\underline{q},\epsilon)$ will have sharp peaks and Rel $\mathcal{T}^c(\underline{q},\epsilon)$ will behave as a rounded off $1/x$ singularity. For this reason while Eq. IV-11 is very hard to evaluate the Brillouin zone integral in Eq. IV-21 can be readily performed The correction terms in Eq. IV-21 do not require a Brillouin zone integration and they can be trivially evaluated once $t_{c,L}$ is known.

Another useful quantity to calculate is the Bloch wave spectral function which we shall now define. The usual spectral function is defined to be $\bar{A}(\underline{q},\epsilon) = -1/\pi\,\mathrm{Im}\,\langle G(\underline{q},\underline{q};\epsilon)\rangle$. As we mentioned in Sec. II this quantity is a set of reasonably well defined peaks at energies equal to the real part of the poles of $\bar{G}(\underline{q},\epsilon)$ in the complex energy plane. Unfortunately it is not possible to calculate this spectral function from the knowledge of the "on the energy shell" t-matrices alone. However, we can define a similar quantity which carries roughly the same kind of information. For a pure system it makes sense to define

$$A_B(\underline{q},\epsilon) = \sum_n \delta(\epsilon - \epsilon_{\underline{q},n}) \qquad \text{IV-22}$$

where n is the band index and call it the Bloch spectral function. Evidently $1/\Omega_{BZ} \cdot \int_{BZ} d^3q\ A_B(\underline{q}, \epsilon) = n(\epsilon)$. Thus $A_B(\underline{q}, \epsilon)$ is the density of states per $\underline{q}$-point in the Brillouin zone. Therefore, in order to find the analogue of Eq. IV-22 we must write Eq. IV-21 as single sum over the Brillouin zone. Using Eq. IV-20 to write the correction terms in Eq. IV-21 in terms of $\mathcal{T}^{c,00}_{L,L'} = \frac{1}{\Omega_{BZ}}\int_{BZ} d^3q\ \mathcal{T}^c(\underline{q}, \epsilon)$ and then differentiating the resulting equation with respect to ϵ leads to

$$\bar{A}_B(\underline{q}, \epsilon) = -\ \mathrm{Im}\frac{1}{\pi}\sum_L \frac{1}{\Omega_{BZ}}\Big[\frac{(t^{-1}_{A,L} - t^{-1}_{c,L})(\frac{\partial}{\partial\epsilon} t^{-1}_{B,L}) - (t^{-1}_{B,L} - t^{-1}_{c,L})(\frac{\partial}{\partial\epsilon} t^{-1}_{A,L})}{t^{-1}_{A,L} - t^{-1}_{B,L}} \times \mathcal{T}^c_{L,L}(\epsilon)\Big] \qquad \text{IV-23}$$

after some straightforward algebra.

By definition Eq. IV-23 is the CPA Bloch spectral function. Its integral over the Brillouin zone is manifestly equal to the density of states and therefore has the same interpretation as Eq. IV-22 for pure systems. Moreover, in the limit of no disorder e. g. $t_{A,L} = t_{B,L} = t_{c,L} = t_L$ Eq. IV-23 reduces to

$$A_B(\underline{q}, \epsilon) = -\frac{1}{\pi}\ \mathrm{Im}\frac{1}{\Omega_{BZ}}\frac{\partial}{\partial\epsilon}\ \ln\,||t^{-1}_L\ \delta_{L,L'} - G_{L,L'}(\underline{q}, \epsilon)|| \qquad \text{IV-24}$$

By recognizing that the above formula is the energy derivative of the phase of the KKR determinant which behaves like a sequence of step functions at the energy eigen values it is easy to see that Eq. IV-24 is the same as Eq. IV-22.

In more physical terms $\bar{A}_B(\underline{q}, \epsilon)$ may be interpreted as the probability that an electron with a Bloch vector $\underline{q}$ has an energy ϵ. Clearly, $\bar{A}_B(\underline{q}, \epsilon)$ is as close as we can get to looking at individual bands in the disordered systems.

We would like to note in passing that the interest in $\bar{A}_B(\underline{q}, \epsilon)$ is not entirely pedagogical. While $\bar{A}_B(\underline{q}, \epsilon)$ like the ϵ vs. $\underline{q}$ curves for pure systems, is not directly observable in any experiment something very close to it is measured in angle resolved photoemission studies. In this experiment, which has become possible only recently, one measures not only the energy but also the wave vectors of the photo-emitted electrons. Hence, if we are willing to disregard the effect of matrix elements and the complications introduced by the surface barrier one can interpret the results in terms of integrals of $\bar{A}_B(\underline{q}, \epsilon)$ along various directions. Unfortunately, we do not have the time here to discuss this very promising field any further. Also, we can only mention the possibility of analysing positron annihilation, and Compton profile studies in terms of $\bar{A}_B(\underline{q}, \epsilon)$ all be it only approximately.

In order to describe the above experiments accurately one has to know the wave function all through the crystal and to calculate these we would need to know more about the electrons than their "on the energy shell" scattering properties, therefore, our on the energy shell KKR-CPA is necessarily an incomplete theory in this regard. However, there are a number local probes of the electronic wave functions for which the calculation within the KKR-CPA can be complete with matrix elements. Examples are the Knight Shift, the Soft X-ray spectra and the Mössbauer Isomer Shift measurements [51].

As an illustration let us consider the soft X-ray spectrum of a binary alloy. If by electron bombardment, or by some other method, we knock out an electron from the core of an A atom then one of the conduction electrons may jump into the hole and emit an X-ray photon in the process. Since there are electrons at all energies over the width of the conduction band there will be a spectrum of X-rays emitted. The same process occurs if the hole is created on a B site. However, the spectrum will appear at different energies because the core state of an A atom is likely to be at an energy different from the corresponding eigen energy of the B atom. A schematic representation of this point is shown in Fig. 17. This separation between the two spectra means that the act of measurement differentiates between A and B sites. This fact must be taken into account when the configuration averages are formed.

Let us just sketch how the calculation goes. The emission intensity will be proportional to the probability per unit time that a conduction electron makes a transition to the empty core hole. Thus using the "Golden Rule" we may write

$$I_A(\omega) = \frac{2\pi}{\hbar} \sum_n |\langle \varphi_c | \underline{p} \cdot \underline{A} | \psi_n \rangle|^2 \delta(\hbar\omega - \epsilon_n + \epsilon_c) \qquad \text{IV-25}$$

where $|\varphi_c\rangle$ is the core state, ϵ_c is the corresponding energy, $|\psi_n\rangle$ and ϵ_n are the corresponding quantities for a conduction electron for a particular alloy configuration. $\underline{A}$ is the vector potential and $\underline{p}$ is the momentum operator. Writing out the square of the matrix element as a double integral in the space of the electron position variable and noting that

$\sum_n \psi_n^*(\underline{r})\,\psi_n(\underline{r}')\delta(\epsilon - \epsilon_n) = -\frac{1}{\pi}$ Im $G(\underline{r},\underline{r}';\epsilon)$ we may rewrite the emission intensity in terms of the Greens function for that configuration as

$$I_A(\omega) \propto \int d^3r \int d^3r'\ \varphi^*_{c,A}(\underline{r})\,\varphi_{c,A}(\underline{r}')\ \underline{\nabla}\cdot\underline{\nabla}'\ \mathrm{Im}\ G(\underline{r},\underline{r}';\epsilon)\Big|_{\epsilon=\hbar\omega+\epsilon_c} \qquad \text{IV-26}$$

We must now average this intensity over the ensemble of all configurations which are consistent with the fact that there is an A atom at the site $\underline{R}_0$. The core state is usually tightly bound and it is safe to assume that $\varphi_{c,A}(\underline{r} - \underline{R}_0)$ falls to zero outside the radius of the muffin tin sphere surrounding the A atom at $\underline{R}_0$. Thus, we only need to know the Greens function within the muffin tin well of the emitting atom. This is what we meant by referring to the soft X-ray experiment as a local probe. Under these circumstances we can use Eq. III-44 to evaluate Im $G(\underline{r},\underline{r}';\epsilon)$. This leads us to an expression for $\overline{I}_A(\omega)$ in terms of the partial average $\langle\mathcal{T}^{00}_{LL'}\rangle_A$ defined earlier. In the CPA $\langle\mathcal{T}^{00}_{LL'}\rangle_A = \mathcal{T}^{A,00}_{LL}$ hence

$$\overline{I}_A(\omega) \propto \sum_L \Big| \int d^3r\ \varphi_{c,A}(\underline{r})\underline{\nabla}\,\Delta^A_L(\underline{r},\hbar\omega+\epsilon_c)\Big|^2 \mathrm{Im}\ \mathcal{T}^{A,00}_{LL}(\hbar\omega+\epsilon_c) \qquad \text{IV-27}$$

Depending on the angular momentum eigenvalue associated with the core state $|\varphi_c\rangle$ the above expression gives the K, $L_{2,3}$, M, N spectra from the A sites in the one electron approximation. Of course a similar expression applies to the spectra from the B sites. If the KKR-CPA equations for the effective scattering amplitudes $t_{c,L}$ have been solved $\mathcal{T}^{A,00}_{LL'}$ can be readily calculated. The matrix elements $\int d^3r\ \varphi_c(\underline{r})\nabla\,\Delta_L(\underline{r})$ are also easily obtained since these depend only on the muffin tin potential at the site where the emission occurs and are independent of the band structure (the multiple scattering effects are completely described by $\mathcal{T}^{A,00}_{L,L'}$). Thus we have a theory where the matrix elements are determined on equal footing with the density of states.

Unfortunately the presence of the core hole will alter the potential seen by the conduction electrons at the emiting site from its usual muffin-tin value. This effect will modify the above theory : However the modification is large only at energies close to the Fermi energy. Nevertheless, we expect that much could be learned about the alloy potential if improved data becomes available and is interpreted in terms of the above CPA theory.

In conclusion we would like to emphasize that, in our

opinion, making progress with the problem of constructing alloy potentials is largely a matter of trial and error and therefore making contact with many different experiments within the context of the same theory is of the greatest importance. Recall that our ideas about the crystal potential for pure systems have been refined, over the past fifteen years, to the point where now we can calculate cohesive energies with some confidence, by calculation of such complicated quantities as the Fermi surface. For random alloys we do not have the very accurate and detailed Fermiology experiments which helped so much in the case of the pure metals. So we have to make do with more crude tools such as Soft X-ray Spectra and Knight Shift and angle resolved photoemission. Nevertheless the hope is that total energy calculations will eventually become feasible.

V. SOLVING THE KKR-CPA EQUATION : THE CASE OF $Ni_c Cu_{1-c}$

1. The local CPA approximation (LCPA)

As we have mentioned in Sec. IV. 3 the main difficulty in solving Eq. IV-13 is in the calculation of $\mathcal{T}^{c,00}_{L,L'}(\epsilon)$ a calculation which must be repeated at each iteration. To find a way around this problem it is useful to think about $\mathcal{T}^{c,00}_{L,L'}(\epsilon)$ in real space instead of the integral representation given in Eq. IV-11. In the space of the lattice points $\mathcal{T}^{c,00}_{L,L'}(\epsilon)$ is given by

$$\mathcal{T}^{c,ij}_{L,L'} = t_{c,L}\,\delta_{L,L'}\,\delta_{i,j} + \sum_{k\neq i}\sum_{L''} t_{c,L}\, G_{L,L''}(\underline{R}_i - \underline{R}_k;\epsilon)\mathcal{T}^{c,kj}_{L'',L'} \qquad \text{V-1}$$

that is to say it may be constructed by adding up the effects of all the individual inelastic scattering processes which begin and end at $\underline{R}_0$. Since the scattering at each centre is inelastic each time the particle scatters there is a finite probability that it will be absorbed. This suggests that $\mathcal{T}^{c,00}_{L,L'}(\epsilon)$ is determined by the scatterings at the neighbouring sites. Since $G_{L,L'}(\underline{R}_i - \underline{R}_j;\epsilon)$ is not attenuated as the function of $\underline{R}_i - \underline{R}_j$ this is not a rigorous argument but it is supported by cluster calculations on pure systems [61,51].

Let us push this argument to its logical limit and assume as a zeroth order approximation that $\mathcal{T}^{c,00}_{L,L}(\epsilon)$ is determined by the scattering at $\underline{R}_0$ alone. This means that we take $\mathcal{T}^{c,00}_{L,L'} = t_{c,L}\,\delta_{L,L'}$. Substituting this expression into Eq. IV-13 it is a matter of simple algebra to show that the resulting

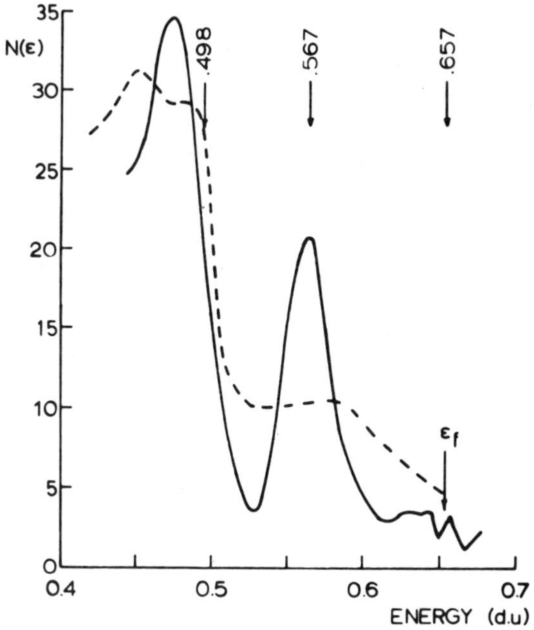

Fig. 18. The full line is the high energy portion of the density of states for $Cu_{87}Ni_{13}$ in the Local Coherent Potential Approximation (LCPA ~ ATA). The energy is measured in dimensionless units (D. U). One D. U = $(2\pi/a)^2$ Ry. Here a = 6.673 and hence $(2\pi/a)^2$ = .886 Ry. The sharp drop in n(ε) at .5 D. U. is the upper edge of the Cu d-band.

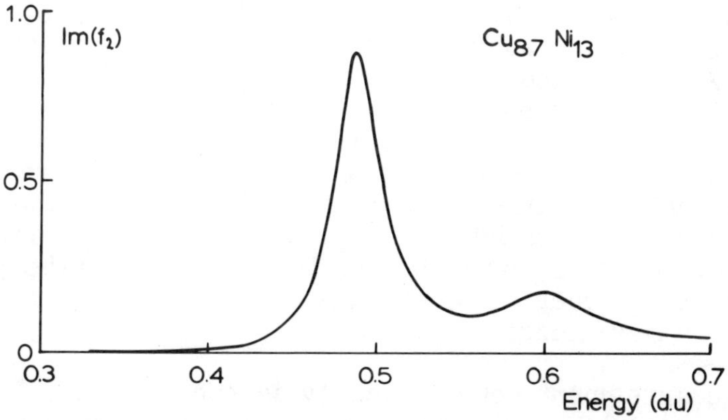

Fig. 19. The imaginary part of the averaged scattering amplitude for $Cu_{87}Ni_{13}$ as a function of energy.

equation is solved by $t_{c,L} = c\, t_{A,L} + (1 - c) t_{B,L}$. Thus the extreme local approximation to the CPA equation (LCPA) is just the ATA for the scattering amplitude $t_{c,L}$. Clearly this is a very attractive starting point for sequence of approximations. However, before describing the other members of this sequence we want to discuss the LCPA in a little more detail.

Having approximated $t_{c,L}$ by $\bar{t}_L$ in the LCPA, we must now evaluate Eq. IV-21 using $\bar{t}_L$. Since Eq. IV-21 was derived using the CPA equations given in Eq. IV-13 it is only valid for $t_{c,L}$ which is the solution of that equation. Consequently the LCPA is a reasonable procedure only if $\bar{t}_L$ is close to $t_{c,L}$. Evidently, the LCPA is a different approximation for the density of states than the ATA discussed in Sec. IV. 2 even though it uses the same scattering amplitude to describe the effective scatterers. Unfortunately at this stage of the development of the theory it is impossible to tell which is better. We shall use the LCPA prescription since it is easier to implement. Yet another approximation results if we use $t_{c,L} = \bar{t}_L(\epsilon)$ in Eq. IV-23 since in going from Eq. IV-21 to Eq. IV-23 we have again used the CPA equation and therefore the two expressions are the same only if $\bar{t}_L$ is a solution of Eq. IV-13. It is lengthy but straightforward to show that this scheme is the same as the one proposed by Schwartz et al[22] as the ATA.

Using the LCPA we have evaluated the density of states for $Cu_{87}Ni_{13}$ for the δ_2^{Ni}, δ_2^{Cu} phase shifts shown in Fig. 15 and the corresponding s and p phase shifts. The q integral in Eq. IV-21 was carried out in the irreducible 1/48-th of the Brillouin zone direction by direction. At each energy the phase of the determinant was evaluated along a series of directions emanating from the zone centre. With each point along the direction we associated an elemental volume chosen so that the sum of the elemental volumes gives the volumes of the Brillouin zone. First, the integral is performed for each direction, these are then summed over to give the required integrated density of states. The density of states was then obtained by numerical differentiation. The high energy portion of the density of states is shown in Fig. 18 together with the photoemission result of Spicer. The two curves were lined up so that the Fermi energies coincide.

The agreement between theory is encouraging. Clearly, the LCPA and presumably any version of the ATA gives an impurity peak above the Cu d-band in agreement with experiments and therefore cures the main failings of the rigid

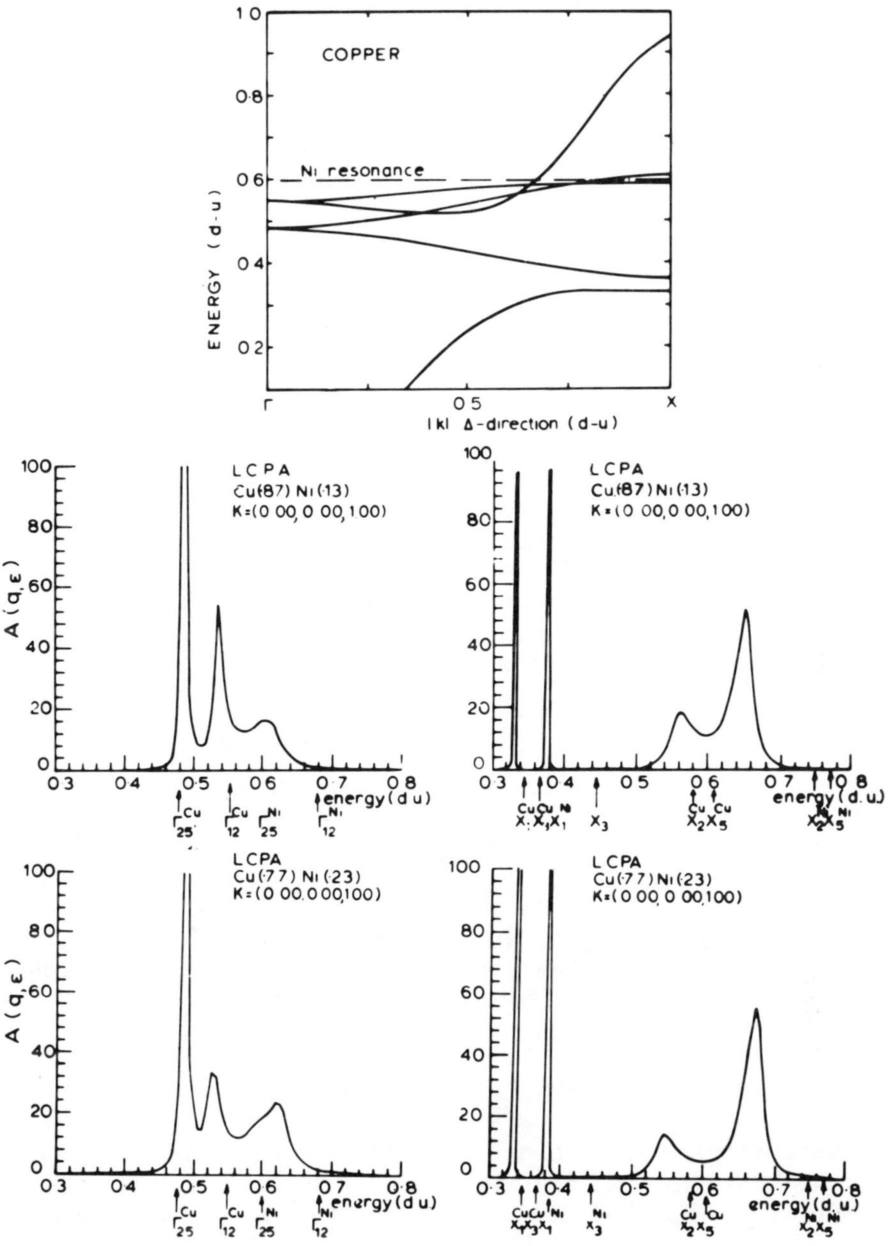

Fig. 20. Energy bands of pure Cu and the Bloch spectral functions for $Cu_{87}Ni_{13}$ and $Cu_{77}Ni_{23}$ in the Δ direction. On the graphs showing the spectral functions at the Γ and X points we mark the positions of the appropriate pure Cu and pure Ni bands for orientation.

band model and the virtual crystal approximation. This is also a sensible result if one bases ones expectations on the virtual bound state model of an isolated impurity [5]. In this connection it is interesting to note that the impurity peak is not exactly at the energy of the Ni resonance where the Argand plot for this concentration shows the second loop. This is more clearly seen if we plot Im f_2 as a function of energy as shown in Fig. 19. Note that the higher Ni resonance peak is slightly lower in energy then its consequence in the Density of state shown in Fig. 18. Presumably this shift is due to the hybridization of the Ni resonance with the Cu s-p band.

Though it appears to be reasonable in its broad outlines the LCPA is still not an adequate theory for the photo emission experiments from Cu rich Ni-Cu alloys. The trouble is that as the concentration changes LCPA predicts that the Ni impurity peak shifts to higher energies with respect to the upper edge of the Cu d-band in contradiction with experiments. To see what is happening we show the spectral density calculated from Eq. IV-23 at various $\underline{q}$ points in the $\Gamma - X$ direction for $Cu_{87}Ni_{13}$ and $Cu_{77}Ni_{23}$, in Fig. 20. From these pictures it is clear that in absolute terms the Fermi energy and the position of the impurity bands remains the same but the peaks which can be associated with the Cu d-band broaden considerably as the concentration increases and, more importantly, move to lower energies. This broadening will give rise to a flattening out of the upper edge of Cu d-band in the density of states which for the purposes of this discussion should be thought of as $\sum_{\underline{q}} A_B(\underline{q}, \epsilon)$ and together with the downward movements of what can be identified as the top of the Cu d-band will have the effect of changing the relative position of the impurity and the Cu d-band. There are a number of other features of the spectral function which are not quite right as given in Fig. 18 but we shall discuss these in the next section.

2. The cluster method for solving the KKR-CPA equations

An obvious way of proceeding further with the argument which leads to the LCPA is to take instead of $\mathcal{T}^{c,00}_{L,L'} = t_{c,L}\,\delta_{L,L'}$ the solution of Eq. V-1 for a finite cluster surrounding the site at $\underline{R}_0$.

To calculate $\mathcal{T}^{c,00}_{L,L'}(\epsilon)$ for a finite number of scatterers one merely has to invert the matrix $t_L^{-1}\delta_{ij}\delta_{L,L'} - G_{L,L'}(\underline{R}_i - \underline{R}_j;\epsilon)$ where the site indices i and j run from 0 to N, the number of scatterers considered, and the angular momentum index L takes all possible values consistent with the largest ℓ for which the scattering amplitude $t_{c,L}$ is taken to be non zero. For example in an f. c. c. lattice an atom and its nearest neighbour shell is a cluster of 13 atoms. If we include a scattering amplitude with $\ell \leq 2$, L can take 9 different values (0,0 1,1 1,0 1,-1 2,2 2,1 2,0 2,-1 2,-2) hence one must invert a 117 by 117 matrix. This can be readily accomplished numerically. If we are willing to treat the s-p component of the effective scattering amplitude in the average t-matrix or virtual crystal approximation on the grounds that they are very much the same any way only the L-index corresponding to the 5 d-states arises and the matrix to be inverted is only 65 by 65. In our actual calculations we shall always be using the d-block only.

With this in mind we propose the following strategy for solving for the effective CPA scattering amplitude $t_{c,L}$:
We solve Eq. IV-13 using a finite cluster centred on the site at $\underline{R}_0$ to evaluate $\mathcal{T}^{c,00}_{L,L'}$ at each iteration. We then repeat the calculation using a larger cluster which includes the next nearest neighbour shell beyond the cluster in the previous calculation. This process will generate a set of scattering amplitudes, one for each cluster size. In principle this sequence will converge to the true solution of Eq. IV-13. However, the physical idea that far away scatterers will have little effect on the very local CPA self-consistency between impurity scattering on a site and its environment suggests that the convergence will be rapid. Indeed, as we shall show presently, already the third nearest neighbour shell has practically no effect on $f_{c,L}$ for the systems we have considered.

In Fig. 21 we show Im $f_{c,L}$ as a function of energy for 0, 1, 2 and three nearest neighbour shell calculations. Evidently Im $f_{c,L}$ changes considerably from its ATA value (0 shell cluster) when the first nearest neighbour shell is introduced. As expected $f_{c,L}$ no longer corresponds to a spherically symmetric scatterer, e. g. it is different for different m. Since the site has cubic symmetry this m dependence is still degenerate and $f_{c,L}$ splits into only two components f_{c,e_g} and $f_{c,t_{2g}}$. When the next shell is introduced the t_{2g} component hardly changes but there is still quite a bit of movement in f_{c,e_g}. This can be understood easily. In an fcc lattice the e_g orbitals about a site $(x^2 - y^2, 2z^2 - x^2 - y^2)$ point towards next nearest neighbours. In a one nearest neighbour shell cluster the central atom sees no scatterers along its e_g orbital lobes. When the second shell is introduced there will be scatterers along these lobes. For

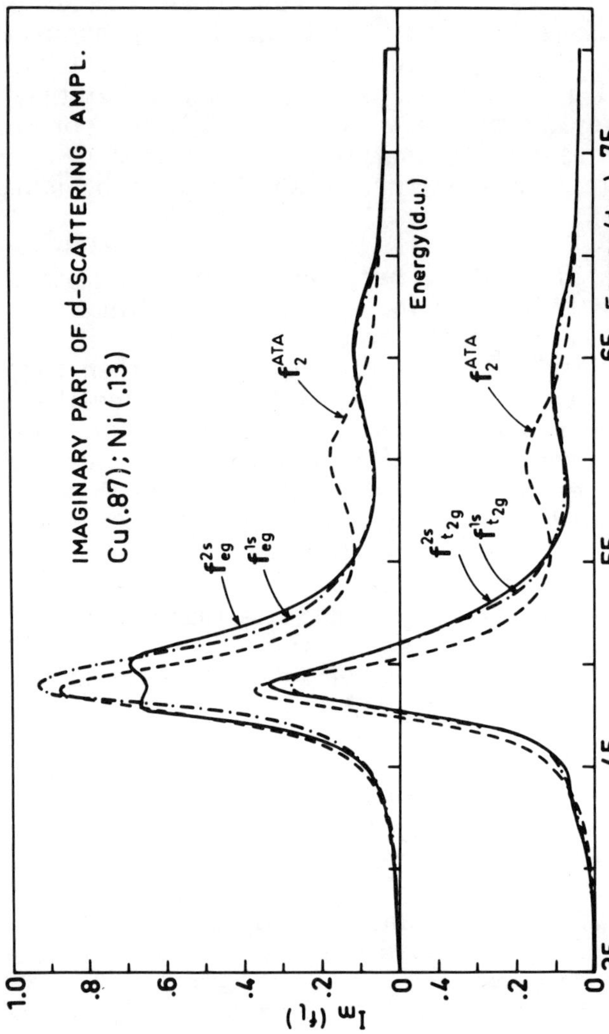

Fig. 21a. The imaginary part of the effective scattering amplitude $f_{c,L}(\epsilon)$ for a 1 shell and 2 shell C. P. A. showing the convergence discussed in the text.

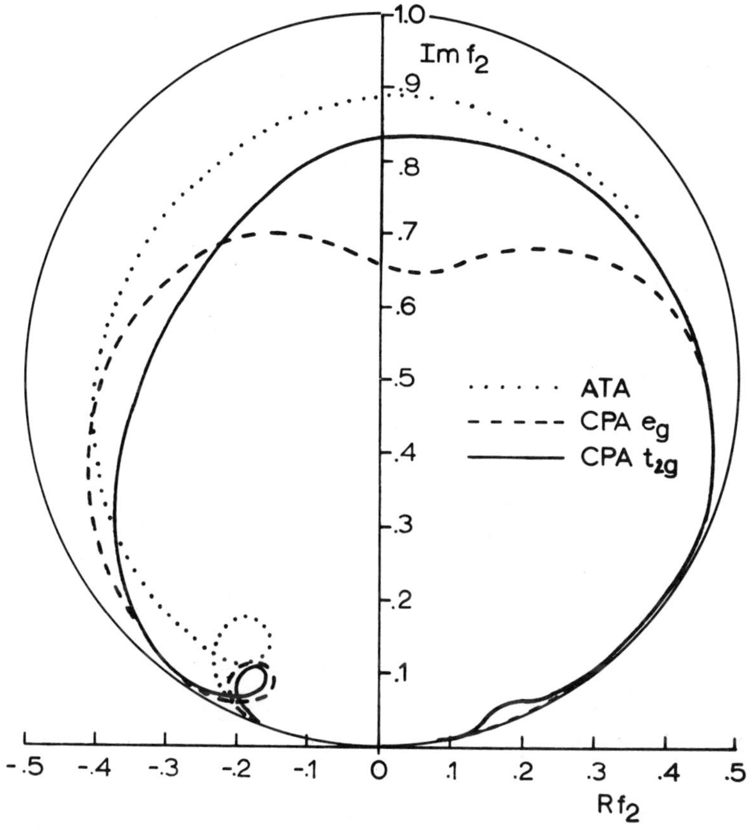

Fig. 21b. The Argand plot of $f_{c,L}(\epsilon)$ as obtained in the 2 shell calculation.

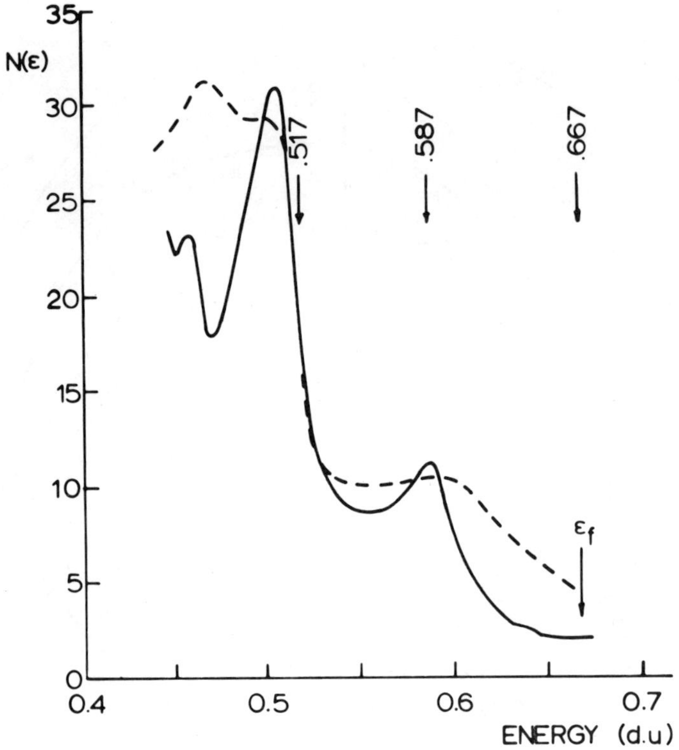

Fig. 22. Comparison of the optical density of states (broken line) and density of states from the KKR-CPA calculation (full line) described in the text for $Cu_{87}Ni_{13}$. The experimental curve is in arbitrary units. The two curves were lined up by setting the respective Fermi energies equal.

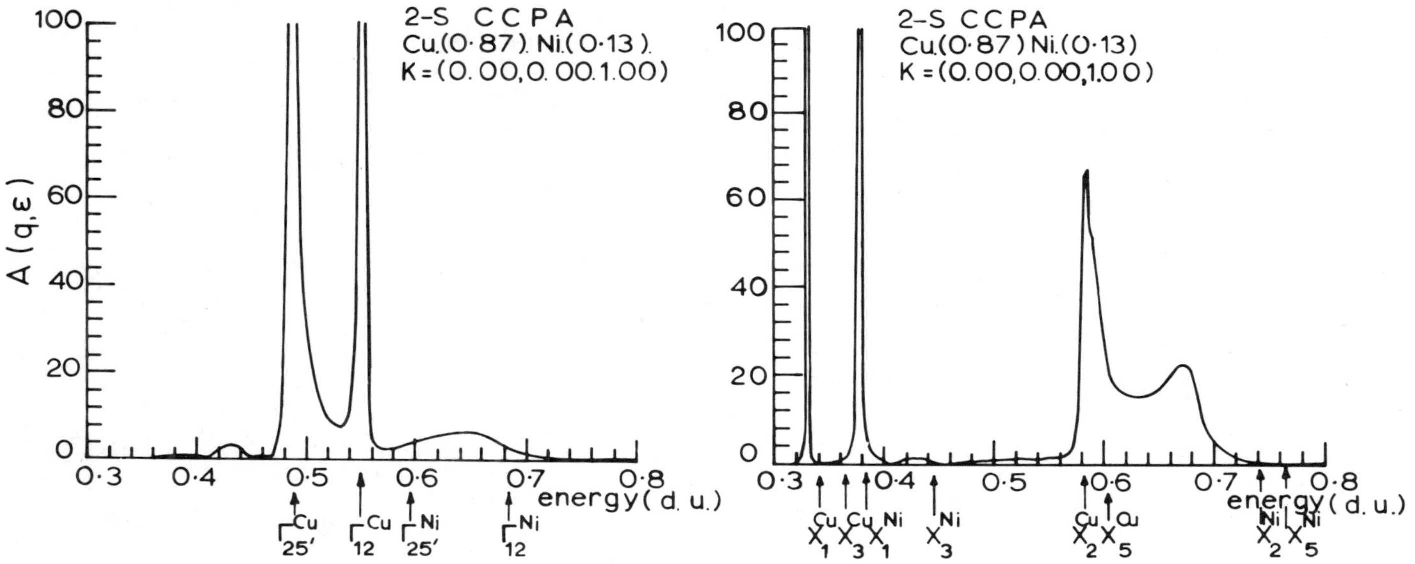

Fig. 23a. The Bloch spectral function at the Γ and X points for the alloy $Cu_{87}Ni_{13}$ as obtained in a 2 shell KKR-CPA calculation. Note that the impurity peak stays roughly at the same position relative to the "Cu like" bands as the concentration increases.

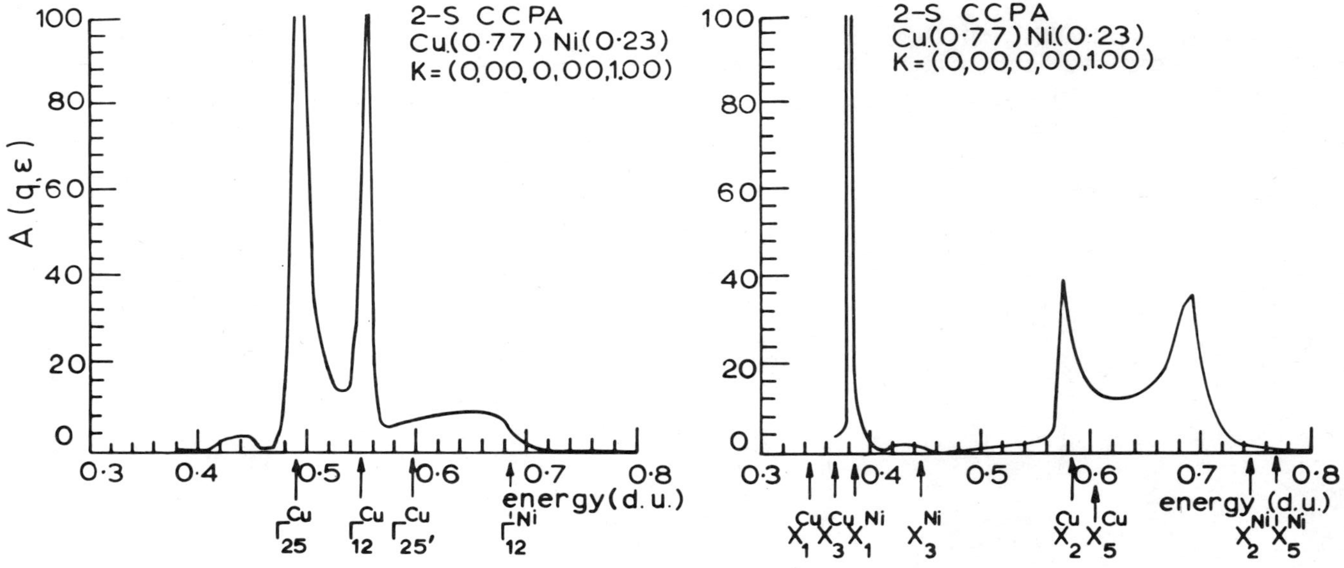

Fig. 23b. The Bloch spectral function at the Γ and X points for the alloy $Cu_{77}Ni_{23}$ as obtained in a 2 shell KKR-CPA calculation. Note that again the impurity peak stays roughly at the same position relative to the "Cu like" bands as the concentration increases.

this reason the CPA self-consistency yields two very different f_{c,e_g}'s. It is reassuring to note that the introduction of the third shell causes changes in both f_{c,e_g} and $f_{c,t_{2g}}$ which are small on the scale of changes in previous steps. Thus we assume that our scheme is reasonably well converged after two shells of neighbours has been included.

Note that in the CPA the Ni peaks of Im $f_{c,2}$ is about .5 eV higher in energy than in the ATA. This is an important physical effect and is the root cause of the concentration dependence of the impurity peak in the density of states. Unlike in the ATA where the two resonances simply add in the CPA they interact through the self-consistency requirement and as it turns out repell each other like interacting bound states (Intersite d-d repulsion). The reason for not moving symmetrically about their centre of mass (only the Ni peak appears to have moved substantially) is that at this concentration the Cu resonance is weighted much more heavily. To put it another way the coupling into the Cu resonance is much stronger as is evident from the Argand plots in Fig. 15.

As a consequence of the fact that the Ni resonance in the scattering amplitude appears at higher energies the impurity peak in the density of states will also appear at higher energies in the CPA than in the LCPA. One might think that this will destroy the good agreement between the theory and experiment as far at the relative distance between the impurity peak and the upper edge of the Cu like d-band is concerned. However, this is not the case as can be seen in Fig. 22 where we show the two shell cluster KKR-CPA density of states for $Cu_{87}Ni_{13}$ together with the photoemission experimental "density of states". Clearly the agreement is very good even quantitatively. This is very satisfying since the phase shifts ($\delta_2^{Ni}(\epsilon)$, $\delta_2^{Cu}(\epsilon)$) were not adjusted to give this agreement and there are no other adjustable parameters in the theory.

The reason why the CPA can agree with the experiments inspite of the higher energy of Ni resonance is that a higher impurity band disturbs the upper Cu d-bands less and therefore the upper edge of the Cu like d-band at the X-point is sharper than in the ATA and is also at a higher energy. This is illustrated in Fig. 23 where we show the Bloch spectral function in the $\Gamma - X$ direction for $Cu_{87}Ni_{13}$ and $Cu_{77}Ni_{23}$. At the Γ and the X points we have marked the energies of the corresponding peaks in pure Cu and pure Ni. As you can see the CPA resolves the Cu like d-bands much better than the ATA.

Detailed comparison with Fig. 21 shows that all the 'Cu' peaks are closer to their corresponding positions in pure Cu for the CPA than for the LCPA. In particular the zone centre $\Gamma_{25'}$ and Γ_{12} states are at exactly the same position as in pure Cu, however the peak which results from the Γ_{12} state in the LCPA is pushed to lower and lower energies as the concentration of Ni is increased. So, in the LCPA the Fermi energy, the impurity peak and the upper edge of the Cu d-band are all lower and this gave for the $Cu_{87}Ni_{13}$ a fortuitous agreement with experiments as far as their relative position is concerned. In the CPA all these quantities appear at higher energies with roughly the same relative position. Moreover, these relative positions do not change as a function of concentration in agreement with experiments. As is clear from comparing the curves the two concentrations in Fig. 23 even at increased concentrations the higher up Ni impurity band leaves the Cu like band pretty much undisturbed, and therefore the relative positions of the peaks in the full density of states remain unchanged.

It is also reassuring to compare the tight-binding calculation of Stocks Williams and Faulkner with our present results. As mentioned earlier our phase shifts have been adjusted so that our calculation would be directly comparable with these tight-binding results. The density of states from these two calculations are compared in Fig. 24. The good agreement suggests that we have indeed solved the CPA.

Another interesting feature of the spectral functions shown in Fig. 23 is the behaviour of the impurity band as a function of $\underline{q}$. Clearly it is a very flat band ; e. g. its position hardly changes from the Γ to the X point. However it sharpens up considerably near the zone boundary. If this behaviour was interpreted in the language of self energy discussed in Sec. II it would mean that we are dealing with a self energy which is $\underline{q}$ dependent. The fact that this situation can arise is the consequence of the fact that we are treating off diagonal randomness on an equal footing with diagonal disorder. Consequently this effect could not appear in the tight binding calculation we have referred to above. In a photo-emission experiment where one can control the direction of the incoming photon as well as measure the momentum of the out going electron one can investigate the band structure $\underline{q}$ point by $\underline{q}$ point [62]. Hopefully, the above $\underline{q}$ dependence of the width of the impurity levels will be studied by this means in the not too distant future.

The Ni rich Cu-Ni alloys are, of course, ferromagnetic

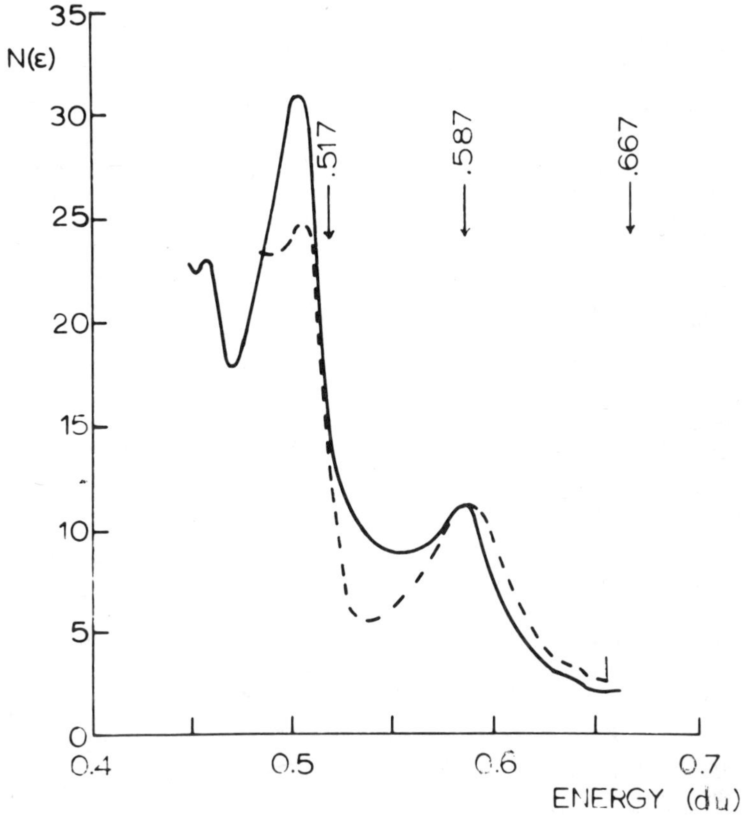

Fig. 24. Comparison of the density of states for $Cu_{87}Ni_{13}$ as calculated in the KKR-CPA (full line) on the one hand and in tight-binding model of Ref. 34 (broken line) on the other. Since in the latter the Fermi energy was not known very accurately we lined up the upper edges of the Cu d-bands for comparison. The end of the broken line marks ϵ_F in the tight binding calculation.

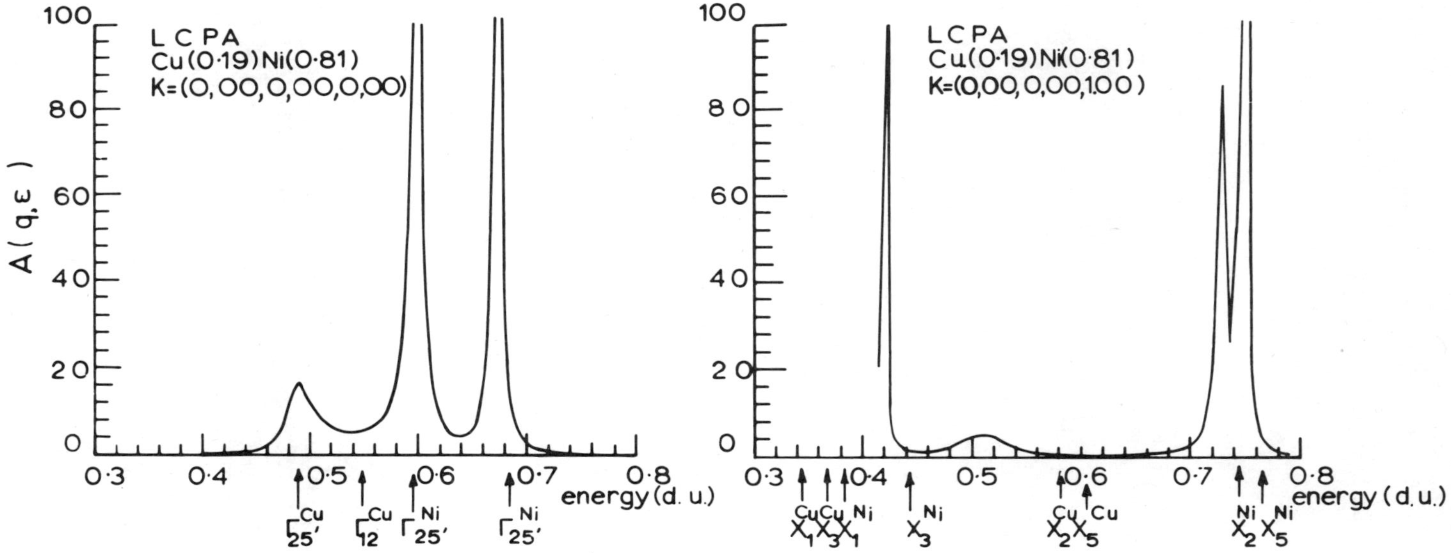

Fig. 25a. Comparison of the Bloch spectral function at the Γ and X points as calculated in the LCPA for $Cu_{19}Ni_{81}$.

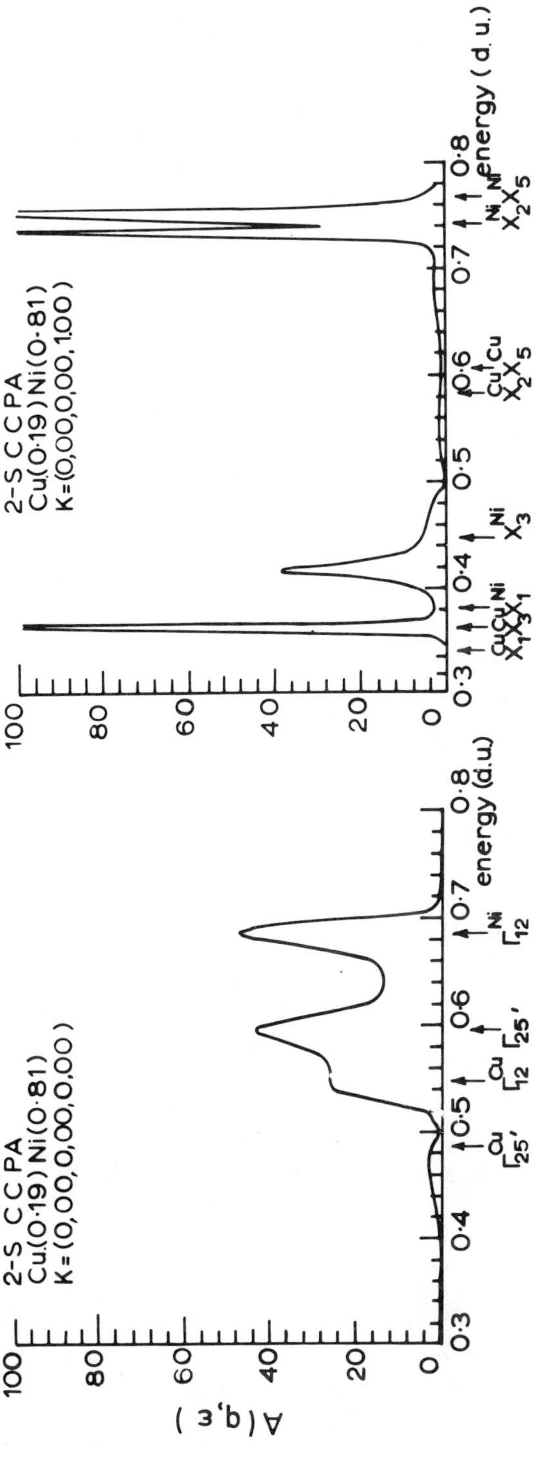

Fig. 25b. Comparison of the Bloch spectral function at the Γ and X points as calculated in the CPA for $Cu_{19}Ni_{81}$.

and the above paramagnetic theory is not directly relevant to many experiments. Nevertheless, it is interesting to consider what happens to the impurity band when it appears below the host band. In Fig. 25 we show a few spectral functions for $Cu_{19}Ni_{81}$ using both the LCPA and a two shell CPA. Obviously, here the two calculations give even more divergent answers than in the Cu rich case. The shoulder at the low energy side of the d-band as marked $\Gamma_{25'}$ in the Γ point spectral function has been identified in the calculation to have Γ_{12} symmetry. Thus, even at this low concentration we see the forming of the upper Cu d-bands instead of a virtual bound state like impurity resonance. Instead of a repulsion between the resonances, if anything, we see an attraction. We do not yet understand this phenomena in detail but it is consistent with the calculations of Stocks et al for the Ni rich alloys[31].

A physically more interesting feature of these Ni rich calculations is the sharpness of the d-states at the top of the d-band. Clearly this has important bearing on the subject of magnetism in this system. In the light of a recent paper by Gunnarson[63] it is an attractive proposition to analyse the condition for magnetism in terms of the paramagnetic bands calculated here. However, even without such detailed work we can see that the rigid band argument which we have referred to in the introduction is not likely to do better for the susceptibility at this end of the concentration range than it did for the specific heat in the Cu rich end. The bands are clearly not rigid. The presence of Cu atoms puts in extra states at low energies. Most probably the extra Cu electrons will occupy these states and the Fermi energy will again not move. Consequently magnetism will disappear not because the Ni holes have been filled with Cu electrons[64] but because the Ni states at the Fermi energy will become too ill defined to be able to benefit energetically from a small exchange splitting. To be sure metallic magnetism is a complex and a still unsolved problem even in pure systems and the above remarks are no substitute for a real theory of magnetism in Ni-Cu alloys. However, we would like to point out that the rigid band idea probably has nothing to do with it.

VI. EPILOGUE

In a structurally disordered system like a liquid metal one does not begin to have a theory until at least the pair wise correlations between sites are included. In amorphous semi-conductors even higher order correlations are necessary to have a qualitatively acceptable theory. However for substitutionally disordered system like the alloys we have been talking about one can go a long way before short range order needs to be considered specifically. All the same the theory we outlined in these lectures is clearly lacking in this respect. A description of the formation of localized moments and of the magnetic state as well as a theory of the order-disorder phase transition or the problem of phase separation clearly requires that we take account of the fluctuations of the environment of a given site about its average. A Ni atom having a Ni rich environment more frequently than one would guess from the concentration alone clearly would help the formation of the magnetic state.

In the tight-binding formulation of our problem there is only one way to proceed. Being a single site theory CPA as it stands is not capable of describing correlations between sites. Therefore the whole scheme needs to be generalized. In the context of electronic states in random alloys the problem of short range order was first considered by Ducastelle and Cyrot-Lackman[65]. Since then a number of attempts, having variable successes, have been made to include short range order[66]. It is probably fair to say to date no scheme has emerged which is as universally accepted as the CPA is for the simpler no correlation problem. The main reason for that is not so much the difficulty of generating sensible looking and tractable approximations[67] but rather the lack of clean cut experiments which would differentiate between them. In these lectures we might have overemphasised its significance but there is little doubt that the photoemission experiments played an important role in establishing the CPA as the most widely used approximation.

Within the theory of alloys we have outlined in these lectures there is an alternative way for treating short range order. This does not require that we should go beyond the CPA in treating a random Hamiltonian, it merely involves a more complicated CPA.

As we mentioned in Sec. IV-1 when generating scattering amplitudes for an A site in a random lattice strictly speaking one obtains a sequence of scattering amplitudes $t^{\nu}_{A,L}$ where ν enumerates the local configurations. Similarly, one obtains a set $t^{\nu}_{B,L}$ for B sites. This suggests that we should consider a 2ν component alloy. The formal generalization of the theory is trivial. The CPA equation for the effective scattering amplitude $t_{c,L}$ is given by

$$\sum_{\nu} C_{\nu}^{A}\, \mathcal{J}_{L,L'}^{A,\nu,00}(\epsilon) + \sum_{\nu} C_{\nu}^{B}\, \mathcal{J}_{L,L'}^{B,\nu,00}(\epsilon) = \mathcal{J}_{L,L'}^{c,00}(\epsilon) \qquad \text{VI-1}$$

where C_{ν}^{A} is the probability for the configuration denoted by ν about an A site and C_{ν}^{B} is the probability that the same configuration occurs about a B site. Clearly, this theory includes short range order to some as yet known extent. Though computationally demanding it should also be tractable for at least a one shell CPA. However, this takes us into the realm of speculations and it is probably time to stop.

In conclusion we would like to emphasize that these lectures did not intend to review all current work on alloys using CPA or its generalizations. Many interesting new developments in the theory of magnetic alloys[67] and significantly in the energetics of metallic alloys[68] were not even mentioned. However these are advanced topics within the program we have been discussing and in the language of these lectures have not yet been seriously considered.

Acknowledgment

The point of view we have expressed in these lectures was developed over the years in long and stimulating discussions with our colleagues in the Theoretical Physics Group at the University of Bristol and the Alloy Theory Group at the Oak Ridge National Laboratory. For their time and valuable comments we are most grateful. In particular we would like to thank Dr. J. S. Faulkner for having introduced us to the alloy problem and scattering theory as well as for many years of fruitful collaborations. The cluster method of solving the CPA equations was developed in a stimulating partnership with Drs S. Julliano and R. Ruggero. We would also like to thank Barbara Gordon, Walter Temmerman and Laszlo Gyorffy for their help in preparing the manuscript and Wendy Redstone for her patience and conscientiousness in typing the text.

APPENDIX

Here, we shall derive an expression for the density of states due to a set of non-overlapping muffin-tin potentials. The final result will be relatively simple and will show that the density of states for such a system depends only on the phase shifts of the individual scatterers and the special arrangement of the scattering centres.

Using the definitions in the text we may begin by writing

$$n(\epsilon) = -\frac{1}{\pi}\,\mathrm{Im}\,\mathrm{tr}\,G(\epsilon) = -\frac{1}{\pi}\,\mathrm{Im}\int d^3k\, G(\underline{k};\epsilon)$$

$$= -\frac{1}{\pi}\,\mathrm{Im}\int d^3k\, G_0(\underline{k};\epsilon) - \frac{1}{\pi}\,\mathrm{Im}\int d^3k\, G_0(\underline{k};\epsilon) T_{\underline{k},\underline{k}}(\epsilon) G_0(\underline{k};\epsilon)$$

$$= -\frac{1}{\pi}\,\mathrm{Im}\int d^3k\, G_0(\underline{k};\epsilon) - \frac{1}{\pi}\,\mathrm{Im}\sum_{i,j}\int d^3k\, G_0(\underline{k};\epsilon)\, \mathcal{T}^{i,j}_{\underline{k},\underline{k}}(\epsilon)\, G_0(\underline{k};\epsilon) \qquad \text{A. 1}$$

where $G_0(\underline{k};\epsilon)$ is the free particle propagator and $\mathcal{T}^{i,j}_{\underline{k},\underline{k}}(\epsilon)$ is the plane wave matrix elements of the scattering path operator $\mathcal{T}^{i,j}(\epsilon)$. Noting that $G_0^2(\underline{k};\epsilon) = -\frac{\partial}{\partial\epsilon} G_0(\underline{k};\epsilon)$ it is straightforward to show that

$$\sum_{i\neq j} \mathrm{Im}\int d^3k\, G_0^2(\underline{k};\epsilon)\, \mathcal{T}^{i,j}_{\underline{k},\underline{k}}(\epsilon) = \sum_{i\neq j}\sum_{L,L'}\left[\frac{\partial}{\partial\epsilon} G_{L,L'}(\underline{R}_i - \underline{R}_j;\epsilon)\right] . \mathcal{T}^{i,j}_{L,L'}(\epsilon) \qquad \text{A. 2}$$

since $\mathcal{T}^{i,j}(\underline{r},\underline{r}';\epsilon) = \mathcal{T}^{i,j}(\underline{r}-\underline{R}_i, \underline{r}'-\underline{R}_j;\epsilon) = 0$ for either $|\underline{r}-\underline{R}_i| > r_{MT}$ or $|\underline{r}'-\underline{R}_j| > r_{MT}$ and therefore one needs the free particle propagator $G_0(\underline{r},\underline{r}';\epsilon)$ only for $\underline{r}$ and $\underline{r}'$ in two different muffin tins (see Eq. III-32). The site diagonal sum $\mathrm{Im}\sum_i \int d^3k\, G_0^2(\underline{k};\epsilon)\, \mathcal{T}^{i,j}_{\underline{k},\underline{k}}(\epsilon)$ is a little more difficult to handle.

Noting that

$$\mathcal{T}^{i,j} = t_i\, \delta_{i,j} + \sum_{\ell\neq i} J^{i,\ell}\, G_0\, t_j \qquad \text{A. 3}$$

is equivalent to Eq. III-30 we can write

$$\mathcal{T}^{i,i} = t_i + \sum_{\ell\neq i} t_i\, G_0\, J^{\ell,i}$$
$$\mathcal{T}^{\ell,i} = \sum_{\ell\neq i} J^{\ell,\ell'}\, G_0\, t_i ; \quad \ell\neq i \qquad \text{A. 4}$$

and therefore

$$\mathcal{J}^{i,i} = t_i + \sum_{\substack{\ell \neq i \\ \ell' \neq i}} t_i G_0 J^{\ell,\ell'} G_0 t_i \qquad \text{A. 5}$$

Consequently

$$\sum_i \text{Im} \int d^3k \, G_0(\underline{k};\epsilon) \mathcal{J}^{i,i}_{\underline{k},\underline{k}} (\epsilon) = \sum_i \text{Im} \int d^3k G_0^2(\underline{k};\epsilon) t_{i;\underline{k},\underline{k}} (\epsilon)$$

$$+ \sum_{\substack{\ell \neq i \\ \ell' \neq i}} \text{Im} \int d^3k \, G_0^2(\underline{k};\epsilon) < \underline{k} | t_i \, G_0 \, \mathcal{J}^{\ell,\ell'} \, G_0 t_i \, | \, \underline{k}> \qquad \text{A. 6}$$

By inserting the operator identity $\int d^3 r |\underline{r}><\underline{r}| = 1$ between all the operators in the sequence $t_i G_0 \mathcal{J}^{\ell,\ell'} G_0 t_i$, and carrying out the spacial integrals with the help of Eq. III-32 one obtains the following identity

$$<t_i G_0 \mathcal{J}^{\ell,\ell'} G_0 t_i> = \sum_{L_1 L_2 L_3 L_4} < \underline{k} | t_i | L_1, \sqrt{\epsilon} > . \qquad \text{A. 7}$$

$$\cdot G_{L_1,L_2}(\underline{R}_i - \underline{R}_\ell;\epsilon) \mathcal{J}^{\ell,\ell'}_{L_2,L_3} (\epsilon) G_{L_3,L_4}(\underline{R}_{\ell'} - \underline{R}_i;\epsilon) < L_4, \sqrt{\epsilon} | t_i \, | \underline{k}>$$

where we have used the notation that

$$< \underline{k} | \, t \, | L_1, \sqrt{\epsilon}> = \int d^3 r \, t(\underline{k},\underline{r};\epsilon) \, Y_L \, (\hat{r}) j_L \, (\sqrt{\epsilon} \, r) \qquad \text{A. 8}$$

The next trick is to notice that

$$\int d^3k <L_4, \sqrt{\epsilon} \, | \, t \, | \underline{k}> G_0^2(\underline{k};\epsilon) < \underline{k} | t | L_1, \sqrt{\epsilon}> = < L_4, \sqrt{\epsilon} | t \; G_0^2 t | L_1, \sqrt{\epsilon}>$$

$$= <L_4, \sqrt{\epsilon} | [\frac{\partial}{\partial \epsilon} (\frac{1}{1 - vG_0} v)] | L_1, \sqrt{\epsilon}> = \frac{\partial}{\partial \epsilon} t_{L_1}(\underline{k},\underline{k};\epsilon)|_{\underline{k} = \sqrt{\epsilon}} \delta_{L_1,L_4} \qquad \text{A. 9}$$

Using this result and the fairly obvious fact that the "on the energy shell" component of Eq. A-5 may be written as

$$\frac{1}{t_{i,L}} (\mathcal{J}^{i,i}_{L,L} - t_{i,L}) \frac{1}{t_{i,L}} = \sum_{\substack{\ell \neq i \\ \ell' \neq i}} \sum_{L_1,L_2} G_{L,L_1}(\underline{R}_i - \underline{R}_\ell;\epsilon) \mathcal{J}^{\ell,\ell'}_{L_1,L_2}(\epsilon)$$

$$G_{L_2,L}(\underline{R}_{\ell'} - \underline{R}_i;\epsilon) \qquad \text{A. 10}$$

it is straightforward to derive the relation

$$\sum_i \text{Im} \int d^3k \, G_0^2(\underline{k};\epsilon) \, \mathcal{T}_{\underline{k},\underline{k}}^{i,i}(\epsilon) = \sum_i \text{Im} \int d^3k \, G_0^2(\underline{k};\epsilon) t_{i;\underline{k},\underline{k}}(\epsilon)$$

$$- \sum_i \sum_L \text{Im}(\frac{1}{t_{i,L}} \frac{\partial}{\partial \epsilon} t_{i,L}) + \sum_{i,L} \text{Im}[\frac{1}{t_{i,L}^2} (\frac{\partial}{\partial \epsilon} t_{i,L}) \mathcal{T}_{L,L}^{i,i}(\epsilon)] \quad \text{A. 11}$$

In the above expression the first term cancels the second since the latter is just the Friedel sum e. g. $\text{Im} \, t_{i,L}^{-1} \frac{\partial}{\partial \epsilon} t_{i,L} =$
$= \text{Im} \, \frac{\partial}{\partial \epsilon} \ell n \, t_{i,L} = \frac{\partial}{\partial \epsilon} \text{Im} \ell n(- \epsilon^{-1/2} \sin \delta_\ell^i e^{i\delta_\ell^i}) = \frac{\partial}{\partial \epsilon} \delta_\ell^i(\epsilon)$ and the former is, by definition, the change in the density of states due to an isolated muffin-tin well at $\underline{R}$;

Substituting Eq. A-2 and Eq. A-11 into Eq. A-1 yields

$$n(\epsilon) = n_0(\epsilon) - \frac{1}{\pi} \text{Im} \sum_{i,j} \sum_{L,L'} [\frac{\partial}{\partial \epsilon} t_{i,L}^{-1} \delta_{i,j} \delta_{L,L'} -$$

$$- \frac{\partial}{\partial \epsilon} G_{L,L'}(\underline{R}_i - \underline{R}_j ; \epsilon)] \mathcal{T}_{L,L'}^{i,j}(\epsilon) \quad \text{A. 12}$$

With the help of the matrix notation introduced in Eqs. III-36, III-37 and III-38 the above formula may be recast as

$$n(\epsilon) = n_0 - \frac{1}{\pi} \text{Im tr} [\frac{\partial}{\partial \epsilon} t^{-1} - \frac{\partial}{\partial \epsilon} G] \mathcal{T}(\epsilon)$$

$$= n_0 - \frac{1}{\pi} \text{Im tr} \frac{\partial}{\partial \epsilon} \ell n \, (t^{-1} - G)$$

$$= n_0 - \frac{1}{\pi} \text{Im} \frac{\partial}{\partial \epsilon} \ell n \, ||t^{-1} - G|| \quad \text{A. 13}$$

where we have made use of the matrix identity that $\text{tr} \ell n A = \ell n \|A\|$

Evidently,

$$N(\epsilon) = N_0(\epsilon) - \frac{1}{\pi} \text{Im} \ell n \, ||t_{i,L}^{-1} \delta_{i,j} \delta_{L,L'} - G_{L,L'}(\underline{R}_i - \underline{R}_j ; \epsilon)||$$
A. 14

This is the Lloyd formula quoted in the text. This derivation of it was first given in Ref. 56.

Instead of the site indices we could use lattice Fourier transform variables. If all the scattering amplitudes are the same e. g. $t_{i,L}(\epsilon) = t_L(\epsilon)$ for all i with respect to the lattice Fourier representation variables $\underline{q}$ the K. K. R. matrix is diagonal.

$$\sum_{i,j} e^{i\underline{q}\cdot\underline{R}_j}\left(t_L^{-1}\delta_{i,j}\,\delta_{L,L'} - G_{L,L'}(\underline{R}_i - \underline{R}_j;\epsilon)\right)e^{-i\underline{q}'\cdot\underline{R}_1} =$$

$$= \delta_{\underline{q},\underline{q}'}\left(t_L^{-1}\,\delta_{L,L'} - G_{L,L'}(\underline{q};\epsilon)\right)$$

Under this circumstance Eq. A-14 reduces to

$$N(\epsilon) = N_0(\epsilon) - \frac{1}{\pi}\sum_{q} \mathrm{Im}\,\ell n\,\left\| t_L^{-1}(\epsilon)\delta_{L,L'} - G_{L,L'}(\underline{q},\epsilon)\right\|$$

where the sum is over the Brillouin zone.

REFERENCES

1. "Thermochemical Data of Alloys" (Pergamon Science Series Vol. 3 1956) by O. Kubaschewski and J. A. Catterall

2. "Constitution of Binary Alloys", First Supplement (McGraw Hill Book Co. 1965) by R. P. Elliott

3. "The Theory and the Properties of Metals and Alloys" N. F. Mott and H. Jones (1936) Oxford University Press

4. N. F. Mott Adv. Phys. 13 326 (1964). See also J. Friedel Nuovo Cimento Sup. 7 287 (1958)

5. M. Shimizu, T. Takahashi and A. Katsuki, J. Phys. Soc. Japan 18 1192. However, Friedel was dissatisfied with this explanation and proposed the "virtual bound state" model as an alternative (J. Friedel Adv. Phys. 3 466 (1954) and Canadian J. Phys. 34 1190 (1956). At low concentration his description is in fact very close to that afforded by the CPA to be discussed in these lectures.

6. D. H. Seib and W. E. Spicer Phys. Rev. B2 1676 (1970) also N. J. Shevchik and C. M. Penchina Phys. Stat. Sol. 6 70 619 (1975)

7. L. Nordheim Ann. Physik 9 607 (1931), 641 (1931) T. Muto Sci. Papers Inst. Phys. Chem. Res. (Tokyo) 34 377 (1938) For actual calculations using the virtual crystal approximation see C. B. Sommers, H. Amar and K. H. Johnson Bull. Am. Phys. Soc. 11 73 (1966)

8. E. A. Stern Phys. Rev. Lett. 26 1630

9. E. G. Brovman and Y. M. Kagan "Dynamical Properties of Solids" Vol. 1 ed. G. K. Horton and A. A. Maradudin (North Holland 1974)

10. "Pseudopotentials in the Theory of Metals" W. A. Harrison (W. A. Benjamin 1966)

11. "Mathematical Physics in One-Dimension" (Academic New York 1966) E. G. Lieb and D. C. Mattis

12. "Theory of Lattice Dynamics in the Harmonic Approximation" (Academic Press 1961) A. A. Maradudin E. W. Montroll and I. P. Ipatova

13. T. Wolfram and J. Callaway Phys. Rev. 130 2207 (1963) also "Amorphous Magnetism" ed. H. O. Hooper and A. M. De Graaf (Plenum, New York 1973)

14. Y. Onodera and Y. Tayarawa J. Phys. Soc. Japan 27 341 (1968)

15. International Conference on Disordered Metallic Systems Strasbourg, France 10-15 Sept. J. Phys. (Paris) 35C

16. R. P. Abou-Chacra, P. W. Anderson and D. J. Thouless J. Phys. C. 6 1934 (1973), 65 (1974)

17. L. Nordheim Ann. Physik 9 607 (1931) 641 (1931)

18. P. Soven Phys. Rev. 156 809 (1967), 178 1136 (1969)

19. D. W. Taylor Phys. Rev. 156 1017 (1968)

20. Y. Onodera and Y. Toyazawa J. Phys. Soc. Japan 24 341 (1968)

21. R. J. Elliott, J. A. Krumhansl and P. L. Leath Rev. Mod. Phys 46 465 (1974)

22. H. Ehrenreich and L. Schwartz "Solid State Physics" ed. H. Ehrenreich, F. Seitz and D. Turnbull (Academic Press) Vol. 31 (1976)

23. B. E. Warren "X-ray diffraction" (Addison-Wesly Publishing Co. Reading, Mass 1969)

24. R. N. Aiyer, R. J. Elliott, J. A. Krumhansl and P. L. Leath Phys. Rev. 1006 (1969)

25. J. Callaway "Quantum Theory of the Solid State" Part B (Academic Press 1974) p. 385

26. B. Velicky, S. Kirkpatrick and H. Ehrenreich Phys. Rev. 175 747 (1968)

27. P. L. Leath J. Phys. C. 6 1559 (1973)

28. L. Schwartz and E. Siggia Phys. Rev. B5 383 (1972)

29. A. R. Bishop Solid State Comm. 17 1405 (1975)

30. R. Alben, M. Blume, H. Krakauer and L. Schwartz Phys. Rev. B12 4090 (1975)

31. G. M. Stocks, R. W. Williams and J. S. Faulkner Phys. Rev. B4 4390 (1971)

32. S. Kirkpatrick, B. Velicky and H. Ehrenreich Phys. Rev. B1 3250 (1970)

33. J. S. Faulkner Phys. Rev. (To be published)

34. G. M. Stocks, R. W. Williams and J. S. Faulkner J. Phys. F. Metal Physics 3 1688 (1973)

35. T. Kaplan and M. Mosstoller Phys. Rev.

36. K. Levin and H. Ehrenreich Phys. Rev. B3 4172 (1971)
F. Brouers and A. V. Vedayev Phys. Rev. B5 348 (1972)

37. "Quantum Theory of the Solid State" (Academic Press 1974) p. 291 J. Callaway

38. J. M. Ziman "Solid State Physics" Ed. H. Ehrenreich, F. Seitz and D. Turnbull (Academic Press 1970) Vol. 26

39. E. Wigner and F. Seitz "Solid State Physics" Ed. F. Seitz and D. Turnbull (Academic Press 1954) vol. I. p. 97

40. "Greens Functions for Solid State Physics" S. Doniach and E. H. Sondheimer (Benjamin Inc. 1974) Chap. 7

41. P. Hohenberg and W. Kohn Phys. Rev. 136B 864 (1964)

42. W. Kohn and L. J. Sham Phys. Rev. 140 A1133 (1965)

43. "Quantum Theory of Molecules and Solids IV (McGraw Hill Book Co. 1974) J. Slater

44. J. Slater "Technical Reports of the Solid State and Molecular Group". M. I. T. 1952-1954

45. "Quantum Theory of Molecules and Solids Vol. IV (McGraw-Hill Book Co. 1974) J. Slater

46. L. Hodges, R. E. Watson and H. Ehrenreich Phys. Rev. B5 3953 (1972)

47. N. D. Mermin Phys. Rev. 137 A 1441 (1965)

48. L. M. Mattheiss Phys. Rev. 133 A1399

49. T. L. Louks "Augmented Plane Wave Method" Benjamin : (New York 1967)

50. Scattering theory has an enormous literature. The most relevant to our lectures is the compendium of results and derivations by P. Lloyd and P. V. Smith Adv. Phys. 21 p. 69 (1972). We shall follow their notation closely. In particular we shall use the same spherical harmonics and spherical Bessel and Neuman functions and units (see their glossary).

51. B. L. Gyorffy and M. J. Stott "Band Structure Spectroscopy of Metals and Alloys" Ed. D. J. Fabian and L. M. Watson (Academic Press 1973)

52. J. Korringa Physica 13 392 (1947), W. Kohn N. Rostoker Phys. Rev. 94, 1111 (1954) see also B. Segall and F. S. Ham "Methods in Computational Physics (Academic Press 1968) Vol. 8 Chap. 7

53. J. S. Faulkner, H. L. Davis and H. W. Joy Phys. Rev. 161 556 (1967)

54. Mittag-Leffler Acta Math. (1884) more usefully "The Theory of the Scattering Matrix" (Macmillan Co. 1967) by A. O. Barut Appendix 6.

55. W. Butler, J. Olson, J. S. Faulkner and B. L. Gyorffy (to be published)

56. B. L. Gyorffy "Fondamenti Di Fisica Dello Stato Solido" Ed. F. Fumi University of Genova

57. J. Korringa J. Phys. Chem. Solids 7 252 (1958)

58. H. Shiba Progr. Theoret. Phys. (Kyoto) 16 77 (1971)

59. P. Soven Phys. Rev. B 2 4715 (1970)

60. B. L. Gyorffy Phys. Rev. 5B, 2382 (1972)

61. B. L. Gyorffy and G.M. Stocks J. de Phys. 5 C4-75 (1974)
D. House, B. L. Gyorffy, G. M. Stocks, J. de Phys. 5 C4-81 (1974)

62. U. Gerhardt and E. Dietz Phys. Rev. Lett. 26 1477 (1971)
P. O. Garland and B. J. Slagsvold Phys. Rev. 12B (1975)

63. O. Gunnarson J. Phys. F : Metal Physics 6 587 (1976)

64. "Introduction to Solid State Physics" (3rd Edition p. 580) (J. Wiley and Sons Inc. 1966) C. Kittel

65. F. Cyrot-Lackman and F. Ducastelle Phys. Rev. Lett. 27 429

66. F. Brouers, F. Ducastelle, F. Gautier, J. van der Rest J. de Phys. 35 C4-89

67. J. Kanamori J. de Phys. 35 C4-131 (1974)

68. J. Giner, F. Brouers, F. Gautier and J. van der Rest J. Phys. F. : Metal Physics 6 1281 (1976)

69. S. B. Woods (Private Communications)

ONE-BODY POTENTIALS IN CRYSTALS

N. H. MARCH

Department of Physics

The Blackett Laboratory, Imperial College
London S. W. 7. 2BZ, England

CONTENTS*

*The lay out follows March (1974). However, additional material is included in almost all sections beyond that discussion.

INTRODUCTION

In these lectures, we shall give an introduction to the basic theory underlying the use of one-body potential theory in crystals. In the course of this discussion, we shall give a few examples to illustrate the theory, and the kind of accuracy one can hope to achieve in particular types of solid, with one-body potential theory. These examples have been chosen from calculations which the author has either been involved in, or of which he has intimate knowledge. In no sense are they claimed to be representative - their role is illustrative.

In spite of the fact that this course is about basic theory underlying the use of one-body potentials, the subject will be taken far enough to show what consequences stem from the applicability of one-body potentials. In particular consequences for interatomic force fields in crystals and for defect calculations will be briefly considered.

Finally, a discussion of spin density descriptions will be given : this should provide an introduction to some of the material in Professor Callaway's lectures.

1. THOMAS-FERMI THEORY OF INHOMOGENEOUS ELECTRON GAS

A valuable starting point illustrating both the philosophy and the practice underlying this course is afforded by the Thomas-Fermi theory (see, for example, March 1975). Here the original idea was to treat the electrons in an atom, molecule or solid as a degenerate electron gas, and to apply free electron relations locally.

Therefore one starts out from an originally uniform electron gas, of density ρ_0, related to maximum momentum $p_{Fermi} \equiv p_f$ by the usual free-electron relation

$$\rho_0 = \frac{8\pi}{3h^3} p_f^3 \tag{1.1}$$

Into this uniform gas, we assume an inhomogeneity is introduced so that $\rho_0 \rightarrow \rho(\underline{r})$, the electron number density at position $\underline{r}$. If the gradients of the electron density are small then eqn. (1.1) can be applied around the point $\underline{r}$ to yield

$$\rho(\underline{r}) = \frac{8\pi}{3h^3} (p_f(\underline{r}))^3 . \tag{1.2}$$

Now we write the classical energy equation of the fastest electron, with energy equal to the Fermi energy E_f as

$$E_f = \frac{p_f^2(\underline{r})}{2m} + V(\underline{r}) \tag{1.3}$$

where we have assumed that the electrons move in a one-body potential $V(\underline{r})$*. Also, in eqn. (1.3), we note that while both the kinetic and potential energies vary with position $\underline{r}$, the Fermi energy does not. For otherwise, if different regions had different Fermi levels electrons could spill over from one region to another to lower the energy.

Combining eqns. (1.2) and (1.3) leads to the Thomas-Fermi relation

$$\rho(\underline{r}) = \frac{8\pi}{3h^3}(2m)^{3/2}\{E_f - V(\underline{r})\}^{3/2} \tag{1.4}$$

between density and one-body potential.

1.1 Use of Measured Density $\rho(\underline{r})$ to Define a One-Body Potential

This relation, though only valid for densities and potentials which vary slowly in space, can be used to illustrate the basic philosophy employed here.

Suppose we measure the density $\rho(\underline{r})$ by X-ray scattering in an electronic system with a slowly varying density. Then, clearly, we could invert eqn. (1.4) to obtain, from the measured density of the interacting electron system a one-body potential $V(\underline{r})$ given by

$$V(\underline{r}) = -\left[\frac{3h^3}{8\pi}\right]^{2/3}\frac{1}{2m}\rho^{2/3}(\underline{r}) + E_f \tag{1.5}$$

This connects $V(\underline{r})$ with the electron density and it is this connection which we shall show below is very deeplying in first principles theory.

---

* Naturally, $V(\underline{r})$ must vary only slowly in space for density gradients to be small. More precisely, $V(\underline{r})$ must change by only a fraction of itself over a distance equal to the de Broglie wavelength of an electron with energy equal to the Fermi energy.

But for the moment, in a slowly varying density $\rho(\underline{r})$, all that we have ensured by (1.5) is that $V(\underline{r})$ will generate the correct density ! There would be no progress here unless we could demonstrate that $V(\underline{r})$ thereby derived is useful for more general purposes than calculating the density.

We stress though that we have reduced the calculation of the interacting electron density to a one-body problem : assuming we know $V(\underline{r})$!

1.2 Exact Examples of Density - Potential Relations

Since eqn. (1.5) is too crude to be really useful in a crystal, let us take two examples, (the first is almost trivial !) in which Thomas-Fermi approximations are not involved, to illustrate the density-potential relationship we have been stressing. The examples are from atomic rather than crystalline solid theory but will suffice for our immediate purposes.

(a) Ground state of hydrogen-like ion, nuclear charge Ze

The ground-state wave function is

$$\psi(\underline{r}) = \left\{\frac{Z^3}{\pi a_0^3}\right\}^{1/2} \exp\left(-\frac{Zr}{a_0}\right) \tag{1.6}$$

From the Schrödinger equation

$$\nabla^2\psi + \frac{2m}{\hbar^2}(E - V(\underline{r})\psi = 0 \tag{1.7}$$

it is clear that we can write, since $V(\underline{r})$ is a scalar (not an operator)

$$V(\underline{r}) = E + \frac{\hbar^2}{2m}\frac{\nabla^2\psi}{\psi} \tag{1.8}$$

and using the explicit form (1.6) we regain

$$V(r) = -\frac{Ze^2}{r} \tag{1.9}$$

apart from a constant. But since in this one-particle case we can clearly write $\rho = \psi^2$ or $\psi = \rho^{1/2}$, eqn. (1.8) can be readily expressed in terms of ρ as

$$V(\underline{r}) = E + \frac{\hbar^2}{2m}\frac{\nabla^2(\rho^{1/2})}{\rho^{1/2}} \tag{1.10}$$

Clearly, the statistical theory of Thomas and Fermi, leading to equation (1.5) would be quite wrong in this case where we have (a) only one electron and (b) a rapidly varying electron density. Nevertheless, eqn. (1.10) demonstrates again that knowledge of $\rho(\underline{r})$ allows $V(\underline{r})$ to be obtained in this elementary example.

(b) Two-electron helium-like ions with nuclear charge Ze

Schwartz (1959) showed how the ground state electron density could be calculated directly to low order in a 1/Z expansion. This density has been used by March and Stoddart (1972) to construct a body potential for this He-like series, for large Z.

Using the density

$$\rho(r) = \frac{2Z^3}{\pi} \exp(-2Zr)G(r) \tag{1.11}$$

with $G(r) = 1 + \dfrac{R(2Zr)}{Z}$

where
$$R(x) = \frac{5x}{8} + \frac{1}{4}\left(3\ln 2 - \frac{23}{4}\right) - \frac{1}{2}e^{-x} - \frac{3}{4x}(e^{-x} - 1) + \frac{3}{4}\int_0^x dt\, \frac{e^{-t} - 1}{t}$$

Writing $\psi(r) = \left(\dfrac{\rho(r)}{2}\right)^{1/2}$ we can again use eqn. (1.10) to determine V(r).

If, however, this potential were used to calculate an excited state, there would seem no physical justification to expect good answers, there being usually a large energy separation between ground and first excited state.

The example discussed by March and Stoddart was, essentially, the use of this V(r) to calculate the momentum density $\rho(p)$ in these systems. They showed in fact that a different one-body potential is required to get the exact form of $\rho(p)$. The difference between these two one-body potentials is then one way to measure the role of electron-electron correlation effects in the ground-state of these two-electron ions*.

What this discussion, so far, has been designed to establish is that, given a density $\rho(r)$ for a many-body system, there is an operational route to the construction of a one-body

*Though, as discussed below, the inversion of a charge density $\rho(\underline{r})$ to get a (Hartree-like) potential $V(\underline{r})$ is unique, there is no assurance that this is so from $\rho(\underline{p})$, except for one and two electron systems.

potential. However, in crystals, the density varies too rapidly in the neighbourhood of the ions to allow the use of eqn. (1.5) We shall next consider therefore how this formula might be corrected to allow for gradient terms. Such an approach has been used with some success for the electron density near a metal surface (see Smith 1969 for solid metal surfaces and Brown and March, 1973 for liquid metal surfaces).

2. IMPROVED APPROXIMATIONS FOR KINETIC ENERGY

2.1 Thomas-Fermi Form

The kinetic energy density $t_r(\rho)$ plays a central role in one-body potential theory. The Thomas-Fermi approximation to this is again got by applying free electron relations locally. Since the mean kinetic energy/particle in a Fermi gas is $\frac{3}{5}$ of the Fermi energy, we evidently have for free electrons that the kinetic energy/unit volume, t_0 say is

$$t_0 = \frac{3}{5} E_f \frac{N}{V} = \frac{3}{5} E_f \rho_0 \tag{2.1}$$

with N electrons in volume V. Using eqn. (1.1) and eqn. (1.3) with $V(\underline{r}) = 0$ in this equation one readily obtains

$$t_0 = C_k \rho_0^{5/3} \qquad C_k = \frac{3}{10m}\left(\frac{3h^3}{8\pi}\right)^{2/3} \tag{2.2}$$

Taking this relation over into an inhomogeneous gas with slowly varying density we find the Thomas-Fermi approximation to the kinetic energy density $t_r[\rho]$, namely

$$t_r(\rho) = C_k \{\rho(\underline{r})\}^{5/3} \tag{2.3}$$

2.2 Kinetic Energy in Terms of Density for One Electron Only

Following the use of the hydrogen atom example in section 1, let us turn briefly to discuss this case in connection with the form of the kinetic energy density $t_r(\rho)$. In fact, this discussion will suggest the type of gradient correction to be made to eqn. (2.3), as we demonstrate below.

Let us write the kinetic energy density for a wave function ψ in one of the well known forms from elementary quantum mechanics, namely

$$t_r[\rho] = \frac{\hbar^2}{2m}(\nabla\psi)^2 \equiv \frac{\hbar^2}{2m}(\nabla\rho^{1/2})^2 = \frac{\hbar^2}{8m}\left(\frac{\nabla\rho}{\rho}\right)^2 \tag{2.4}$$

von Weizsäcker suggested that one should simply add the

contribution (2. 4) as it stands to eqn. (2. 3) in order to account for gradient corrections. However, as demonstrated by Kirsnitz (1957), there is a very substantial correction factor (1/9 in fact) multiplying eqn. (2. 4) when the gradient corrections are developed systematically. Thus according to Kirsnitz, the kinetic energy density should be modified from the form (2. 3) to read

$$t_r[\rho] = C_k \{\rho(\underline{r})\}^{5/3} + \frac{\hbar^2}{72m}\frac{(\nabla\rho)^2}{\rho} + \ldots\ldots\ldots \qquad (2.5)$$

eqn. (2. 5) displaying only the leading gradient term in what must be (presumably) an infinite series. Though eqn. (2. 5) is still simple and explicit, it already shows us that the local relation between ρ and $t_r[\rho]$ implied by eqn. (2. 3)* is too restricted ; we must have a non-local relation**.

In general, we must caution the reader that, except in special cases like the metal surface electron density referred to above, the development (2. 5) is sometimes disappointing when used in practice, (however, Parr and his colleagues get good results in atoms, including the next term also (Hodges 1973)). Fortunately, solid-state physics now has its own ways of dealing with the single-particle kinetic energy $t_r[\rho]$, so this no longer constitutes a problem (other than large computations!) in practice.

Because of this, we shall turn, without more ado, to the heart of the problem of constructing the one-body potential $V(\underline{r})$ in a real crystal : namely the contribution from electron-electron interactions, other than the classical Hartree part. This is the problem of exchange and correlation energy. To introduce it we shall discuss next Dirac's introduction of exchange into the Thomas-Fermi theory.

3. INTRODUCTION OF EXCHANGE INTO THOMAS-FERMI THEORY

We note first of all that the Thomas-Fermi relation (1. 4) between density and potential is readily derived from a varia-

* Local implies that knowledge of the density at $\underline{r}$ is sufficient to calculate the kinetic energy density at $\underline{r}$.

** Since the gradient of ρ is involved in eqn. (2. 5), we need to know the density slightly away from $\underline{r}$ to get the kinetic energy density at $\underline{r}$. A more sophisticated non-local relation will be given below.

tional principle. Thus, we use the form (2. 3) for the kinetic energy density to construct the total energy **E** in the Thomas-Fermi approximation as

$$E = C_k \int \{\rho(\underline{r})\}^{5/3} d\underline{r} + \int \rho V_N(\underline{r}) d\underline{r} + \frac{1}{2} e^2 \int \frac{\rho(\underline{r})\rho(\underline{r}')}{|\underline{r} - \underline{r}'|} d\underline{r}\, d\underline{r}' \qquad (3.1)$$

Here V_N is the potential energy of the bare nuclei in the system considered, the first potential energy term then being evidently the electron-nuclear interaction. The second term in the potential energy is evidently the classical electrostatic energy of the electronic charge cloud.

If we minimize E with respect to ρ, subject to the normalization condition $\int \rho(\underline{r})\, d\underline{r} = N$, the total number of electrons in the system, we find after a short calculation

$$0 = \frac{5}{3} C_k \rho^{2/3} + V_N + V_e + \lambda \qquad (3.2)$$

where λ plays the role of a Lagrangian multiplier taking care of the normalization of ρ (it is the negative of the chemical potential $\equiv E_f$). V_e is simply given by

$$V_e(\underline{r}) = e^2 \int \frac{\rho(\underline{r}')}{|\underline{r} - \underline{r}'|} d\underline{r}' \qquad (3.3)$$

and $V_N + V_e = V(\underline{r})$ is the total (Hartree-like) one-body potential. Thus we regain eqn. (1. 4) but now with a specific prescription for $V(\underline{r})$. We have here carried out a semi-classical equivalent of a one-electron Hartree calculation.

3. 1 Free-Electron Exchange Energy

As Dirac (1930) was the first to demonstrate, it is important to include exchange in the Thomas-Fermi theory. To do so, we start again from the exchange energy per particle of a uniform electron gas. This can be written in terms of the usual interelectronic spacing r_s as

$$\text{Exchange energy/particle} = -\frac{0.916e^2}{r_s} \qquad (3.4)$$

and hence the exchange energy density, $\varepsilon_x[\rho]$ say, is given by

$$\varepsilon_x[\rho] = -C_e \rho^{4/3} \; * \quad C_e = \left(\frac{3}{\pi}\right)^{1/3} \frac{3}{4} e^2 \qquad (3.5)$$

---

* It is a simple matter of dimensional analysis to show that, once $C_e \propto e^2$ as follows from averaging e^2/r_{ij}, the Coulomb interaction between electrons, the power of ρ is 4/3.

We now add, to get the Thomas-Fermi-Dirac theory, the total exchange energy A given by

$$A = -C_e \int \rho^{4/3}\, d\underline{r} \tag{3.6}$$

to the right-hand-side of eqn. (3.1) and repeat the minimization. Evidently eqn. (3.2) is modified to read

$$0 = \frac{5}{3} C_k \rho^{2/3} + V_N + V_e - \frac{4}{3} C_e \rho^{1/3} + \lambda \tag{3.7}$$

and evidently if we wish to define a new one-body potential including exchange we must add to $V_N + V_e$ an exchange potential $-\frac{4}{3} C_e \rho^{1/3}$. This is a central result of the theory. However, it is clearly only correct if the density gradients are small.

We shall refer to this term below as the Dirac-Slater $\rho^{1/3}$ exchange potential. The point to be made is that if we want to construct a one-body potential incorporating exchange we must write

$$V(\underline{r}) = V_N(\underline{r}) + e^2 \int \frac{\rho(\underline{r}')}{|\underline{r} - \underline{r}'|} d\underline{r}' - \frac{4}{3} c_e \rho^{1/3} \tag{3.8}$$

3.2 Inclusion of Correlation

The same philosophy of using uniform gas theory locally immediately suggests that we take the best jellium model calculation for the exchange and correlation energy density ε^0_{xc} and then use this locally to obtain the total exchange and correlation energy as

$$E_{xc} = \int \varepsilon^0_{xc}(\rho)\, d\underline{r} \tag{3.9}$$

Evidently, eqn. (3.8) is then modified to read

$$V(\underline{r}) = V_{Hartree} + \frac{\delta \varepsilon^0_{xc}(\rho)}{\delta \rho} \tag{3.10}$$

and we regain the Dirac-Slater exchange potential if we approximate $\varepsilon^0_{xc}(\rho)$ by $-c_e\rho^{4/3}$. In general, we can include correlation through approximate formulae such as those of Wigner (1938) or of Nozières and Pines (1958).

3.3 Gradient Correction Terms to Exchange and Correlation Energy

Just as we corrected the Thomas-Fermi theory by including gradient terms in the kinetic energy density, so we ought to contemplate refining the exchange and correlation energy $\varepsilon^0_{xc}(\rho)$ by adding gradient corrections.

Herman et al (1969) first pointed out that the leading term must take the form $e^2(\nabla\rho)^2/\rho^{4/3}$, again on dimensional grounds*. However, care is needed, for if we treat exchange and correlation energies separately, neither will have a gradient expansion (see Beattie, Stoddart and March 1971 ; also Geldart and Rasolt, 1976 and other references given there : Sjölander et al, 1975).

Actually Geldart and Rasolt write

$$\varepsilon_{xc}[\rho] = \varepsilon^0_{xc}[\rho] + e^2 C_{xc} \frac{(\nabla\rho)^2}{\rho^{4/3}} + \ldots\ldots \qquad (3.11)$$

and calculating C_{xc} by writing

$$C_{xc} = C(r_s) \qquad (3.12)$$

they find that $C(r_s)$ varies from $\sim 2.5 \times 10^{-3}$ at $r_s = 0$ to around 1.7×10^{-3} at $r_s = 6$.

There is some evidence from the metal surface problem referred to above that the correction given by eqns. (3.11) and (3.12) is improving the theory significantly (see Geldart and Rasolt, 1976 for further details).

4. DIRAC DENSITY MATRIX AND GRADIENT EXPANSIONS

Having established one of the major results of one-body potential theory , namely eqn. (3.10), which immediately focusses on approximations to the exchange and correlation energy, it is still worth emphasising that the goal of finding $t_r[\rho]$ is eminently worthwhile. Thus, if we generalize the single-particle kinetic energy from the Thomas-Fermi plus gradient corrections level and write

$$T = \int t_r[\rho]\, d\underline{r} \qquad (4.1)$$

we can again formally derive the Euler equation of the variation problem as

---

* Thus $(\nabla\rho)^2 \sim (\text{length})^{-8}$ and $\rho^{4/3} \sim (\text{length})^{-4}$ yielding correctly $\varepsilon_{xc}(\rho) \sim e^2\rho^{4/3} \sim e^2 (\text{length})^{-4}$

$$\frac{\delta t_r[\rho]}{\delta \rho} + V + \lambda = 0 \qquad (4.2)$$

from which we regain eqn. (3.2) if we put $t_r[\rho] \cong c_k \rho^{5/3}$

It is important to note that eqn. (4.2) would allow us to derive $V(\underline{r})$ if we knew :

(a) The functional $t_r[\rho]$
(b) The exact density.

We stressed earlier that, in principle, the electron density is observable by X-ray scattering experiments. There would clearly be rich returns therefore if we could get the single-particle kinetic energy functional.

We shall sketch below a route by which this aim may eventually be achieved. The essential point is that in a one-body problem, both the electron density $\rho(\underline{r})$ given in terms of the single-particle wave functions $\psi_i(\underline{r})$ generated by the one-body potential $V(\underline{r})$ as

$$\rho(\underline{r}) = \sum_{\substack{\text{occupied}\\ \text{states}}} \psi_i^*(\underline{r})\, \psi_i(\underline{r}) \qquad (4.3)$$

and the Dirac density matrix $\rho(\underline{r}\ \underline{r}')$ are functionals of the one-body potential $V(\underline{r})$. If, as we have argued above, it is possible to get $V(\underline{r})$ as a functional of ρ, then, in principle we have a route by which to find the density matrix as a functional of its diagonal element $\rho(\underline{r})$. In principle then, we can calculate the single-particle kinetic energy density $t_r[\rho]$ from the Dirac density matrix, as a functional of ρ.

To date, this programme can only be carried through by:
(a) Perturbative methods (for a summary, see Jones and March '73)
(b) Gradient expansions.

It turns out that the two approaches (a) and (b) are closely related. Thus, from the perturbation theory of the density matrix, Stoddart, Beattie and March (1971) have derived a gradient expansion of the Dirac density matrix as

$$\rho(\underline{r} + \underline{x}, \underline{r}, E_f) = \rho_0(\underline{x}B) + \Big[\frac{1}{72} \int_0^1 da\ \rho^{-2}(\underline{r} + a\underline{x})(\nabla_r \rho(\underline{r} + a\underline{x}))^2$$

$$- \frac{1}{36} \int_0^1 da\ \rho^{-1}(\underline{r} + a\underline{x})\ \nabla_r^2\ \rho(\underline{r} + a\underline{x})\Big]\rho_0'(x, B)$$

$$+ F_1(\underline{r}\,\underline{x})\,\rho_0''(\underline{x}B) + F_2(\underline{r}\,\underline{x})\,\rho_0'''(\underline{x}B) + \ldots\ldots$$

with

$$F_1(\underline{r}\,\underline{x}) = -\int_0^1 da\frac{a(1-a)}{2}\frac{(3\pi^2)^{2/3}}{2}\left\{\frac{2}{9}\rho^{-4/3}(\underline{r}+a\underline{x})(\nabla_r\rho(\underline{r}+a\underline{x}))^2 - \frac{2}{3}\rho^{-1/3}(\underline{r}+a\underline{x})\,\nabla_r^2\rho(\underline{r}+a\underline{x})\right\}$$

$$F_2(\underline{rx}) = \frac{(3\pi^2)^{4/3}}{8}\int_0^1 da\,a^2\int_0^1 db_1\int_0^1 db_2\,b_1b_2\,\frac{4}{9}$$

$$\rho^{-1/3}(\underline{r}+ab_1\underline{x})\,\rho^{-1/3}(\underline{r}+ab_2\underline{x})\times(\nabla_r\rho(\underline{r}+ab_1\underline{x})).(\nabla_r\rho(\underline{r}+ab_2\underline{x})).$$

Here

$$B = \frac{(3\pi^2)^{2/3}}{2}\int_0^1 da\,\rho^{2/3}(\underline{r}+a\underline{x})$$

and

$$\rho_0(x,E) = \frac{(2E)^{3/2}}{\pi^2}\,\frac{j_1\{(2E)^{1/2}x\}}{(2E)^{1/2}x} \qquad (4.4)$$

In concluding this section, we want to refer to the interesting work of Light and his colleagues, which is closely related to the above discussion (Yuan, Lee and Light, 1974 ; Light 1974 ; Light 1973 and other references given there).

5. PARTIAL SUMMATIONS OF GRADIENT SERIES

The limitations of the above approach reside in the fact that it is unlikely that a rapidly convergent series for the kinetic energy density $t_r[\rho]$ could be obtained solely from knowledge of the density in the close vicinity of $\underline{r}$. And this is, essentially, what is implied when we carry out gradient expansions.

In fact, a theory which is non-local at a deeper level can be constructed. This, in essence is equivalent to partial summations of gradient series. This seems to us to be a basically sounder approach. But unfortunately, it is quite a bit more complicated to work with than a gradient expansion.

Though again, in practice, the useful route is to focus on the exchange and correlation contribution to the one-body potential, and not on $t_r[\rho]$, it will afford a useful starting point to the discussion of ε_{xc} to start out with the kinetic energy density $t_r[\rho]$

Hohenberg and Kohn (1964; see also Kohn and Sham, 1965) pointed out that the gradient series for the kinetic energy density could be partially summed (with $\hbar$ = m = 1) as

$$t_r[\rho] = \frac{3(3\pi^2)^{2/3}}{10}\{\rho(\underline{r})\}^{5/3} + \frac{1}{8}\int d\underline{r}'\, K(\underline{r}',\rho(\underline{r}))\,[\rho(\underline{r}+\tfrac{1}{2}\underline{r}') - \rho(\underline{r}-\tfrac{1}{2}\underline{r}')]^2 \qquad (5.1)$$

where K is a known function recorded in their papers.

The corresponding form for the exchange and correlation energy is given by

$$\varepsilon_{xc}[\rho] = \varepsilon^0_{xc}[\rho] - \frac{1}{2}\int d\underline{r}'\, B(\underline{r}',\rho(\underline{r}))[\rho(\underline{r}+\tfrac{1}{2}\underline{r}') - \rho(\underline{r}-\tfrac{1}{2}\underline{r}')]^2 \qquad (5.2)$$

where the form of B is discussed by Stoddart, Beattie and March (1971).

This result has been used to obtain the exchange and correlation potential in eqn. (5. 3) below.

5.1 One-body potential in metallic Be

We refer here briefly to the use of an experimentally determined electron density for Be (Brown, 1972) to calculate the one-body potential (Stoddart et al, 1974).

Explicitly, the one-body exchange and correlation potential can be written in terms of the kernel B as

$$V_{xc}(\underline{r}) = V^0_{xc}(\underline{r}) - \frac{1}{2}\int d\underline{r}'\,\frac{\delta B(\underline{r}',\rho(\underline{r}))}{\delta\rho(\underline{r})}[\rho(\underline{r}+\tfrac{1}{2}\underline{r}') - \rho(\underline{r}-\tfrac{1}{2}\underline{r}')]^2$$
$$- 2\int d\underline{r}'\, B(\underline{r}-\underline{r}',\rho(\tfrac{\underline{r}+\underline{r}'}{2}))[\rho(\underline{r}) - \rho(\underline{r}')] \qquad (5.3)$$

Using the form of B(q) given by Singwi et al (1970), and the measured density of Brown (1972), Stoddart et al find that, at any point in the unit cell, the correction to the local density result V^0_{xc} (r) is less than 30%, and usually substantially less. This is encouraging, and suggests that the calculation of the exchange and correlation potential via gradient expansions is a useful practical approach in Be.

6. DENSITY FUNCTIONAL THEORY OF MANY-BODY PROBLEM

Early work using density descriptions was, at least in principle, based on a single Slater determinant as the approximation to the many-body wave function.

The question arises as to whether correlation effects can be properly treated within such a density description. The answer is yes, as demonstrated by Hohenberg and Kohn (1964). Various workers, and explicitly Gombás and his school, had anticipated this result and had already generalized the Thomas-Fermi-Dirac theory discussed above to include correlation effects. These attempts to include correlation were almost always based on the Wigner (1938) interpolation formula for the correlation energy of a uniform electron gas, namely

$$\varepsilon_c(\rho) = -\frac{0.056\rho^{4/3}}{0.079+\rho^{1/3}} \qquad (6.1)$$

It turns out that such a philosophy is well based, though because of basic limitations in the Thomas-Fermi treatment of kinetic energy, it was not apparent from the early work that the approach could be successful quantitatively.

Proof of unique charge density for each external potential

Following Hohenberg and Kohn (1964), we first prove that a unique charge density exists for each external potential.

Suppose that two external potentials V and V_1 generate two ground-state wave functions ψ and ψ_1. If the corresponding Hamiltonians are H and H_1, we can write for the ground-state energy E of H (we assume a non-degenerate ground state)

$$E = \int \psi^* H \psi \, d\underline{r} < \int \psi_1^* H \psi_1 \, d\underline{r} \qquad (6.2)$$

Now suppose the electron densities associated with ψ and ψ_1 are the same. Then we can write

$$\int \psi^*[V - V_1] \psi \, d\underline{r} = \int \psi_1^* [V - V_1] \psi_1 \, d\underline{r} = \int [V - V_1] \rho(\underline{r}) \, d\underline{r} \qquad (6.3)$$

in this case. Also

$$H = H_1 + (V - V_1) \tag{6.4}$$

and hence we find

$$E < \int \psi_1^* H_1 \psi_1 \, d\underline{r} + \int [V - V_1] \, \rho(\underline{r}) d\underline{r} \tag{6.5}$$

Using the variational principle for H_1 with energy E_1,

$$E_1 < \int \psi^* H \psi \, d\underline{r} + \int [V_1 - V] \, \rho(\underline{r}) d\underline{r} \tag{6.6}$$

By addition these yield $E + E_1 < E + E_1$ and the conclusion is that two different external potentials cannot generate the same charge density. But E is uniquely determined by the external potential and hence we deduce that the ground-state energy is a unique functional of the electron density $\rho(\underline{r})$.

General Euler equation, including exchange and correlation

The total electronic energy can now be written as

$$E = \int t_r[\rho] \, d\underline{r} + \int \rho V_N \, d\underline{r} + \frac{e^2}{2} \int \frac{\rho(\underline{r})\rho(\underline{r}')}{|\underline{r} - \underline{r}'|} d\underline{r} \, d\underline{r}' + \int \varepsilon_{xc}[\rho] \, d\underline{r} \tag{6.7}$$

Here $t_r[\rho]$ is a single-particle kinetic energy associated with density ρ. Thus the exchange and correlation energy $\varepsilon_{xc}[\rho]$ includes correlation kinetic energy. Provided only that we assume the existence of the functional derivative $\delta\varepsilon_{xc}[\rho]/\delta\rho$, the Euler equation now reads

$$\mu = \frac{\delta t_r[\rho]}{\delta\rho} + V_{\text{Hartree}} + \frac{\delta\varepsilon_{xc}[\rho]}{\delta\rho} \tag{6.8}$$

7. GRADIENT EXPANSION OF MA AND BRUECKNER

We merely record here that Ma and Brueckner (1968) essentially combine the density functional philosophy outlined above with the use of diagrammatic techniques in many-body theory. We cannot give the detailed diagrammatic arguments here (see, for example, Jones and March 1973). They arrive by such arguments at an energy density of the form $(\nabla\rho)^2/\rho^{4/3}$ proposed for exchange by Herman et al, with a calculated value of the coefficient. Some difficulties remain in this theory, since, as remarked earlier, gradient expansions exist only if exchange and correlation effects are treated together. Nevertheless the Ma-Brueckner calculation points the way to possible future quantitative progress in the gradient expansion of exchange plus correlation energy.

8. APPLICATIONS OF DENSITY FUNCTIONAL THEORY

From the density functional theory discussed above, we are left in pure crystals with a periodic potential $V(\underline{r})$ given by

$$V(\underline{r}) = V_{\text{Hartree}}(\underline{r}) + \frac{\delta\varepsilon_{xc}[\rho]}{\delta\rho} \tag{8.1}$$

which we can use to calculate energy band structure. It should be now be clear that this potential enters the expression for the chemical potential or Fermi energy. But no basic justification for using such a potential away from the Fermi energy has been given. Thus we have really no right to suppose that this $V(\underline{r})$ can be used to calculate the local density $\rho(\underline{r}E)$ of electrons lying below energy E, except of course when $E = E_f = \mu$.

Actually therefore, even if we could get $\varepsilon_{xc}[\rho]$ exactly, which we manifestly cannot at present, we ought really only to calculate $\rho(\underline{r}E_f)$ which anyway we can measure by X-ray scattering experiments. Nevertheless, the first principles work of S. Lundquist and Hedin (see, for example, their review ; 1969) on the one hand and the success of Dirac-Slater exchange in band structure calculations on the other, show that the above potential is very useful in many cases, away from the Fermi level.

However, we now outline below some other applications of such an approach, based on the density, in the physics of crystals. These applications, however, have primarily to do with perturbing the system, say by introducing a phonon, or by an impurity atom.

8.1 Phonon Theory and Electron Density

As a first application, the one-body potential theory discussed above can be used to formulate the lattice dynamical problem, within the Born-Oppenheimer and the harmonic approximations.

Then, the essential quantity one must obtain is the change in the electron density due to small displacements of the ions from their perfectly periodic lattice sites.

Before dealing with the calculation of the density change, let us summarize the way the lattice dynamics can be handled, (see Born and Huang, 1954 ; also Vosko, Taylor and Keech, 1965). The phonon frequencies ω are determined by the eigenvalue equation

$$\sum_{\beta} D_{\alpha\beta}\ (\underline{k})\ \varepsilon_k^{\beta} = \omega^2\ \varepsilon_k^{\alpha} \tag{8.2}$$

where ε_k is a polarization vector. D is the usual dynamical matrix given in the harmonic approximation by

$$D_{\alpha\beta}\ (\underline{k}) = \frac{1}{M}\ \sum \Phi_{\alpha\beta}(\underline{l})\,[\ \exp(-i\underline{k}.\underline{l}\,) - 1] \tag{8.3}$$

where M is the ionic mass, and

$$\Phi_{\alpha\beta}(\underline{l}' - \underline{l}\,) = \left(\frac{\partial^2 \Phi}{\partial u_{\ell'}^{\alpha}\, 1\ \partial u_{\ell}^{\beta}}\right)_0 \tag{8.4}$$

Φ being the total potential energy governing the motion of the ions. The vectors $\underline{l}$ denote the equilibrium lattice sites and the derivative in the above equation for $\Phi_{\alpha\beta}$ is evaluated at these equilibrium positions.

Next the electronic contribution to $\Phi_{\alpha\beta}$ can be calculated from Feynman's theorem relating the Hamiltonian H to the energy E when a parameter λ is present in the Hamiltonian by

$$\frac{\partial E}{\partial \lambda} = <\psi_\lambda \mid \frac{\partial H}{\partial \lambda} \mid \psi_\lambda> \tag{8.5}$$

in an obvious notation.

If $V(\underline{r})$ is the potential due to an ion, this equation yields

$$\frac{\partial E}{\partial u_{\ell'}^{\beta}} = \int \rho(\underline{r})\ \frac{\partial V(\underline{r} - \underline{l}\,' - \underline{u}_{\ell'})}{\partial u_{\ell}^{\beta}}\ d\underline{r} \tag{8.6}$$

and therefore

$$\frac{\partial^2 E}{\partial u_{\ell}^{\alpha}\, u_{\ell'}^{\beta}} = \int \frac{\partial \rho(\underline{r})}{\partial u_{\ell}^{\alpha}}\ \frac{\partial V(\underline{r} - \underline{l}\,' - \underline{u}_{\ell'})}{\partial u_{\ell'}^{\beta}}\ d\underline{r} + \delta_{\ell'\ell} \int \rho(\underline{r})\ \frac{\partial^2 V(\underline{r} - \underline{l} - \underline{u}_{\ell})}{\partial u_{\ell}^{\alpha}\, \partial u_{\ell}^{\beta}}\ d\underline{r} \tag{8.7}$$

The second term on the right-hand side, involving the second derivative of the ionic potential (and representing in fact intrinsic two-phonon processes, which we need not go into further) is not needed for the determination of the phonon frequencies.

The essential point which allows further progress to be made with the lattice dynamics is that the density involved in the above equation can be calculated, in principle exactly, from one-body potential theory.

8.2 Generalized Rigid Ion Model

A model that has been very useful in earlier work on lattice dynamics is to assume that an electronic charge cloud could be assigned to each ion in the crystal and then, at least for the small displacements we are concerned with, this charge cloud moved rigidly with the nucleus when it was displaced. Thus, the periodic density, $\rho(\underline{r})$ was written as a sum of 'rigid-ion' densities $\sigma(\underline{r})$, centred on the lattice sites $\underline{l}$, namely

$$\rho(\underline{r}) = \sum \sigma(\underline{r} - \underline{l}). \tag{8.8}$$

Such a decomposition of a periodic density into localized distributions does not involve approximation, but unfortunately is far from unique. Thus, $\rho(\underline{r})$ is characterized by its Fourier components ρ_{K_n} at the reciprocal lattice vectors $\underline{K}_n$ whereas to know $\sigma(\underline{r})$ we must know its Fourier transform $\sigma(\underline{k})$ at all $\underline{k}$, not just at the $\underline{K}_n$'s.

However, as indicated above, the early workers in lattice dynamics were essentially assuming that there is a choice of $\sigma(\underline{r})$ which not only allows $\rho(\underline{r})$ in the perfectly periodic crystal to be built up, but which also permits, via the rigid-ion model, the density to be constructed when the ions are displaced to new sites $\underline{l} + \underline{u}$. It is already physically clear that if such a rigid-ion model is valid then one can define pairwise forces between the ions. We shall see from one-body potential theory that, in general, this is not possible, but that a suitable exact generalization of the rigid-ion model in fact exists.

Let us now proceed to calculate the first-order density change $\rho_1(\underline{r})$ when the ions are moved through small displacements $\underline{u}$ from the sites l (Jones and March, 1970). This density change can be written

$$\rho_1(\underline{r}) = \sum_l \underline{u}_l \cdot \frac{\partial \rho(\underline{r})}{\partial \underline{u}_l} \tag{8.9}$$

and alternatively can be expressed via linear response theory as

$$\rho_1(\underline{r}E) = \int \Delta V^{(1)}(\underline{r}') \, F(\underline{r}\ \underline{r}'\ E)\, d\underline{r}' \tag{8.10}$$

Here F can be expressed through the result (see, for example, Stoddart, March and Stott, 1969)

$$\frac{\partial F}{\partial E} = 2\mathrm{Re}\ \ G_0(\underline{r}\ \underline{r}'\ E+) \frac{\partial \rho_0(\underline{r}'\ \underline{r}\ E)}{\partial E} \tag{8.11}$$

where G_0 is the perfect lattice Green function and $\rho_0(\underline{r}\ \underline{r}'\ E)$ is a Dirac density matrix whose diagonal element $\rho_0(\underline{r})$ is the exact crystal density. We want to emphasize that this response function F is determined solely by one-body band theory and does not require a many-body calculation. The local one-body potential which generates the exact density completely determines F. Of course, the price we have to pay is that in the equation for the potential $\Delta V^{(1)}(\underline{r})$ is the change due to ionic displacements in the one-body potential incorporating electron-electron interactions. Since, for a given ionic configuration, the one-body potential always takes the form

$$V(\underline{r}) = V_{\mathrm{electrostatic}} + V_{\mathrm{exchange}} + V_{\mathrm{correlation}} \tag{8.12}$$

where these are solely functionals of the electron density, $\Delta V^{(1)}$ can be expressed, to first order in ρ as

$$\Delta V^{(1)} = \Delta V^{(1)}_{\mathrm{electrostatic}} + \int U(\underline{r}\ \underline{r}')\rho_1(\underline{r}')d\underline{r}' \tag{8.13}$$

where

$$\Delta V^{(1)}_{\mathrm{electrostatic}} = \int \frac{\rho_1(\underline{r}')}{|\underline{r} - \underline{r}'|}\, d\underline{r}' \tag{8.14}$$

$$+ \sum_l \left[\frac{Ze}{|\underline{r} - \underline{l} - \underline{u}|} - \frac{Ze}{|\underline{r} - \underline{l}|}\right]$$

and Ze represents the charge on each ion. More generally, we could, when it proves necessary, take a potential describing the nucleus plus core electrons, instead of Ze/r.

We can now solve by writing

$$\rho_1(\underline{r}) = \sum_l \underline{u}_l \cdot \underline{R}_l(\underline{r}) \equiv \sum_l \underline{u}_l \cdot \underline{R}(\underline{r} - \underline{l}) \tag{8.15}$$

and

$$\Delta V^{(1)}(\underline{r}) = \sum_{\ell} \underline{u}_{\ell} \cdot \underline{P}(\underline{r} - \underline{l}), \tag{8.16}$$

when we find

$$\underline{P}(\underline{r}) = \int \frac{\underline{R}(\underline{r}')d\underline{r}'}{|\underline{r} - \underline{r}'|} - \frac{Ze\underline{r}}{r^2} + \int U(\underline{r}\ \underline{r}')\ \underline{R}(\underline{r}')d\underline{r}' \tag{8.17}$$

Also $\underline{R}$ and $\underline{P}$ are related to F via

$$\underline{R}(\underline{r}) = \int \underline{P}(\underline{r}')\, F(\underline{r}\ \underline{r}')d\underline{r}' \tag{8.18}$$

If we knew the exchange and correlation function $U(\underline{r}\ \underline{r}')$, then these two equations could be solved to yield $\underline{R}(\underline{r})$.

8.3 Approximations leading to rigid-ion model

To establish contact with the rigid-ion model, let us put all the $\underline{u}$'s equal, that is make a uniform translation of the lattice. Then if ρ_0 is the periodic perfect crystal density, we have

$$\nabla \rho_0(\underline{r}) \doteqdot \sum \underline{R}(\underline{r} - \underline{l}) \tag{8.19}$$

whereas the rigid-ion model gives

$$\nabla \rho_0(\underline{r}) \doteqdot \sum \nabla \sigma (\underline{r} - \underline{l}) \tag{8.20}$$

In other words, if R can be written as the gradient of a scalar, then the rigid-ion model is regained. From the first eqn. for $\nabla \rho_0(\underline{r})$ it is clear that

$$\mathrm{curl}\, \nabla \rho_0 = \mathrm{curl} \sum \underline{R}(\underline{r} - \underline{l}) = 0 \tag{8.21}$$

but the rigid-ion model insists on the stronger constraint

$$\mathrm{curl}\ \underline{R} = 0 \tag{8.22}$$

This latter eqn is not generally true, of course.

If we Fourier transform $\underline{R}(\underline{r})$ then the Fourier components $\underline{R}(\underline{k})$ have the property that

$$\underline{R}(\underline{K}_n) = i\, \underline{K}_n\, \rho_{\underline{K}_n} \tag{8.23}$$

which is an exact result at the reciprocal lattice vectors K_n.

Whereas the rigid-ion model gives $\underline{R}(\underline{k})$ in the direction of $\underline{k}$ for all $\underline{k}$, this is not generally true unless $\underline{k}$ is a reciprocal lattice vector. Deviations of $R(\underline{k})$ from the direction of k reflect the existence of many-body forces in the ion-ion interaction. The one-body potential theory presented here affords a systematic basis for their inclusion in lattice dynamics. No division into two-body, three-body, etc. forces is necessary if we work in terms of the vector $R(\underline{r})$. But at present, in practice, combining this formally exact theory with pseudopotentials, one will then classify corrections to the pair force model in terms of 3, 4 etc body forces.

9. ELEMENTARY THEORY OF PAIRWISE INTERACTIONS

As an elementary example, let us use the response function F as calculated for a uniform electron gas to derive the corresponding pair forces (Jones and March, 1970). To do so, we write for the pair potential ϕ (X) in terms of the localized density $\sigma(\underline{r})$ and the bare-ion potential V_b

$$\phi(X) = \int \sigma(\underline{r})\, V_b\, (\underline{r} - \underline{X})\, d\underline{r} \tag{9.1}$$

when we find

$$\frac{\delta^2 \phi\,(X)}{\delta X_\alpha \delta X_\beta} = \int \frac{\delta\sigma(r)}{\delta x_\alpha} \times \frac{\delta V_b\,(\underline{r} - \underline{X})}{\delta X_\beta}\, d\underline{r} \tag{9.2}$$

and it is clear that

$$R_\alpha\,(\underline{r}) = \frac{\delta\sigma}{\delta x_\alpha} \tag{9.3}$$

which is equivalent to assuming rigid ions. For a uniform electron gas we have

$$F(\underline{r}, \underline{r}') = F(\underline{r} - \underline{r}') \tag{9.4}$$

and

$$U(\underline{r}, \underline{r}') = U(\underline{r} - \underline{r}') \tag{9.5}$$

Hence

$$V(\underline{r}) = \int U(\underline{r} - \underline{r}')\,\Delta\rho\,(\underline{r}')d\underline{r}' + V_b\,(\underline{r}) + \int \frac{\Delta\rho\,(\underline{r}')d\underline{r}'}{|\underline{r} - \underline{r}'|} \tag{9.6}$$

or in Fourier transform

$$V(k) = U(k)\ \Delta\sigma_k + V_b(k) + \Delta\sigma_k k^{-2} = U'(k)\ \Delta\sigma_k + V_b(k) \tag{9.7}$$

where $U(k) + k^{-2} = U'(k)$. But we also have

$$\Delta\sigma_k = F(k)V(k) \tag{9.8}$$

which can be written as

$$\Delta\sigma_k = F(k)U'(k)\ \Delta\sigma_k + F(k)V_b(k) \tag{9.9}$$

Representing the relation between $\Delta\sigma$ and V_b in the conventional way by a dielectric function $\varepsilon(k)$ we find

$$\Delta\sigma_k = k^2 V_b(k)[\varepsilon^{-1}(k) - 1] \tag{9.10}$$

and hence we find

$$\varepsilon^{-1}(k) = \frac{1 - F(k)U(k)}{1 - F(k)U'(k)} \tag{9.11}$$

Explicitly

$$F(\underline{r}\ \underline{r}'\ E) = \frac{-k^2}{2\pi^3}\ \frac{j_1(2k|\underline{r} - \underline{r}'|)}{|\underline{r} - \underline{r}'|^2}\ :\ E = \frac{k^2}{2} \tag{9.12}$$

where $j_1(x) = (\sin x - x\cos x)/x^2$ and Fourier transforming this we regain the Lindhard theory for a uniform electron gas when $U'(k)$ is replaced by k^{-2}.

Finally, by adding the ion-ion interaction ϕ_{ii} we obtain the total pair potential as

$$\begin{aligned}\phi_{total}(\underline{X}) &= \int\int \Delta\sigma_k \exp(i\underline{k}.\underline{r})d\underline{k}\ V_b(\underline{r} - \underline{X})d\underline{r} + \phi_{ii} \\ &= \int\int \Delta\sigma_k V_b(k)\ \exp(i\ \underline{k}.\underline{X})d\underline{k} + \phi_{ii} \\ &= \int \frac{k^2 V_b^2(k)}{\varepsilon(k)}\ e^{i\underline{k}\cdot\underline{X}}\ d\underline{k}\end{aligned} \tag{9.13}$$

This reduces to the $\underline{r}$-space result of Corless and March (1961) for point ions, whereas for ions with structure the result is the same as that given, for example, by Ziman (1964), with $V_b(k)$ appropriately interpreted as a pseudopotential, though in the present case the dielectric function incorporates exchange and correlation. A very useful approximate representation of the interaction U(k) is afforded by the work of Singwi, Tosi and Sjölander (1968, 1970), using an effective-field approach.

10. DEFECTS IN CRYSTALS

Defects will be discussed in another lecture course in detail. We shall therefore restrict ourselves to some brief comments on the usefulness of one-body potentials in this field. First of all, electron theory classifies a defect rather basically by the charge it displaces. Thus, if ρ_0 is the periodic charge density, we wish to calculate the new charge density $\rho(\underline{r})$ in the defect lattice, the displaced charge $\Delta(\underline{r})$ being defined by $\Delta(\underline{r}) = \rho(\underline{r}) - \rho_0(\underline{r})$. If the defect represented a genuinely small perturbation on the perfect lattice, then we could express $\Delta(\underline{r})$ in linear response terms. Then, knowledge of the one-body potential of the perfect lattice would enable F to be calculated. This, plus a model for ΔV would allow one to map out the charge displaced round the impurity or defect.

Some prescriptions have been developed for estimating ΔV. For example, in a free-electron metallic matrix, the defect potential must satisfy the Friedel sum rule (cf Kittel, 1963)

$$Z = \frac{2}{\pi} \sum_{\ell} (2l + 1)\, \eta_{\ell}\, (k_f)$$

where Z is the excess charge on the impurity centre, while $\eta_{\ell}(k_f)$ is the phase shift of the l th partial wave due to the presence of the defect potential, calculated at the Fermi wave number k_f. Obviously, if the response function F were known from a Bloch wave calculation, the angularity of the charge density round a defect could be studied.

We must stress though that whereas for a 'defect' which is a phonon, the perturbation is small because the displacements u are small, impurities in crystals are often strong perturbations, and then cannot be treated by linear response theory. Nevertheless, we also want to emphasize that, for properties which can be described in terms of the displaced charge, one-body potential theory is again appropriate.

11. FERMI SURFACE AND ONE-BODY POTENTIAL THEORY

Let us briefly summarize what the potential (8.1) can do. It can generate the exact charge density $\rho(\underline{r})$ in an N-electron system from the sum of the squares of N one-body wave functions. Secondly, knowledge of $\rho(\underline{r})$ can be employed to calculate the ground-state energy, using the approximate exchange and correlation energy functionals.

Thirdly, we can, in principle, from a knowledge of exchange and correlation, calculate the phonon dispersion relations. We now want to emphasize that one of the methods which can map out the Fermi surface in a metal is from the so-called Kohn anomaly in the lattice dispersion relations. This arises from the 'kink' in the Lindhard dielectric function at $k = 2k_f$, which in turn, is a consequence of the sharp Fermi surface. And since the phonon dispersion curves can be got from the density, in essence, it seems clear that this Fermi surface must be correctly generated by the one-body potential (8.1). We have not seen how to prove that the Fermi surfaces which are measured by different techniques, such as the de Haas-van Alphen effect or cyclotron resonance must be precisely the same as that determined from the Kohn anomaly (Kohn 1959), but the experimental evidence available tells us that any differences must be pretty small.

There is a further point which must be made. If the electron density is high, the Landau quasi-particle picture (see, for example, Jones and March, 1973) is valid and a Fermi surface is well defined. However, if we could lower the density of a metal sufficiently, we could at some stage pass through a transition to an insulating phase, after which the concept of a Fermi surface would no longer be useful. It is not presently clear to us how one-body potential theory would deal with this, even though it might be possible to characterize the new phase by an electron density with broken symmetry. This aspect of one-body potential theory needs further investigation.

12. K-DEPENDENT POTENTIALS AND ENERGY BANDS

It is often argued in metals that the concept of one-electron states is only useful relatively near to the Fermi surface, at which the quasi-particle states have infinite lifetimes. It might seem therefore that it is not very meaningful to calculate energy bands over a wide range of energies away from the Fermi surface, because of the short lifetimes of quasi-particle states well away from this surface. This does not, however, appear to be the case, for band structure information over wide energy bands appears to be very helpful, for instance in interpreting experiments on optical properties of crystals.

Adopting such a pragmatic point of view, it is then natural to ask whether the potential in eqn (8.1), which is appropriate at the Fermi energy, can be used to calculate energy bands over a wide range of energy (as, say, reported for lithium metal immediately below). We certainly see no basic reason why this must be so in general, though it seems reasonable to expect that properties near the Fermi surface might be usefully described in this manner.

One reason for fearing that very different potentials might have to be constructed as we move away from the Fermi surface is that Hartree-Fock theory leads to a potential which varies strongly with E or $\underline{k}$. However, in metals, such variation leads to qualitative disagreement with experiment (e.g. zero density of states at the Fermi level). It is quite clear that the effect of correlations is to weaken the energy dependence.

Nevertheless, numerous prescriptions exist for setting up $\underline{k}$ dependent potentials. We can only mention the fundamental work of Hedin and Lundqvist (1969), together with the work on non-local effects in Fermi surface studies on Li, by Vosko and Rasolt (1974).

As an example of the use of the above procedure for calculating energy bands, we refer here to the work of Perrin et al (1975) on body-centred cubic Li metal. They used the Nozières and Pines (1958) expression for the exchange and correlation energy of the electron gas, yielding

$$\frac{\delta\,\varepsilon_{xc}(\rho)}{\delta\rho} = \frac{-2}{\pi}\left([3\pi^2\rho(r)]^{1/3} - [3\pi^2\rho_0]^{1/3}\right) - \frac{0.031}{3}\ln\frac{\rho(r)}{\rho_0} \qquad (12.1)$$

Perrin et al show that a sensible account of the Fermi surface is thereby given, by comparison with the experiment of Randles and Springford (1973 ; also Springford 1976, private communication).

Furthermore, the electron density $\rho(\underline{r}, E_f)$ determined with the above one-body potential is in pretty good agreement with that given by the superposition of pseudoatoms, using the pseudoatom density calculated by Dagens et al (1975). This fact is very useful in calculating phonon spectra and interionic forces.

13. SPIN DENSITY DESCRIPTIONS

We shall conclude with a brief discussion of the extension of the density functional theory to include situations (e. g. ferromagnetism) where the total electron density must be decomposed into two parts, an upward spin contribution $\rho_+(\underline{r})$ and a downward spin density $\rho_-(\underline{r})$ with $\rho_+(\underline{r}) \neq \rho_-(\underline{r})$ and

$$\rho(\underline{r}) = \rho_+(\underline{r}) + \rho_-(\underline{r}) \tag{13.1}$$

Stoddart and March (1971) gave a discussion of the spin density functional theory and applied it to the problem of magnetic instabilities in metals. Their approach has been developed later by von Barth and Hedin (1972) and by Callaway and Rajogopal (1973). The discussion below will be posed in fairly simple terms, and in particular we give as a starting (purely illustrative) example the case of a uniform electron gas described in the Hartree-Fock approximation. We know now that Hartree-Fock theory overemphasizes the tendency to get a ferromagnetic ground state, since there is a (largely correct) Fermi hole correlation between parallel spin electrons, but no correlation between antiparallel spins. This tells us that it is essential to include correlation in a realistic discussion of the criteria for ferromagnetism.

13.1 Uniform gas ferromagnetism in Hartree-Fock approximation

As Bloch (1928) was the first to recognize, if we consider the jellium model of electrons moving in a uniform neutralizing density of positive ions, then at sufficiently low density Hartree-Fock theory shows that the ferromagnetic ground state must become stable, instead of the normal paramagnetic state. Thus, from eqn (2.2) the kinetic energy density is proportional to $\rho_0^{5/3}$, ρ_0 being the positive ion density ($\rho(\underline{r}) \equiv \rho_0$ in jellium) or in terms of the usual electron gas parameter r_s ($\rho_0 = 3/4\pi r_s^3$), the kinetic energy/particle is $\propto r_s^{-2}$ wheras from eqn (3.4) the exchange energy per particle is proportional to r_s^{-1}. And now, if we flip the spins so that all of them are parallel (completely saturated ferromagnetism), though we double the volume of the occupied Fermi sphere, and hence increase the kinetic energy, for sufficiently large r_s we gain more exchange energy and the ferromagnetic state is stabilized*.

---

* The nature of the exact ground state of low density jellium is still not finally settled. At very large r_s it is probably antiferromagnetic.

Let us turn these qualitative statements into a calculation posed in the spirit of the spin density functional approach. The spin densities ρ and ρ are constants in the simple magnetic state considered here. Then it is a simple matter to write down :

The kinetic energy

This is

$$T(\rho_+ \rho_-) = \text{constant } \rho_+^{5/3} + \text{constant } \rho_-^{5/3} \tag{13.2}$$

and introducing (for convenience) the relative magnetization τ defined by

$$\tau = \frac{\rho_+ - \rho_-}{\rho_+ + \rho_-} \tag{13.3}$$

we can write the total ground state energy $E(\rho_+ \rho_-)$ as

$$\frac{E(\tau)}{N E_f} = \frac{3}{10}[(1+\tau)^{5/3} + (1-\tau)^{5/3}] - \frac{3}{8}\frac{\varepsilon_x}{E_f}[(1+\tau)^{4/3} + (1-\tau)^{4/3}] \tag{13.4}$$

where E_f is the usual Fermi energy of N electrons in volume V, i. e.

$$E_f = \frac{h^2}{8m}\left(\frac{3N}{\pi V}\right)^{2/3} \quad \text{and} \quad \varepsilon_x = e^2\left(\frac{3N}{\pi V}\right)^{1/3} = \frac{4}{3} c_e \rho_0^{1/3},$$

in terms of the constant c_e introduced in eqn (3. 5).

The above result is Bloch's Hartree-Fock energy. In the spirit of the spin density approach, the magnetic state is obtained for a given density of jellium obtained by finding τ such that $E(\tau)$ is a minimum.

We shall simply refer to the curves of Lidiard (1951) who has plotted (his Figure 3) $[E(\tau) - E(0)]/N$ against the relative magnetization for particular values of ε_x/E_f. We simply quote the result that in the Hartree-Fock theory of jellium the simple ferromagnetic state $\tau = 1$ is stable for $\varepsilon_x/E_f > 4(2^{1/3} + 1)/5$ while the paramagnetic (HF) state $\tau = 0$ is stable relative to the ferromagnetic state for $\varepsilon_x/E_f < 4(2^{1/3} + 1)/5$.

No intermediate values of τ are possible in this model in contrast to the (more realistic) Stoner collective electron model (1938, 1939).

This model, in fact, for uniform electrons, could be regarded as related to eqn (13.4) if we argue, motivated by the (relative) success of the Weiss theory of ferromagnetism, that the effect of including correlation is to restore an interaction energy in eqn (13.4) whose dependence on the relative magnetization τ is simply quadratic. Then the character of the solution is altered drastically and intermediate values of τ are possible.

In addition to this merit of the Stoner model, one can insert energy bands (one could also do this in eqn (13.4) above, and the criterion that ferromagnetism will occur takes the form.

$$I\,N(E_f) > 1 \tag{13.5}$$

where I measures the strength of the exchange and correlation interaction, while $N(E_f)$ is the one-electron density of states at the Fermi surface. Stoddart and March (1971) have discussed modifications to this criterion for the existence of ferromagnetism from their formulation of spin-density functional theory, but we shall not give the details here. Finally, in connection with eqn (13.5) we must also refer to the recent discussion of Gunnarsson (1976) who has used the same spin-density functional theory to calculate magnetic properties of the transition metals. Gunnarsson derives an approximate formula for the strength of the interaction I in terms of the electron density and the radial wave functions of the d electrons in the transition metal under consideration. Using a local spin density approximation, in conjuction with paramagnetic band structure calculations, Gunnarsson's results for $IN(E_f)$ are shown in the Table below

Table 13.1

Transition metal	V	Fe	Co	Ni	Pd	Pt
$IN(E_f)$	0.8 - 0.9	1.5 - 1.7	1.6 - 1.8	2.1	0.8	0.5

It will be seen that the ferromagnetic metals Fe, Co, and Ni do indeed satisfy the criterion (13.5) while the other three paramagnetic metals do not. This circumstance would not hold with a straightforward application of Dirac-Slater theory which gives larger values of $I\,N(E_f)$ than in the local spin density theory used by Gunnarsson.

13.2 Magnetic susceptibility

We shall conclude the lectures by giving a brief summary

of the way the spin-density functional formalism can be used to calculate the spin paramagnetic susceptibility in metals.

One can easily calculate the spin susceptibility χ from eqn (13.4) in the paramagnetic region both at $T = 0$ and also (approximately) at elevated temperatures (March and Donovan, 1954). If μ is the Bohr magneton, and N the total number of electrons then eqn (13.4) yields for the susceptibility.

$$\frac{\mu^2 N}{\chi E_f} = \frac{2}{3} - \frac{1}{3}\frac{\varepsilon_x}{E_f} \tag{13.6}$$

yielding the usual Pauli result for free electrons if we put $\varepsilon_x = 0$. The interaction (exchange) term enhances the susceptibility ; a well known result.

Actually, in early work, Sampson and Seitz (1940) calculated the spin susceptibility by a density approach. They essentially corrected equation (13.4) by adding correlation energy, which they noted to be principally dependent on the number of electrons with opposite spin. Their treatment used Wigner's (1938) formula for the correlation energy and their computed susceptibility for sodium was close to the observed value.

In connection with the Stoner theory, the same enhancement effects occur in the susceptibility, the non-interacting electron susceptibility χ_0 being increased according to

$$\chi = \frac{\chi_0}{1 - I\,N(E_f)} \tag{13.7}$$

the connection of eqn (13.7) with the criterion (13.5) being evident.

13.3 Wave-number dependent magnetic susceptibility

Having given this elementary introduction, we shall conclude by referring briefly to the work of Stoddart (1975), in which the wave-number (q) dependent magnetic susceptibility is expressed in terms of the spin-density formalism.

Stoddart's work essentially effects the generalization of the relation between perturbative methods and gradient expansions referred to in section 4 above to the spin-polarized case. We shall confine ourselves to the essential steps in his argument. Working in terms of $\rho(\underline{r})$ in eqn (13.1) plus the magnetization

$$m(\underline{r}) = \frac{1}{2}[\rho_-(\underline{r}) - \rho_+(\underline{r})] \tag{13.8}$$

Stoddart writes formally the total ground-state energy as

$$E[\rho, m] = G[\rho, m] + \frac{1}{2}\iint \frac{\rho(\underline{r})\rho(\underline{r}')}{|\underline{r} - \underline{r}'|} d\underline{r}\, d\underline{r}' + E_{ext} \tag{13.9}$$

where E_{ext} represents the energy of interaction with an external magnetic field (say). Then we can write

$$E_{ext} = -\int m(\underline{r})H(\underline{r})d\underline{r} \tag{13.10}$$

As we discussed above, in the absence of spin polarization the most useful way of proceeding is by density-gradient expansions. Stoddart therefore writes

$$\begin{aligned} G[\rho, m] &= G_0(\rho, m) + \int g_1(\underline{r}, m(\underline{r})[\nabla_r \rho(\underline{r})]^2 d\underline{r} \\ &+ \int g_2(\rho, m)\, [\nabla_r \rho(\underline{r}) . \nabla_r\, m(\underline{r})]\, d\underline{r} \\ &+ \int g_3(\rho, m)\, [\nabla_r\, m(\underline{r})]^2 d\underline{r} + \ldots . \end{aligned} \tag{13.11}$$

where the spin-polarized system is assumed throughout to have a single magnetization direction*.

In eqn (13.11)

$$G_0(\rho, m) = \int d\underline{r}\; g_0(\rho(\underline{r}), m(\underline{r})) \tag{13.12}$$

where $g_0(\rho, m)$ is the sum of kinetic, exchange and correlation energies per unit volume of a uniform electron gas with density ρ and magnetization m (compare eqn (13.4) for this result in the Dirac-Slater exchange approximation). More general energy functionalsand a corresponding spin-dependent one-body potential have been proposed by von Barth and Hedin (1972).

Stoddart then effects the generalization of the perturbation expansion of $G(\rho, m)$ to read for the magnetic case

$$\begin{aligned} G[\rho, m] &= G_0(\rho_0, m_0) + \int G_1(\underline{r} - \underline{s})\,\Delta(\underline{r})\,\Delta(\underline{s})d\underline{r}\, d\underline{s} \\ &+ \int [G_2(\underline{r} - \underline{s})\,\Delta(\underline{r})\delta(\underline{s}) + G_3(\underline{r} - \underline{s})\delta(\underline{r})\,\delta(\underline{s})]\, d\underline{r}\, d\underline{s} \end{aligned} \tag{13.13}$$

* Obviously, the theory would have to be generalised to treat spiral spin arrangements (e. g. some rare earth metals)

where $\Delta(\underline{r})$ is the displaced charge defined by

$$\Delta(\underline{r}) = \rho(\underline{r}) - \rho_0 \qquad (13.14)$$

while

$$\delta(\underline{r}) = m(\underline{r}) - m_0 \qquad (13.15)$$

Following the method used by Stoddart et al (1971) for the non-magnetic case, the g_i in the gradient development (13.11) can be related to the kernels G_i in the perturbative expansion (13.13).

Finally, defining the spin response $\chi(q)$* to the external magnetic field through

$$\chi(q) = \frac{-\delta(q)}{H(q)} \qquad (13.16)$$

Stoddart obtains the result

$$\chi(q) = \frac{1}{2}\left[G_3(q, \rho_0, m_0) - \frac{G_2^2(q, \rho_0 m_0)}{4[(2\pi/q^2) + G_1]}\right]^{-1} \qquad (13.17)$$

where G_i is defined in terms of eqn (13.13) as

$$G_i(q, \rho_0, m_0) = \int d\underline{r}\; G_i(\underline{r}) \exp(i\, \underline{q}.\underline{r}) \qquad (13.18)$$

If we (too drastically) illustrate eqn (13.17) by neglecting exchange and correlation, we find (with $m_0 = 0$)

$$\begin{aligned} G_1(q, n_0, 0) &= -\frac{1}{4} K \\ G_2 &= 0 \\ G_3 &= -K \end{aligned} \qquad (13.19)$$

where $K = K(q, \frac{1}{2}\rho_0)$ and $K(q,\rho_0)$ is the well known (Lindhard-type) response function (compare Stoddart and March)

$$K(q,\rho_0) = \frac{-2\pi^2}{k_f}\left(\frac{1}{2} + \frac{1-n^2}{4n} \ln\left|\frac{1+n}{1-n}\right|\right)^{-1} \qquad (13.20)$$

with $n = q/2k_f$ and $\rho_0 = k_f^3/3\pi^2$ as usual.

*We shall not consider the most general susceptibility here, which for a periodic lattice involves, of course, reciprocal lattice vectors.

Then $\chi(q)$ reduces to the free-electron result (cf Hebborn and March, 1970)

$$\chi(q) = \chi_{\text{Pauli}} \left[\frac{1}{2} + \frac{1 - n^2}{4n} \ln \left| \frac{1 + n}{1 - n} \right| \right] \tag{13.21}$$

Extensive band-structure calculations, based on the functionals and corresponding spin-dependent one-body potentials discussed here have been carried out, for example, by Langlinais and Callaway (1972) on Ni using Dirac-Slater exchange and Professor Callaway will, no doubt, discuss these in his lectures.

APPENDIX I

Relation Between Ground-State Energy and Sum of Eigenvalues One-Body Potential Theory

In practice, for a chosen approximation to $\varepsilon_{xc}(\rho)$, it is often not convenient to go back to the basic total energy functional E, even though this is obviously a quite fundamental route to the energy.

Instead, following the procedure in Hartree theory, we construct E from the sum of the one-electron eigenvalues $\sum\varepsilon_i$, the sum being taken over all states occupied by the N electrons. In Hartree theory, it is well known that this procedure leads to the ground state energy when we correct for the fact that in $\sum\varepsilon_i$ we have counted the electron-electron contribution twice over. Thus, if we make the Hartree approximation $\varepsilon_{xc} = 0$, then we can write

$$E_{\text{Hartree}} = \sum \varepsilon_i - \frac{1}{2} \int \rho V_e \, d\underline{r}.$$

Now we turn to the exact one-body potential theory based on eqn. (3.10). If we could use this form exactly, then clearly we can write for the sum of the one-electron eigenvalues derived from this potential $V(\underline{r})$

$$\sum\varepsilon_i = \int t_r[\rho] \, d\underline{r} + \int \rho V(\underline{r}) \, d\underline{r}$$

and if we now insert eqn. (3.10) into this equation we find the formula

$$\sum\varepsilon_i = \int t_r[\rho] \, d\underline{r} + \int \rho (V_e + V_N) \, d\underline{r} + \int \rho \frac{\delta\varepsilon_{xc}[\rho]}{\delta\rho} \, d\underline{r}$$

where as usual V_N is the bare nuclear potential energy and V_e that created by the electron cloud $\rho(\underline{r})$.

From the (formally) exact total energy

$$E = \int t_r[\rho] \, d\underline{r} + \int \rho V_N \, d\underline{r} + \frac{1}{2} \int \rho V_e \, d\underline{r} + \int \varepsilon_{xc}[\rho] d\underline{r}$$

we see that

$$E = \sum \varepsilon_i - \frac{1}{2} \int \rho V_e \, d\underline{r} - \int \left\{\rho \frac{\delta\varepsilon_{xc}(\rho)}{\delta\rho} - \varepsilon_{xc}(\rho)\right\} d\underline{r}$$

and this is the exact generalization of the Hartree result given above. It should occasion no surprise that we have a correction to make to the exchange and correlation term, since, as with the electrostatic electron term corrected in $E_{Hartree}$, the term $\varepsilon_{xc}(\rho)$ is a consequence of electronic pair interactions and we are over counting in the eigenvalue sum.

If one is working simply within the Dirac-Slater framework then of course we can readily evaluate the total energy to obtain

$$E_{Dirac\text{-}Slater} = \sum \varepsilon_i - \frac{1}{2} \int \rho V_e \, d\underline{r} + \frac{1}{3} C_e \int \rho^{4/3} d\underline{r}.$$

Finally we wish to caution the reader that while there is a formally exact expression for the total energy here, once an approximation is made for $\varepsilon_{xc}[\rho]$ (such as the Dirac-Slater treatment above) there is no assurance that one could not go below the exact ground-state energy somewhat. Nevertheless, existing calculations strongly point to the fact that good energies can be obtained from the one-body potential approach in numerous (though by no means all !) cases with an approximation to $\varepsilon_{xc}[\rho]$ as simple as the Dirac-Slater form.

APPENDIX 2

Example of Kinetic Energy Functional for an Electron Gas with a Surface

As an example of the construction of the kinetic energy functional, we summarize here the approximate theory of Peuckert (1976).

Peuckert shows that the total single-particle kinetic energy $T_s[\rho]$ may be written as :

$$T_s[\rho] = A \int dz \left[\rho(z) \{\mu - V_H(z)\} - \lim_{\varepsilon \to 0} \int_{-\infty}^{\mu} d\mu' \rho_\varepsilon(z, \mu')\right]$$

In this equation, T_s is to be regarded as a functional of $\rho(z, \mu')$: ie ρ must be known for all z and μ'. What we really need though is T_s as a functional of $\rho(z)$, where μ is fixed by the condition of charge neutrality. Therefore, to turn this statement into a useful functional, we must perform the μ' integration. To date, this can only be achieved approximately. In the above equation A is the surface area of the system. μ is the chemical potential while $V_H(z)$ is the Hartree potential.

Approximate Evaluation of μ' Integration

Peuckert now performs the μ' integration by making use of the gradient expansion of the density. His result is then that $T_s[\rho]$ can be written approximately as the sum of three parts :

$$T_s[\rho] = F_1 + F_2 + F_3$$

$F_1[\rho]$ is a frank extension of Kirzhnits work, and yields

$$F_1[\rho] = A \int dz \Big[\frac{1}{10\pi^2}(3\pi^2\rho)^{5/3} + \frac{1}{72}\frac{\rho'^2}{\rho} + \frac{1}{12960}(3\pi^2)^{-2/3}$$

$$\left(8\frac{\rho'^4}{\rho^{11/3}} - 27\frac{\rho''(\rho')^2}{\rho^{8/3}} + \frac{24(\rho'')^2}{\rho^{5/3}}\right) + \ldots\ldots\ldots\Big]$$

Clearly, the first two terms in the square bracket are the Thomas-Fermi plus Kirzhnits result, discussed in the body of the text.

F_2 and F_3 eventually, making again gradient type

approximations can be written in lowest order (Peuckert, 1976) as

$$F_2 + F_3 \cong - \frac{1}{160\pi^2} \int dz \left\{\frac{\partial}{\partial z} \frac{V_s''(z)V_s'(z)}{u(z)^{3/2}} + \ldots\ldots.\right\}$$

where V_s is the slowly varying part of the Hartree potential V_H while

$$u(z) = 2(\mu - V_s(z))$$

Using finally the Thomas-Fermi approximation in the form

$$V_s'(z) = -(3\pi^2\rho_s)^{-1/3} \cdot \rho_s',$$

the Hartree functional takes the final form (Peuckert, 1976)

$$E_H[\rho] = F_1[\rho] - 2\pi e^2 \int dzdz' \, |z - z'| \, \{\rho(z) - \rho_+(z)\}$$

$$\{\rho(z') - \rho_+(z')\} - \frac{A}{864\pi^2} (3\pi^2)^{-2/3} \int dz \frac{d}{dz} \frac{1}{\rho} \frac{d}{dz} \frac{(\rho')^2}{\rho^{2/3}}$$

$$- \frac{1}{4} A \int dz \, (\rho(z) - \rho_+(z')) \times \left\{2\pi e^2 \int dz' \, z' \{\rho(z') - \rho_+(z')\}\right.$$

$$\left. - (3\pi^2 \rho_+(\infty))^{2/3}\right\}$$

This gives the Hartree energy for $\rho = \rho_s$, ρ_s being the smooth, slowly varying part of the density. ρ_+ is the positive background in the semi-infinite jellium model of the surface under discussion, i. e. $\rho_+(z) = \rho_0 \, \theta(-z)$, where $\rho_0 = k_f^3/3\pi^2$, k_f being the Fermi wave number.

Though numerous approximations have been made in getting to the above result for $E_H[\rho]$, it is a concrete example which well illustrates the practice behind the density functional approach : the dominant theme of the lectures.

REFERENCES

Beattie, A. M., Stoddart, J. C. and March, N. H., 1971, Proc. Roy. Soc. A326, 97.
Bloch, F., 1928, Zeits. fur Physik 57, 545.
Born, M. and Huang, K., 1954, Dynamical Theory of Crystal Lattices (Clarendon Press : Oxford).
Brown, P. J., 1972, Phil. Mag., 26, 1377.
Brown, R. C. and March, N. H., 1973, J. Phys. C6, L363.
Dirac, P. A. M., 1930, Proc. Camb. Phil. Soc. 26, 376.
Donovan, B. and March, N. H., 1953, Proc. Phys. Soc. A66, 1104.
Geldart, D. J. W. and Rasolt, M., 1976, Phys. Rev. B13, 1477.
Gunnarsson, O., 1976, J. Phys. F. 6, 587.
Hebborn, J. E. and March, N. H., 1970, Adv. Phys. 19, 175.
Hedin, L. and Lundqvist, S., 1969, Solid State Physics (eds. F. Seitz, D. Turnbull and H. Ehrenreich) Vol. 23, p1 (Academic Press : New York).
Herman, F., van Dyke, J. P. and Ortenburger, I. B., 1969, Phys. Rev. Letts. 22, 807.
Hodges, C. H., 1973, Can. J. Phys. 51, 1428.
Hohenberg, P. and Kohn, W., 1964, Phys. Rev. 136, B864.
Jones, W. and March, N. H., 1970, Proc. Roy. Soc. A317, 359, 1973, Theoretical Solid-State Physics, Wiley-Interscience, London, Volumes 1 and 2.
Kirsnitz, D. A., 1957, Sov. Phys. JETP, 5, 64.
Kittel, C., 1963, Quantum Theory of Solids (Wiley : New York).
Kohn, W. and Sham, L. J., 1965, Phys. Rev. 140A, 1133.
Langlinais, J. and Callaway, J., 1972, Phys. Rev. B5,124
Lidiard, A. B., 1951, Proc. Phys. Soc. A64, 814.
Light, J. C., 1973, J. Chem. Phys. 58, 660.
1974, J. Chem. Phys. 61, 3417.
Ma, S. K. and Brueckner, K. A., 1968, Phys. Rev. 165, 18.
March, N. H., 1974, Orbital Theories of Molecules and Solids (Clarendon Press : Oxford).
1975, Self-consistent Fields in Atoms (Pergamon : Oxford).
March, N. H. and Stoddart, J. C., 1972, in Computational Solid State Physics : (eds. Herman et al) (Plenum : New York).
Nozières, P. and Pines, D., 1958, Phys. Rev. 111, 442.
Perrin, R. C., Taylor, R. and March, N. H., 1975, J. Phys. F5, 1490.
Peuckert, V., 1976, J. Phys. C9, 809.
Rajagopal, A. K. and Callaway, J., 1973, Phys. Rev. B7, 1912.
Randles, D. L. and Springford, M., 1973, J. Phys. F3, L185.
Sampson, J. B. and Seitz, F., 1940, Phys. Rev. 58, 633.
Singwi, K. S., Sjölander, A., Tosi, M. P. and Land, R. H., 1970, Phys. Rev. B1, 1044.

Sjölander, A., Niklasson G. and Singwi K. S., 1975, Phys. Rev. B11, 113.
Slater, J. C., 1951, Phys. Rev. 81, 385.
Smith, J. R., 1969, Phys. Rev. 181, 522.
Stoddart, J. C., 1975, J. Phys. C8, 3391.
Stoddart, J. C., Beattie, A. M. and March, N. H., 1971, Int. J. Quant. Chem. 4, 35.
Stoddart, J. C. and March, N. H. 1967, Proc. Roy. Soc. A299, 279 ; 1971, Annals of Physics, 64, 174.
Stoddart, J. C., March, N. H. and Stott, M. J., 1969, Phys. Rev. 186, 683.
Stoddart, J. C., Stoney, P., March, N. H. and Ortenburger, I. B., 1974, Nuovo Cimento, 23B, 15.
Stoner, E. C., 1938, Proc. Roy. Soc. A165, 372.
1939, Proc. Roy. Soc. A169, 339.
von Barth, U. and Hedin, L., 1972, J. Phys. C 5, 1629.
Vosko, S. H. and Rasolt, M., 1974, Phys. Rev. Letts., 32, 297.
Vosko, S. H., Taylor, R. and Keech, G. H., 1965, Can. J. Phys. 43, 1187.
Wigner, E. P., 1938, Trans. Far. Soc. 34, 678.
Yuan, J. M., Lee, S. Y. and Light, J. C., 1974, J. Chem. Phys. 61, 3394.
Ziman, J. M., 1964, Adv. Phys. 13, 89.

AB INITIO METHODS FOR ELECTRONIC STRUCTURES OF CRYSTALLINE SOLIDS

Frank E. Harris

Department of Physics, University of Utah

Salt Lake City, Utah 84112

Two approaches have been dominant in theoretical studies of the electronic structures and properties of crystalline solids. The first and older approach involves the use of an independent-electron formalism, but with a semi-empirically determined effective Hamiltonian designed to reproduce the effects of electron exchange and correlation. A variety of effective Hamiltonians have been used, ranging from pseudopotentials, chosen mainly on the basis of agreement with experiment, to *a priori* choices such as the highly popular Xα method (Slater, 1970, 1971; Johnson and Smith, 1971), where the exchange-correlation potential is based on studies of the uniform electron gas (Slater, 1951; Gaspar, 1954; Kohn and Sham, 1965). This approach, although highly successful for many problems, leaves certain questions unanswered; among these are detailed assessments of the quality of the Hartree-Fock wavefunction, the quantitative roles of exchange and correlation, and the elucidation of collective phenomena inherently ruled out by an independent-electron model.

The second approach, which forms the subject of this paper, is that of *ab initio* description of the electronic energy bands and properties, using antisymmetrized many-electron wavefunctions built from an appropriate set of crystal orbitals. Initially, such wavefunctions may be constructed in the independent-electron approximation, possibly as solutions to the Hartree-Fock equations, but it is also possible to proceed beyond this approximation to obtain a description of electron correlation and its effect upon the energy bands and other properties.

A number of investigations (e.g., Harris, Kumar and Monkhorst,

1973) have shown that the inhomogeneity of the electron distributions of solids at normal densities is comparable with that of the atoms of which they are composed, so that it is natural to consider constructing crystal orbitals from appropriate atomic orbitals. In fact, studies such as those of our group (Kumar and Monkhorst, 1974) indicate that substantial portions of some discrepancies in properties predictions based on plane waves may be due, not to electron correlation, but to inhomogeneities in the electron distribution which are neglected in the plane-wave description. Accordingly, we present here a discussion of calculation methods using crystal orbitals built from atomic wavefunctions.

One of the reasons for the earlier development of the semi-empirical approach to solid-state electronic-structure problems was the presumed difficulty of the *ab initio* calculations. This difficulty is now circumvented by the use of Fourier representation methods, which greatly reduce the geometrical complexity of the calculations without unduly obscuring their physical basis (Harris and Monkhorst, 1969). The use of these methods was probably delayed by the belief that they would not exhibit sufficiently rapid convergence; the convergence has in fact turned out to be adequate if viewed in terms of sums of symmetry-related quantities. For example, the sum of orbital products of a central orbital with the star of up to 48 orbitals equivalently symmetry-related thereto is, in direct space, a far more diffuse and symmetrical object than any of the individual orbital products of which it is composed; as a consequence, the Fourier transform of this symmetry sum is both symmetric and localized near the origin in reciprocal space. From this localization come the convergence properties.

In addition to the reduction of geometrical complexity, the Fourier representation method has several further advantages. One such advantage is the fact that the Fourier transform of a Slater-type orbital (STO) is a simple algebraic expression, and all quantities needed for energy calculations take convenient forms involving the transforms of individual orbitals. The result is that, in contrast to atomic and molecular studies, calculations based on STO's are comparable in convenience to those based on Gaussian orbitals. Since the STO's provide a far superior description of atomic electron distributions than do Gaussians, it becomes more convenient to use wavefunctions of higher quality. A second advantage is in the treatment of the long-range cancellations of Coulombic interactions of opposite signs, leading to a physically relevant partitioning of the various quantities entering the expressions for the electronic energy. It turns out that the long-range cancellation can be directly related to the scale and symmetry of the crystal lattice, and that free-electron and inhomogeneity effects are conveniently separated. Finally, we find that the Fourier representation method eliminates completely the explicit occurrence of molecular integrals, recasting them as sums on the relevant reciprocal lattice.

Thus far, almost all the *ab initio* solid-state calculations have been in the Hartree-Fock approximation. While Hartree-Fock results have the drawback of indicating an unphysical vanishing of the density of states at the Fermi surface, they nevertheless correspond to a well-defined starting point, and one from which perturbation-theoretic formulations are in simplest form. *Ab initio* Hartree-Fock calculations have by now been reported for at least the following crystalline systems: atomic hydrogen (Harris and Monkhorst, 1969; Harris, Kumar and Monkhorst, 1973), H_2 (Ramaker, Kumar and Harris, 1975), Li (Calais and Sperber, 1972, 1973; Kumar, Monkhorst and Harris, 1974; Kumar and Monkhorst, 1974), diamond (Euwema, Wilhite and Surratt, 1973), and Ne and LiF (Euwema *et al*, 1974). While a number of investigators have developed approximate schemes for going beyond the Hartree-Fock approximation by including an estimate of the correlation energy, only a very few correlation-energy calculations have actually been carried out starting from reasonably good Hartree-Fock states. Examples of such studies are in the work of Monkhorst and Oddershede (1973) and of Brener (1975a, 1975b). None of the investigations thus far reported constitute initial steps of a systematic method which can be expected to converge toward quantitative correlation-energy values.

As will be discussed in more detail later, the configuration-interaction methods which are so popular for small atomic and molecular systems cannot usefully be applied to crystalline solids. This fact has necessitated the use of more sophisticated many-body techniques to obtain even preliminary estimates of correlation effects. The methods of many-body perturbation theory have been used in this context, usually in partial summations of terms corresponding to variants of the random phase approximation. These methods have not been wholly satisfactory, in part because the order in perturbation theory seems not to be an appropriate significance parameter for determining which terms to retain. In this paper we present an alternative many-body method, namely a coupled-cluster expansion of the type described by Cizek and Paldus (Cizek, 1966; Cizek and Paldus, 1971). This method appears to provide the significance ordering lacking in other approaches, and preliminary calculations indicate its feasibility.

The material to follow is for the most part a presentation of the methods in use in the author's laboratory. It is organized into three major sections, of which the first is a detailed discussion of the classical Madelung problem. This discussion is required, both to motivate the procedure to be used in quantum-mechanical calculations, and to elucidate the cancellation of individually divergent long-range contributions to the Coulomb energy and the role of surface effects. The second section develops the Hartree-Fock formalism and introduces the Fourier representation technique. This section includes a discussion of the key quantities entering the calculations. A final major section discusses the correlation problem, developing the coupled-cluster method and relating it to more familiar perturbation-theoretic approaches.

Some of the mathematical and computational techniques referred to in this presentation have been described in more detail elsewhere (Harris and Monkhorst, 1971; Harris, 1975). The reader is referred to these sources for further information.

MADELUNG SUMS

The Madelung summations with which we shall be concerned here are over macroscopic periodic lattices each interior cell of which is electrically neutral. Let $\vec{r}_i$ denote the position of the origin of Cell i, and let $\vec{s}_m, m=1,\ldots,d$, denote the respective locations, relative to the cell origin, of charges q_m in each cell. If there is a charge at the cell origin, we assign it the symbol q_o. Then the Madelung sum, whose value is the electrostatic potential at $\vec{r}_i$ due to ions at all other positions in the lattice, has the form

$$V(\vec{r}_i) = q_o \sum_{j \neq i} |\vec{r}_j - \vec{r}_i|^{-1} + \sum_j \sum_{m=1}^{d} q_m |\vec{r}_j - \vec{r}_i + \vec{s}_m|^{-1} \quad . \tag{1}$$

In this and all equations to follow, sums over lattice cells will include all cells except as explicitly indicated otherwise. As written, the j summations in Eq. (1) are individually divergent, but it is to be understood that the two summations are to be combined to produce a convergent result. Taken together, the summations thereby represent a conditionally convergent series. Because Madelung sums are conditionally convergent, care must be taken in defining the procedure for evaluating the summations. Assuming the intent to be the description of energy quantities within a macroscopic but finite crystal, it will be necessary to specify the characteristics of the crystal surfaces and then to interpret the Madelung sums consistently therewith.

The classical methods for the evaluation of Madelung sums are those of Evjen (1932) and Ewald (1921). These sums are in Evjen's method analyzed by grouping contributions into shells with vanishing net charge. In the Ewald method, the sum is described as an integral representation which is then broken into two parts, each of which can be represented as a summation which is more rapidly convergent than the original sum. Refinements and modifications of the Evjen and Ewald procedures have also been investigated; among these are the contributions of Emersleben (1923, 1950) and of Bertaut (1952).

Surface Effects

As already suggested, the result of a Madelung summation will depend upon the detailed specification of the boundary surfaces of the crystalline sample within which energy quantities are to be calculated. Since the summations are to be performed by analytical processes appropriate to infinite series, the surface specifications will have to be used to define limit processes fixing definite values for the Madelung sums. It will not always turn out that a natural sample specification and associated limit process will be those implied by any particular method of evaluating a Madelung sum.

For simplicity, let us restrict consideration to crystal samples consisting of an integral number of neutral microscopic repeating units. A crystal sample whose surfaces cannot be represented by translation of the chosen repeating unit can be characterized as a sample which can be so represented plus such additional charges near the boundary surfaces as are necessary. The repeating unit need not contain charges restricted to lie within a single crystal unit cell, but we assume all charges of a repeating unit to be separated by distances which are microscopic compared to the overall crystal dimensions. The repeating unit will be assigned a volume defined by the translation vectors describing its repeat distances. We note that the same charge distribution in the interior of a crystal can be described by many different choices of the repeating unit; however, different repeating units may lead to different specifications of the crystal surfaces. This observation is illustrated in Figure 1. We further restrict consideration to

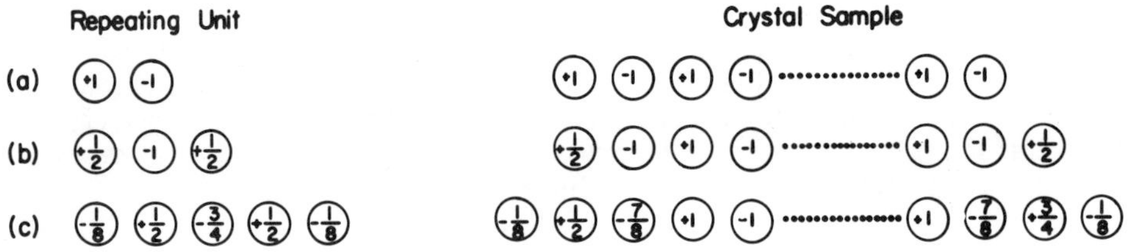

Figure 1. Different repeating units leading to the same interior charge distribution for a one-dimensional crystal, and crystal samples they generate: (a) simple unit with nonzero dipole moment; (b) unit with zero dipole moment and nonzero quadrupole moment; (c) unit with zero dipole and quadrupole moments.

Madelung sums describing the electrostatic potential within a crystal at points macroscopically distant from all crystal surfaces.

Even under the restrictions of the preceding paragraph, a Madelung sum will not be unambiguously defined unless either the repeating unit has vanishing dipole and quadrupole moments, or the sample shape is specified and the summation over repeating units is carried out over shells of increasing size which preserve the specified shape. We restrict consideration here to repeating units with vanishing dipole and quadrupole moments. Such a unit can always be constructed by moving some of the charge within a unit cell to equivalent points in adjoining cells. A one-dimensional example of a construction leading to a zero dipole moment is shown in Figure 1(b), while in Figure 1(c) we show a construction which also causes the quadrupole moment to vanish.

A sample built from repeating units with vanishing dipole and quadrupole moments will, as implied above, have potentials at macroscopically interior points which are independent of the shape of the sample. However, because different repeating units lead to different surface specifications, the potential at a macroscopically interior point of a crystal will, in general, depend upon the choice of the repeating unit, even when that unit has vanishing dipole and quadrupole moments. As Euwema and Surratt (1975) have shown, the surface effects by which samples differ can be related to the traces of the second-moment tensors of their repeating units. This trace need not vanish even if the quadrupole moment does, as for example in systems which possess overall neutrality but consist of concentric spherical shells of charge.

Samples whose repeating units have different second-moment traces per unit volume will have potentials differing at interior points by a position-independent amount. Let T denote this trace per unit volume,

$$T = \frac{1}{2v_r} \sum_{\text{unit}} q_i r_i^2 \quad , \tag{2}$$

where v_r is the repeating-unit volume, the q_i are charges, the r_i are distances from an arbitrary origin, and the sum is over all charges of the repeating unit. Then if the two samples have respective T values T_1 and T_2, their respective potentials V_1 and V_2 at any interior point will differ by

$$V_2 - V_1 = -\frac{4\pi}{3}(T_2 - T_1) \quad . \tag{3}$$

It has been shown (Harris, 1975) that the Ewald method yields a result corresponding to a choice of repeating unit whose entire second-moment tensor vanishes. This is also true of the Fourier representation method. When applied to atomic charge distributions as indicated in Figure 2, these methods thereby carry the implicit assumption that the crystal surfaces are as represented in Figure 2(a), rather than the more physically realistic situation pictured in Figure 2(b). If we wish to obtain results corresponding to Figure 2(b), we must identify a repeating unit leading to that situation, following which we may use Eq. (3) to determine the shift in potential from the Ewald result.

Finally, we note that samples containing a net dipole or quadrupole moment can be characterized as an assembly of repeating units without such moments, plus a suitable distribution of surface charge (or dipole moment). The potential at any interior point of such a sample will be that calculated as described above, plus the potential from the surface charge distribution.

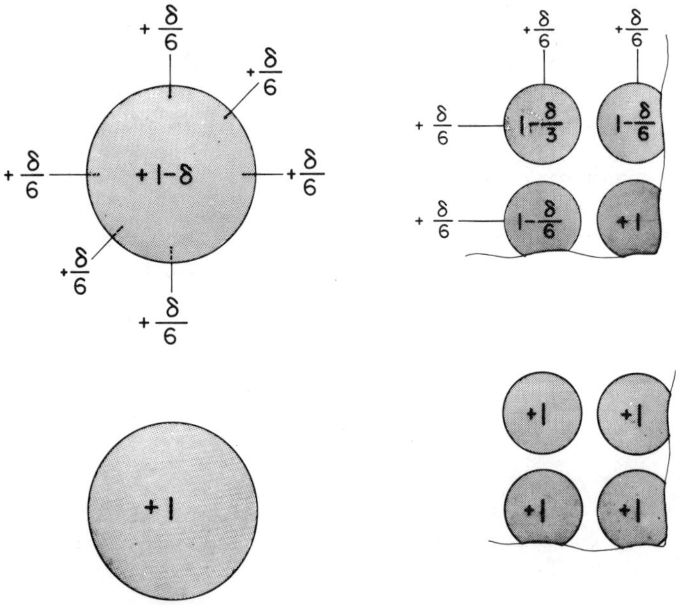

Figure 2. Repeating units with vanishing dipole and quadrupole moments for a simple cubic atomic crystal, and crystal samples they generate: (a) unit with zero second moment; (b) usual physical unit with nonzero second moment. The orbital sphere (shaded) has one unit of negative charge.

Ewald Method

We now consider the application of the Ewald method to the Madelung sums of Eq. (1), to be interpreted as described above. We start by introducing the integral transform

$$\frac{1}{r} = \frac{2}{\sqrt{\pi}} \int_0^\infty e^{-r^2 z^2} dz \tag{4}$$

for the potential due to each charge. We then have

$$V(\vec{r}_i) = \frac{2q_o}{\sqrt{\pi}} \sum_{j \neq i} \int_0^\infty e^{-|\vec{r}_j - \vec{r}_i|^2 z^2} dz + \sum_j \sum_{m=1}^{d} \frac{2q_m}{\sqrt{\pi}} \int_0^\infty e^{-|\vec{r}_j - \vec{r}_i + \vec{s}_m|^2 z^2} dz. \tag{5}$$

The essence of the Ewald method consists of two further steps, of which the first is the separation of the transform integral into the two ranges $(0,Z)$ and (Z,∞), where Z is arbitrary and may be chosen to cause rapid convergence of the final formulas. The second key step is the interchange of the j summation and the integration over the range $(0,Z)$, followed by application of the formula (see for example, Tosi, 1964)

$$\sum_j e^{-|\vec{r}_j + \vec{s}|^2 z^2} = \frac{\pi^{3/2}}{v_o} \sum_\mu z^{-3} e^{-k_\mu^2/4z^2 + i\vec{k}_\mu \cdot \vec{s}} , \tag{6}$$

where $\vec{k}_\mu$ are points of the lattice reciprocal to that of the points $\vec{r}_j$, and v_o is the unit cell volume of the $\vec{r}_j$ lattice.

If we examine carefully the steps proposed in the preceding paragraph, we will note that in the neighborhood of z=0 it is not permitted to interchange the order of the j summation and the z integration, because in that neighborhood the summation is not uniformly convergent. We therefore modify the classical Ewald procedure by separating the z integration into the three ranges $(0,\varepsilon)$, (ε,Z), and (Z,∞), following which we interchange operations and apply Eq. (6) only to the range (ε,Z). We then may consider the limit $\varepsilon \to 0$ under conditions maintaining uniform convergence for $z \geq \varepsilon$. For finite ε the procedure just outlined leads to

$$V(\vec{r}_i) = q_o \sum_j{}' \frac{2}{\sqrt{\pi}} \int_0^\varepsilon e^{-r_j^2 z^2} dz + \int_\varepsilon^Z \left(\frac{2\pi}{v_o} \sum_\mu z^{-3} e^{-k_\mu^2/4z^2} - \frac{2}{\sqrt{\pi}}\right) dz$$

$$+\sum_j{}' \frac{2}{\sqrt{\pi}} \int_Z^\infty e^{-r_j^2 z^2}\, dz$$

$$+\sum_{m=1}^{d} q_m \Bigg[\sum_j \frac{2}{\sqrt{\pi}} \int_0^\varepsilon e^{-|\vec{r}_j+\vec{s}_m|^2 z^2}\, dz + \frac{2\pi}{v_o} \sum_\mu \int_\varepsilon^Z z^{-3}\, e^{-k_\mu^2/4z^2 + i\vec{k}_\mu\cdot\vec{s}_m} dz$$

$$+ \sum_j \frac{2}{\sqrt{\pi}} \int_Z^\infty e^{-|\vec{r}_j+\vec{s}_m|^2 z^2}\, dz \Bigg] . \tag{7}$$

In obtaining Eq. (7) we have used the fact that after summing, a sum over $\vec{r}_j-\vec{r}_i$ is equivalent to one over $\vec{r}_j$; we have subtracted unity from both sides of Eq. (6) when the term j=i is to be eliminated in Eq. (5), and we have introduced the convention that a primed summation indicates omission of a point for which $\vec{r}_j$ or $\vec{k}_\mu$ is zero.

In Eq. (7) the integrations over the range (ε, Z) are elementary. Evaluating these integrals, introducing also the definition

$$\mathrm{erfc}(x) = \frac{2}{\sqrt{\pi}} \int_x^\infty e^{-z^2}\, dz \tag{8}$$

and segregating all terms containing ε, we have

$$V(\vec{r}_i) = \sum_{m=0}^{d} q_m \overline{V}_m + \Delta V , \tag{9}$$

with

$$\overline{V}_o = \frac{4\pi}{v_o} \sum_\mu{}' k_\mu^{-2}\, e^{-k_\mu^2/4Z^2} + \sum_j{}' r_j^{-1}\, \mathrm{erfc}(Zr_j) - \frac{\pi}{v_o Z^2} - \frac{2Z}{\sqrt{\pi}} , \tag{10}$$

$$\overline{V}_m = \frac{4\pi}{v_o} \sum_\mu{}' k_\mu^{-2} e^{-k_\mu^2/4Z^2 + i\vec{k}_\mu \cdot \vec{s}_m} + \sum_j |\vec{r}_j + \vec{s}_m|^{-1} \operatorname{erfc}(Z|\vec{r}_j + \vec{s}_m|)$$

$$- \frac{\pi}{v_o Z^2} , \quad m=1,\ldots,d \quad , \tag{11}$$

$$\Delta V = \sum_{m=o}^{d} q_m \left[\frac{\pi}{v_o \varepsilon^2} - \frac{4\pi}{v_o} \sum_\mu{}' k_\mu^{-2} e^{-k^2/4\varepsilon^2 + i\vec{k}_\mu \cdot \vec{s}_m} \right.$$

$$\left. + \sum_j \frac{2}{\sqrt{\pi}} \int_o^\varepsilon e^{-|\vec{r}_j + \vec{s}_m|^2 z^2} dz \right] \quad . \tag{12}$$

Equations (9)-(11) describe the usual evaluation prescribed in the Ewald method if ΔV can be caused to vanish.

We are now ready to consider the limit $\varepsilon \to 0$. Looking once again at Eq. (6), we note that its use implies the convergence of the series of its left-hand side, which in turn implies that z must be restricted to values greater than M/r_{max}, where r_{max} is the shortest distance from $\vec{r}_i$ to the surface of the sample under study, and M is independent of r_{max} and large compared to unity. We thus may not require ε to tend to zero more rapidly than M/r_{max}. If we also let r_{max} be large enough that $r_{max}/v_o^{1/3} >> M$, we will have $k_\mu r_{max} >> M$ for all nonzero k_μ, and therefore concomitantly $k_\mu^2/4\varepsilon^2 >> 1$. Under these conditions, the μ summation of Eq. (12) will become negligible, while the term involving $\pi/v_o\varepsilon^2$ will vanish by virtue of charge neutrality. We are left in the limit of small ε with

$$\Delta V = \lim_{\varepsilon \to 0} \sum_{m=0}^{d} q_m \sum_j \int_o^\varepsilon e^{-|\vec{r}_j + \vec{s}_m|^2 z^2} dz \quad . \tag{13}$$

It is not even obvious *a priori* that the limit indicated in Eq. (13) actually exists, as the j summation contains of the order of ${r_{max}}^3$ terms, many of which will be of magnitude M/r_{max}. It has been pointed out (Harris, 1975) that the limit giving ΔV exists and is independent of the sample shape if the sample consists of repeating units with vanishing net charge, dipole moment and quadrupole moment. However, ΔV will vanish only if, in additon, the repeating unit has a vanishing second-moment trace. This is a significant requirement implicit in the Ewald method.

Before leaving the Ewald method, let us note that after setting $\Delta V=0$ it is possible to attribute contributions $\overline{V}_m$ in Eq. (9) to lattices of unit positive charges defined by displacements $\vec{s}_m$. Of course, $\overline{V}_m$ is not the actual potential produced by such a lattice, as further contributions therefrom have been placed in ΔV where they have been caused to cancel against contributions from other charges. However, it can be shown (cf. Harris, 1975) that if $\vec{s}_m$ is varied over a unit cell the average value of $\overline{V}_m$ vanishes, implying that $\overline{V}_m$ can be interpreted as the potential of a lattice of unit positive charges defined by $\vec{s}_m$, plus a compensating uniform background. Note, however, that the conventional Ewald result implies the use of a repeating unit with a vanishing second-moment tensor, and this is not the repeating unit most intuitively visualized for the problem under discussion. An appropriate repeating unit and corresponding crystal sample for this problem are illustrated in Figure 3.

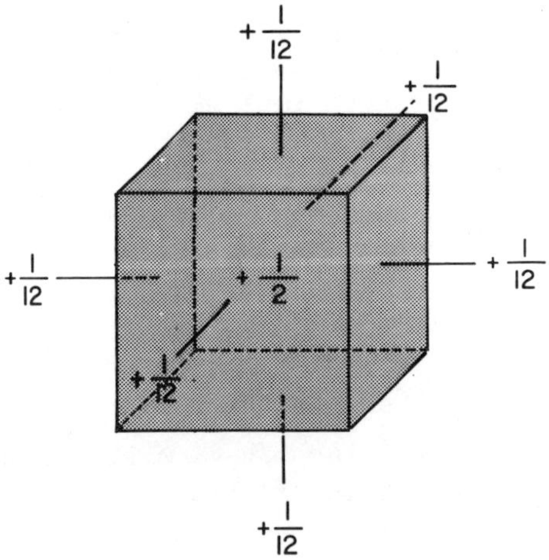

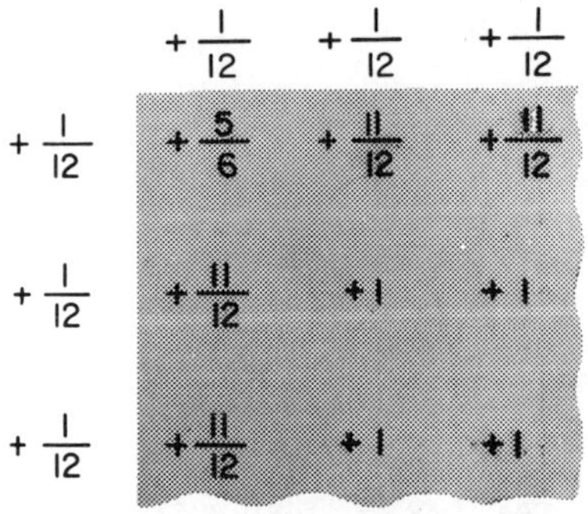

Figure 3. Repeating unit and crystal sample appropriate to Ewald and Fourier representation methods of Madelung summation for lattice of point charges with uniform compensating background. The shading indicates a uniform negative charge density of one unit per unit cell.

Fourier Representation Method

In this section we develop a method for the evaluation of Madelung sums based on the use of Fourier integral representations for the potentials of the charges of a crystalline system. The representation we shall use is of the form

$$r^{-1} = \frac{1}{2\pi^2} \int \frac{d\vec{k}}{k^2} e^{-i\vec{k}\cdot\vec{r}} \quad . \tag{14}$$

We note for future reference that the integral in Eq. (14) is conditionally convergent and that in spherical polar coordinates it may be interpreted as the result obtained when the angular integration is performed first. In cylindrical coordinates the appropriate result is obtained if the axial variable is integrated first.

Inserting Eq. (14) into the Madelung sum of Eq. (1), we have

$$V(\vec{r}_i) = \frac{q_o}{2\pi^2} \sum_{j \neq i} \int \frac{d\vec{k}}{k^2} e^{-i\vec{k}\cdot(\vec{r}_j - \vec{r}_i)}$$

$$+ \sum_j \sum_{m=1}^{d} \frac{q_m}{2\pi^2} \int \frac{d\vec{k}}{k^2} e^{-i\vec{k}\cdot(\vec{r}_j - \vec{r}_i + \vec{s}_m)} \quad . \tag{15}$$

Our next step is to interchange the order of the summation and integration in each term of Eq. (15), but we note that the uniform convergence necessary to permit the interchange is lacking in the neighborhood of $\vec{k}=0$. We therefore divide the $\vec{k}$ integrations into two regions, the first being a sphere of radius ε about $\vec{k}=0$, the second region being the remainder of $\vec{k}$ space. It has been shown (Harris, 1975) that in the limit of small ε, if the summation is interpreted as described earlier, and if the crystal sample is built from a repeating unit with vanishing net charge, dipole moment, and second-moment tensor, the first integration region makes no contribution to $V(\vec{r}_i)$. In the second integration region, we interchange summation and integration, reaching

$$V(\vec{r}_i) = \frac{1}{2\pi^2} \int' \frac{d\vec{k}}{k^2} \left[\sum_j e^{-i\vec{k}\cdot(\vec{r}_j - \vec{r}_i)} \left(q_o + \sum_{m=1}^{d} q_m e^{-i\vec{k}\cdot\vec{s}_m}\right) - q_o \right] . \tag{16}$$

The prime indicates the exclusion of a sphere of radius ε about $\vec{k}=0$. In writing Eq. (16), we have added the term j=i to the summation involving q_o and then explicitly subtracted it again.

Our next step is to invoke the use of lattice orthogonality relations. The summation over j in Eq. (16) consists of exponentials of imaginary argument whose values will be complex numbers of unit magnitude. If these complex numbers are distributed in phase, the summation will lead to a negligible result, while if they are of the same phase, they will reinforce to give an infinite result. Reinforcement will occur when $\vec{k}\cdot\vec{r}_j$ is a multiple of 2π, which is equivalent to the requirement that $\vec{k}$ be a point of the lattice reciprocal to that of the points $\vec{r}_j$. After considering questions of scale, we have the specific formula:

$$\sum_j e^{-i\vec{k}\cdot(\vec{r}_j-\vec{r}_i)} = \frac{8\pi^3}{v_o}\sum_\mu \delta(\vec{k}-\vec{k}_\mu) \quad , \tag{17}$$

where the summation on the right-hand side is over all points $\vec{k}_\mu$ of the reciprocal lattice, and δ is the Dirac delta function. Introducing Eq. (17), two of the terms in the $\vec{k}$ integration of Eq. (16) reduce to reciprocal lattice sums, and

$$V(\vec{r}_i) = \frac{q_o}{2\pi^2}\left[\frac{8\pi^3}{v_o}\sum_\mu{}' k_\mu^{-2} - \int{}' k^{-2}d\vec{k}\right] + \frac{4\pi}{v_o}\sum_{m=1}^{d} q_m \sum_\mu{}' k_\mu^{-2} e^{-i\vec{k}_\mu\cdot\vec{s}_m} \quad . \tag{18}$$

In Eq. (18) the summations exclude the point $\vec{k}_\mu=0$ because the region including that point was absent from the integration region in Eq. (16).

The portion of Eq. (18) in square brackets consists of two individually divergent terms; by examining the manner in which Eq. (18) was derived from Eq. (16), it is clear that the square bracket is to be interpreted as the limit reached when the integration and summation are over identical regions of $\vec{k}$ space, with the included region of $\vec{k}$ space extended infinitely in one of the fashions specified following Eq. (14). This limit, to which we assign the symbol C, depends upon the lattice structure. It is clear that C need not be zero, as it is a close three-dimensional analog of the one-dimensional limit leading to Euler's constant. We therefore write Eq. (18) in the final form

$$V(\vec{r}_i) = \frac{q_o C}{2\pi^2} + \frac{4\pi}{v_o}\sum_{m=1}^{d} q_m \sum_\mu{}' k_\mu^{-2} e^{-i\vec{k}_\mu\cdot\vec{s}_m} \quad . \tag{19}$$

The evaluation of C has been discussed elsewhere (Harris, 1975); values for several lattices are presented in Table I.

HARTREE-FOCK CALCULATIONS

Let us consider a three-dimensional periodic lattice consisting of N unit cells, each of which contains $\underline{d}$ nuclei at positions $\vec{s}_u$ (u=1,...,d) relative to the cell origins $\vec{R}_\mu$, with $s_1=0$. For simplicity we assume that all nuclei are crystallographically equivalent and of atomic number Z. The lattice will then also contain ZNd electrons. In atomic units (distances in bohrs, energies in hartrees; 1 bohr = 0.5292 A, 1 hartree = 2 Rydbergs = 27.210 eV), the Hamiltonian for this system, excluding relativistic and magnetic effects and nuclear motion is

$$H = \sum_{i=1}^{ZNd} (-\tfrac{1}{2}\nabla_i^2) + \sum_{1\leq i<j\leq ZNd} h(\vec{r}_i,\vec{r}_j) \ , \tag{20}$$

where $\vec{r}_i$ and ∇_i^2 refer to the position and Laplace operator for Electron i, and

$$h(\vec{r}_1,\vec{r}_2) = r_{12}^{-1} - \frac{1}{Nd}\sum_{\mu}\sum_{u=1}|\vec{r}_1-\vec{R}_\mu-\vec{s}_u|^{-1}$$

$$- \frac{1}{Nd}\sum_{\mu}\sum_{u=1}^{d}|\vec{r}_2-\vec{R}_\mu-\vec{s}_u|^{-1} + \frac{1}{N^2d^2}\sum_{\mu\mu'}{}^{''}\sum_{u,u'=1}^{d}|\vec{R}_\mu-\vec{R}_{\mu'}+\vec{s}_u-\vec{s}_{u'}|^{-1} \ . \tag{21}$$

Table I. Values of lattice structure constants C [Eq. (19)]

symmetry	$\frac{b}{a}$	$\frac{c}{a}$	C(units $2\pi/a$)
cubic	1	1	-8.913633
tetragonal	1	2	-5.673219
orthorhombic	$\sqrt{8}/3$	$\sqrt{3}$	-5.688760
orthorhombic	$\sqrt{3}$	$\sqrt{6}$	-4.170255
hexagonal	1	$\sqrt{8}/3$	-7.033153

In Eq. (21) the double prime indicates omission of the term $\mu=\mu'$ when $u=u'$. The Hamiltonian of Eq. (20) is inexact by terms which are of order less than N.

The Hamiltonian will be applied to crystal wavefunctions which are built from basis Bloch functions, i.e., functions possessing the periodicity of the crystal. There are several ways to define such Bloch functions; for simplicity we consider here one possibility based on atomic orbitals ϕ_i:

$$|\vec{k}_i> = e^{i\vec{k}\cdot\vec{r}} \sum_{\mu} \sum_{u=1}^{d} \phi_i(\vec{r}-\vec{R}_\mu-\vec{s}_u) \quad . \tag{22}$$

In this equation $\vec{k}$ is the Bloch wave vector associated with $|\vec{k}_i>$, and the subscript i identifies the functional form of the atomic orbital (e.g., its quantum numbers and screening parameter). Equation (22) describes Bloch functions which may be referred to as modulated plane waves; these orbitals are not periodic in $\vec{k}$ but nevertheless possess the periodicity of the $\vec{R}_\mu$ lattice.

The orbitals $|\vec{k}_i>$ are orthogonal except between $\vec{k}$ values which differ by reciprocal-space lattice vectors. When a crystal contains several equivalent atoms per unit cell, there may also be orthogonality between functions of $\vec{k}$ values differing by certain reciprocal lattice vectors. For macroscopic samples, Eq. (22) defines $Nv_0/(2\pi)^3$ basis orbitals per i value per unit volume in $\vec{k}$ space, where v_o is the unit cell volume of the $\vec{R}_\mu$ lattice.

From sets of the basis Bloch functions, we may construct crystal orbitals as linear combinations according to the formula

$$|\vec{k}>_\alpha = \sum_i c_{i\alpha}(\vec{k})\,|\vec{k}_i> \quad . \tag{23}$$

The subscript α, which we shall often suppress, indicates that for any $\vec{k}$ value we may make as many linearly independent orbitals $|\vec{k}>_\alpha$ as we have basis Bloch functions $|\vec{k}_i>$. Variation of the coefficients $c_i(\vec{k})$, in conjunction with a sufficient set of $|\vec{k}_i>$, enables the construction of orbitals $|\vec{k}>$ of arbitrary $\vec{k}$ dependence and of an $\vec{r}$ dependence which, except for periodicity, is also arbitrary.

The crystal orbitals $|\vec{k}>$ can be combined into many-electron antisymmetrized wavefunctions to which we may apply H.

We start by making a Hartree-Fock calculation to define a set of crystal orbitals which may also be used for correlation energy studies. For the Hartree-Fock calculation, we consider a single-determinant wavefunction containing only doubly-occupied crystal orbitals. We classify the $|\vec{k}>$ obtained from the Hartree-Fock method by their one-electron energies, and doubly occupy the $\frac{1}{2}ZNd$ orbitals $|\vec{k}>$ of lowest energy. Temporarily restricting the discussion to situations in which only one set of crystal orbitals $|\vec{k}>_\alpha$ is occupied in the wavefunction obtained by the Hartree-Fock procedure, we note that the occupied crystal orbitals will correspond to some region of $\vec{k}$ space. The surface bounding this region is known as the Fermi surface.

If we use the fact that all the occupied crystal orbitals $|\vec{k}>$ are orthogonal, the orbital occupancy we have chosen causes the expectation value of the total energy to take the form

$$<E> = \frac{2Nv_o}{(2\pi)^3} \int d\vec{k} \frac{<\vec{k}|-\frac{1}{2}\nabla^2|\vec{k}>}{<\vec{k}|\vec{k}>}$$

$$+ \frac{2Nv_o^2}{(2\pi)^6} \int d\vec{k}d\vec{k}' \frac{<\vec{k}\vec{k}'|h|\vec{k}\vec{k}'> - \frac{1}{2}<\vec{k}\vec{k}'|h|\vec{k}'\vec{k}>}{<\vec{k}|\vec{k}><\vec{k}'|\vec{k}'>} \quad . \tag{24}$$

In Eq. (24), and henceforth, the integrals of $\vec{k}$ and $\vec{k}'$ are over the reciprocal-lattice space enclosed by the Fermi surface, of volume $(2\pi)^3 d/(2v_o)$. Except when $\vec{k}=\vec{k}'$, the only part of h contributing to $<\vec{k}\vec{k}'|h|\vec{k}'\vec{k}>$ is the term r_{12}^{-1}, and with negligible error we may therefore replace this matrix element by $<\vec{k}\vec{k}'|r_{12}^{-1}|\vec{k}'\vec{k}>$. If the crystal sample is specified in a manner which would yield a convergent result in a classical calculation, the application of Eq. (24) will give a value for <E> which depends linearly on N. Equation (24) is independent of the scale chosen for the $|\vec{k}>$. It is convenient to normalize $|\vec{k}>$ proportionally to the sample size; specifically, we use

$$<\vec{k}|\vec{k}> = Nd \quad . \tag{25}$$

We may use Eq. (23) to express <E> in terms of the coefficients $c_i(\vec{k})$ and integrals over the basis Bloch functions. With the above normalization, the result is

$$<E> = \frac{2v_o}{(2\pi)^3 d} \int d\vec{k} \sum_{ij} c_i^*(\vec{k})<\vec{k}_i|-\frac{1}{2}\nabla^2|\vec{k}_j>c_j(\vec{k})$$

$$+ \frac{2v_o}{(2\pi)^6 d^2} \int d\vec{k}d\vec{k}' \sum_{ijmn} c_i^*(\vec{k})c_j(\vec{k})c_m^*(\vec{k}')c_n(\vec{k}')$$

$$\times (<\vec{k}_i\vec{k}'_m|h|\vec{k}_j\vec{k}'_n> - \frac{1}{2} <\vec{k}_i\vec{k}'_m|r_{12}^{-1}|\vec{k}'_n\vec{k}_j>) \quad . \tag{26}$$

Upon introduction of Eq. (23), the normalization condition becomes

$$<\vec{k}|\vec{k}> = \sum_{ij} c_i^*(\vec{k})<\vec{k}_i|\vec{k}_j>c_j(\vec{k}) = Nd \quad . \tag{27}$$

The Hartree-Fock equations are now obtained by minimizing <E> by variation of the $c_i(\vec{k})$ and of the form of the Fermi surface, subject to the normalization condition as expressed in Eq. (27) and to the requirement that the Fermi surface enclose a volume $(2\pi)^3 d/(2v_o)$. Using Lagrangian-multiplier methods, we consider the variation $\delta[<E> - (2/d) \int d\vec{k}\ \varepsilon(\vec{k})<\vec{k}|\vec{k}>]$, with $\varepsilon(\vec{k})$ the Lagrangian multiplier. The result is

$$\sum_j F_{ij}(\vec{k})c_j(\vec{k}) = \varepsilon(\vec{k}) \sum_j S_{ij}(\vec{k})c_j(\vec{k}) \quad , \tag{28}$$

$$F_{ij}(\vec{k}) = (\frac{1}{Nd}) \Big[<\vec{k}_i| - \frac{1}{2}\nabla^2|\vec{k}_j>$$

$$+ (\frac{2}{d}) \sum_{mn} \int d\vec{k}' c_m^*(\vec{k}')c_n(\vec{k}')(<\vec{k}_i\vec{k}'_m|h|\vec{k}_j\vec{k}'_n> - \frac{1}{2} <\vec{k}_i\vec{k}'_m|r_{12}^{-1}|\vec{k}_n\vec{k}_j>) \Big] , \tag{29}$$

$$S_{ij}(\vec{k}) = (1/Nd)<\vec{k}_i|\vec{k}_j> \quad . \tag{30}$$

If Eq. (28) is multiplied by $c_i^*(\vec{k})$ and summed over i, we obtain, analogously to atomic and molecular Hartree-Fock calculations,

$$\varepsilon(\vec{k}) = \sum_{ij} c_i^*(\vec{k})\ F_{ij}(\vec{k})c_j(\vec{k}) \quad . \tag{31}$$

An examination of the form given for $F_{ij}(\vec{k})$ in Eq. (29) leads naturally to the expected interpretation of $\varepsilon(\vec{k})$ as a one-electron energy to be associated with crystal orbital $|\vec{k}\rangle$ in the presence of nuclei and the remaining electrons of the crystal. Under ordinary circumstances $\varepsilon(\vec{k})$ will be expected to increase with the magnitude of $\vec{k}$, and the occupied orbitals which minimize the total energy will then be located within a Fermi surface for which

$$\varepsilon(\vec{k}_F) = \varepsilon_F, \quad \vec{k}_F \text{ on the Fermi surface} \quad . \tag{32}$$

The value of the constant ε_F is to be chosen so that the Fermi surface encloses the appropriate volume. Equation (32) corresponds to Koopman's theorem but it is here exact because of the continuous variations in $\vec{k}$ as N approaches infinity.

Crystal Integrals

As a preliminary to a detailed discussion of all the integrals needed in energy calculations, let us start by considering the overlap integrals among the basis orbitals $|\vec{k}_i\rangle$. We have

$$\langle\vec{k}_i|\vec{k}_j'\rangle = \sum_{\mu\mu'} \sum_{u,u'=1}^{d}$$

$$\langle\phi_i(\vec{r}-\vec{R}_\mu-\vec{s}_u)|e^{i(\vec{k}'-\vec{k})\cdot\vec{r}}|\phi_j(\vec{r}-\vec{R}_{\mu'}-\vec{s}_{u'})\rangle \quad . \tag{33}$$

Changing the integration variable $\vec{r}$ to $\vec{r} + \vec{R}_\mu$, and regrouping terms, Eq. (33) becomes

$$\langle\vec{k}_i|\vec{k}_j'\rangle = \sum_{\mu\mu'} \sum_{u,u'=1}^{d} e^{i(\vec{k}'-\vec{k})\cdot\vec{R}}$$

$$\times \langle\phi_i(\vec{r}-\vec{s}_u)|e^{i(\vec{k}'-\vec{k})\cdot\vec{r}}|\phi_j(\vec{r}+\vec{R}_\mu-\vec{R}_{\mu'}-\vec{s}_{u'})\rangle \quad . \tag{34}$$

Since μ is summed over the entire $\vec{R}_\mu$ lattice, we may replace $\vec{R}_\mu$ by $\vec{R}_\mu + \vec{R}_{\mu'}$, following which we may carry out the μ' summation. Using Eq. (17), we have

$$\langle\vec{k}_i|\vec{k}_j'\rangle = \frac{8\pi^3}{v_o} \sum_\lambda \delta(\vec{k}'-\vec{k}-\vec{K}_\lambda) \sum_\mu \sum_{u,u'=1}^{d} e^{i(\vec{k}'-\vec{k})\cdot\vec{R}_\mu}$$

$$\times \langle \phi_i(\vec{r}-\vec{s}_u)|e^{i(\vec{k}'-\vec{k})\cdot r}|\phi_j(\vec{r}+\vec{R}_\mu-\vec{s}_{u'})\rangle \quad , \tag{35}$$

where $\vec{K}_\lambda$ is a vector of the lattice reciprocal to the $\vec{R}_\mu$. Equation (35) shows that $|\vec{k}_i\rangle$ and $|\vec{k}'_j\rangle$ are orthogonal unless $\vec{k}$ and $\vec{k}'$ differ by a reciprocal lattice vector.

Integrating Eq. (35) over a small volume of $\vec{k}'$ about the point $\vec{k}$, and using the fact that there are $Nv_o/8\pi^3$ orbitals $|\vec{k}'_j\rangle$ per unit volume, we obtain

$$\langle\vec{k}_i|\vec{k}_j\rangle = N \sum_\mu \sum_{u,u'=1}^{d} \langle\phi_i(\vec{r}-\vec{s}_u)|\phi_j(\vec{r}+\vec{R}_\mu-\vec{s}_{u'})\rangle \quad . \tag{36}$$

Introducing the definition

$$\Phi_{ij}(\vec{q}) = \sum_\mu \sum_{u,u'=1}^{d} \langle\phi_i(\vec{r}-\vec{s}_u)|e^{i\vec{q}\cdot\vec{r}}|\phi_j(\vec{r}+\vec{R}_\mu-\vec{s}_{u'})\rangle \quad , \tag{37}$$

Eq. (36) may be written

$$\langle\vec{k}_i|\vec{k}_j\rangle = N\,\Phi_{ij}(0) \quad . \tag{38}$$

We note in passing that $\Phi_{ij}(\vec{q})$ is a weighted lattice sum of Fourier transforms of orbital products, with $\vec{q}$ being the transform variable. This fact will make Φ_{ij} a useful quantity in our later analysis.

We next continue to the kinetic energy integrals $\langle\vec{k}_i|-\frac{1}{2}\nabla^2|\vec{k}_j\rangle$. Applying the operator $-\frac{1}{2}\nabla^2$ to the form given in Eq. (22) and following steps similar to those used for the overlap integral, we may obtain

$$\langle\vec{k}_i|-\tfrac{1}{2}\nabla^2|\vec{k}_j\rangle = N \sum_\mu \sum_{u,u'=1}^{d} \Bigg(\tfrac{1}{2}k^2\langle\vec{k}_i|\vec{k}_j\rangle$$

$$- i\langle\phi_i(\vec{r}-\vec{s}_u)|\vec{k}\cdot\nabla|\phi_j(\vec{r}+\vec{R}_\mu-\vec{s}_{u'})\rangle$$

$$+ \langle\phi_i(\vec{r}-\vec{s}_u)|(-\tfrac{1}{2}\nabla^2)|\phi_j(\vec{r}+\vec{R}_\mu-\vec{s}_{u'})\rangle\Bigg) . \qquad (39)$$

The general plan for the evaluation of the remaining integrals over Bloch functions is to use Fourier representation methods to avoid explicit consideration of the "multicenter" integrals involving products of atomic orbitals at as many as four different spatial locations. Central to this plan is an identification of the cancellation of energy contributions analogous to the cancellations encountered in the evaluation of classical Madelung sums.

Looking first at the "Coulomb" integral $\langle\vec{k}_i\vec{k}'_m|h|\vec{k}_j\vec{k}'_n\rangle$, we see that this integral describes the classically calculated electrostatic energy of two charges distributions $k_i^*k_j$ and $k'^*_m k'_n$ interacting with each other and with the Nd nuclei, together with the nuclear-nuclear repulsion energy. If we introduce the Fourier representations for these energy contributions, we will find, as in classical calculations, that we may without error omit contributions from the neighborhood of the origin in the transform variable, and that a lattice of point charges (the nuclei) will generate a potential at a lattice point, due to charges on the other lattice points, which can be expressed as a limit of the type of C in Eq. (19), plus further contributions which cancel. These features, as in the classical case, depend upon the assumption that the crystal is built from a repeating unit possessing zero net charge, dipole moment, and second-moment tensor. With the understanding that all lattice sums involved in the evaluation of the Coulomb integral are to be interpreted as provided above, we may decompose it into separate terms and for each develop its Fourier representation. When finite contributions would be found to arise from the neighborhood of the origin in the transform space, such contributions must be understood to be suppressed.

We shall find that no significant convergence problems are associated with the "exchange" integral $\langle\vec{k}_i\vec{k}'_m|r_{12}^{-1}|\vec{k}'_n\vec{k}'_j\rangle$. Integrals of this type for different $\vec{k}$ and $\vec{k}'$ do not accumulate to yield a potentially divergent result, and the contributions from the region $\vec{k}\approx\vec{k}'$ are not large enough to overcome the fact that this region is of infinitesimal volume in $\vec{k}$ space. However, we shall find that the Fourier representation methods appropriate for the Coulomb integrals are also applicable to the exchange integrals and may there be used to advantage.

We now examine the terms obtained when the Coulomb integral $\langle\vec{k}_i\vec{k}'_m|h|\vec{k}_j\vec{k}'_n\rangle$ is expanded by inserting Eq. (21) for h. The Fourier representation formula needed to reduce the first term of this integral is of the general form (Bonham, Peacher, and Cox, 1964)

$$\int f(\vec{r}_1) r_{12}^{-1} g(\vec{r}_2) d\vec{r}_1 d\vec{r}_2 = \frac{1}{2\pi^2} \int \frac{d\vec{q}}{q^2} f^T(\vec{q}) g^T(-\vec{q}) \quad , \tag{40}$$

where f^T and g^T are the Fourier transforms respectively of f and g. We obtain

$$\langle \vec{k}_i \vec{k}'_m | r_{12}^{-1} | \vec{k}_j \vec{k}'_n \rangle = \sum_{\mu\mu'\nu\nu'} \sum_{u,u'=1}^{d} \sum_{v,v'=1}^{d}$$

$$\times \int \frac{d\vec{q}}{q^2} [\phi_i^*(\vec{r}-\vec{R}_\mu-\vec{s}_u)\phi_j(\vec{r}-\vec{R}_{\mu'}-\vec{s}_{u'})]^T(\vec{q})$$

$$\times [\phi_m^*(\vec{r}-\vec{R}_\nu-\vec{s}_v)\phi_n(\vec{r}-\vec{R}_{\nu'}-\vec{s}_{v'})]^T(-\vec{q}) \quad . \tag{41}$$

Changing $\vec{r}$ to $\vec{r}+\vec{R}_\mu$ in the first Fourier transform, using the relation

$$[f(\vec{r})]^T(\vec{q}) = e^{i\vec{q}\cdot\vec{R}} [f(\vec{r}+\vec{R}_\mu)]^T(\vec{q}) \tag{42}$$

and similarly changing $\vec{r}$ to $\vec{r}+\vec{R}_\nu$ in the second transform, then changing $\vec{R}_\mu-\vec{R}_{\mu'}$ to $\vec{R}_{\mu'}$ and $\vec{R}_\nu-\vec{R}_{\nu'}$ to $\vec{R}_{\nu'}$, we reduce Eq. (41) to the form

$$\langle \vec{k}_i \vec{k}'_m | r_{12}^{-1} | \vec{k}_j \vec{k}'_n \rangle = \sum_{\mu\nu} \frac{1}{2\pi^2} \int \frac{d\vec{q}}{q^2}$$

$$\times e^{i\vec{q}\cdot(\vec{R}_\mu-\vec{R}_\nu)} \Phi_{ij}(\vec{q}) \Phi_{mn}(-\vec{q}) \quad , \tag{43}$$

where we have introduced the notation Φ_{ij} defined in Eq. (37). Finally, using Eq. (17), the sum over μ becomes a sum of delta functions centered at the points $\vec{q}_\mu$ of the reciprocal lattice and the ν sum simply yields a factor N:

$$<\vec{k}_i\vec{k}'_m|r_{12}^{-1}|\vec{k}_j\vec{k}'_n> = \frac{N}{2\pi^2}\frac{(2\pi)^3}{v_o}$$

$$\times \int \frac{dq}{q^2} \sum_{\mu} \delta(\vec{q}-\vec{q}_\mu)\Phi_{ij}(\vec{q})\Phi_{mn}(-\vec{q}) \quad . \tag{44}$$

Remembering that the $\vec{q}$ integration is supposed to avoid the vicinity of $\vec{q}=0$, Eq. (44) can be reduced to the final form

$$<\vec{k}_i\vec{k}'_m|r_{12}^{-1}|\vec{k}_j\vec{k}'_n>$$

$$= \frac{4\pi N}{v_o} \sum_{\mu}{}' \frac{1}{q_\mu^2} \Phi_{ij}(\vec{q}_\mu)\Phi_{mn}(-\vec{q}_\mu) \quad . \tag{45}$$

The second and third terms of the Coulomb integral $<\vec{k}_i\vec{k}'_m|h|\vec{k}_j\vec{k}'_n>$ are reduced using a Fourier representation formula analogous to Eq. (40):

$$\int f(\vec{r})\,|\vec{r}-\vec{R}|^{-1}d\vec{r} = \frac{1}{2\pi^2}\int \frac{d\vec{q}}{q^2} f^T(\vec{q})\; e^{-i\vec{q}\cdot\vec{R}} \quad . \tag{46}$$

These terms become

$$\frac{1}{Nd} <\vec{k}_i\vec{k}'_m| \sum_{\mu}\sum_{u=1}^{d} |\vec{r}_1-\vec{R}_\mu-\vec{s}_u|^{-1} |\vec{k}_j\vec{k}'_n>$$

$$= \frac{4\pi N}{v_o d} \sum_{\mu}{}' \frac{S(-\vec{q}_\mu)}{q_\mu^2} \Phi_{ij}(\vec{q}_\mu)\Phi_{mn}(0) \quad , \tag{47}$$

$$\frac{1}{Nd} <\vec{k}_i\vec{k}'_m| \sum_{\mu}\sum_{u=1}^{d} |\vec{r}_2-\vec{R}_\mu-\vec{s}_u|^{-1} |\vec{k}_j\vec{k}'_n>$$

$$= \frac{4\pi N}{v_o d} \sum_{\mu}{}' \frac{S(\vec{q}_\mu)}{q_\mu^2} \Phi_{ij}(0)\Phi_{mn}(-\vec{q}_\mu) \quad , \tag{48}$$

where $S(\vec{q})$ is the structure factor, defined as

$$S(\vec{q}) = \sum_{m=1}^{d} e^{i\vec{q}\cdot\vec{s}_m} \quad . \tag{49}$$

We now turn to the last term of the Coulomb integral, for which we use the Fourier representation formula given in Eq. (14). We have

$$\langle\vec{k}_i\vec{k}'_m|\frac{1}{N^2d^2}\sum_{\mu\mu'}{}''\sum_{u,u'=1}^{d}|\vec{R}_\mu-\vec{R}_{\mu'}+\vec{s}_u-\vec{s}_{u'}|^{-1}|\vec{k}_j\vec{k}'_n\rangle$$

$$= \frac{1}{N^2d^2}\langle\vec{k}_i|\vec{k}_j\rangle\langle\vec{k}'_m|\vec{k}'_n\rangle$$

$$\times\frac{1}{2\pi^2}\int\frac{d\vec{q}}{q^2}\left[\sum_{\mu\mu'}\sum_{u,u'=1}^{d}e^{-i\vec{q}\cdot(\vec{R}_\mu-\vec{R}_{\mu'}+\vec{s}_u-\vec{s}_{u'})}-Nd\right] \quad , \tag{50}$$

where the term -Nd arises because we removed the restrictive condition from the summations. Proceeding by our usual methods, and grouping terms suitably, we find that the $\vec{q}$ integration reduces to

$$\frac{1}{2\pi^2}\int\frac{d\vec{q}}{q^2}\left[\;\right] = \frac{4\pi N}{v_o}\sum_{\mu}{}'\frac{|S(\vec{q}_\mu)|^2-d}{q_\mu^2}$$

$$+\frac{Nd}{2\pi^2}\left(\frac{8\pi^3}{v_o}\sum_{\mu}{}'q_\mu^{-2}-\int q^{-2}d\vec{q}\right) \quad . \tag{51}$$

The last parentheses in Eq. (51) contain the quantity C appearing in Eq. (19). As pointed out there, C is a constant dependent upon the lattice structure and scale, but not upon the charge distribution. We therefore write

$$\frac{1}{2\pi^2}\int\frac{d\vec{q}}{q^2}\left[\;\right] = \frac{4\pi ND}{v_o} \quad , \tag{52}$$

with

$$D = \sum_{\mu}{}' \frac{|S(\vec{q}_\mu)|^2 - d}{q_\mu^2} + \frac{v_o d}{8\pi^3} C \quad . \tag{53}$$

In terms of D,

$$<\vec{k}_i \vec{k}'_m | \frac{1}{N^2 d^2} \sum_{\mu\mu'} \sum_{u,u'=1}^{d} |\vec{R}_\mu - \vec{R}_{\mu'} + \vec{s}_u - \vec{s}_{u'}|^{-1} | \vec{k}_j \vec{k}'_n>$$

$$= \frac{4\pi N D}{v_o d^2} \Phi_{ij}(0) \; \Phi_{mn}(0) \quad . \tag{54}$$

Finally we consider the exchange integral $<\vec{k}_i \vec{k}'_m | r_{12}^{-1} | \vec{k}'_n \vec{k}_j>$. Writing out the Fourier representation for this integral and manipulating it so as to be able to introduce the definition in Eq. (37), we obtain

$$<\vec{k}_i \vec{k}'_m | r_{12}^{-1} | \vec{k}'_n k_j> = \frac{1}{2\pi^2} \sum_{\mu\nu} \int \frac{d\vec{q}}{q^2}$$

$$\times \; e^{i(\vec{k}' - \vec{k} + \vec{q}) \cdot (\vec{R}_\mu - \vec{R}_\nu)} \Phi_{in}(\vec{q} + \vec{k}' - \vec{k}) \; \Phi_{mj}(-\vec{q} - \vec{k}' + \vec{k}) \quad . \tag{55}$$

The summations over μ and ν lead to delta functions of argument $\vec{q} + \vec{k}' - \vec{k} - \vec{q}_\mu$, and Eq. (55) reduces to

$$<\vec{k}_i \vec{k}'_m | r_{12}^{-1} | \vec{k}'_n \vec{k}_j> = \frac{4\pi N}{v_o} \sum_{\mu} \frac{1}{|\vec{q}_\mu - \vec{k}' + \vec{k}|^2}$$

$$\times \; \Phi_{in}(\vec{q}_\mu) \; \Phi_{mj}(-\vec{q}_\mu) \quad . \tag{56}$$

Now that we have completed the somewhat tedious task of finding computationally convenient forms for the integrals needed in energy calculations, let us summarize the results we have obtained both as to their computational features and as to their physical interpretation. The computational features of our formulation are fairly obvious. All integrals needed for energy calculations have been reduced to lattice sums, and the key quantity appearing in the sums, Φ_{ij}, is essentially the Fourier transform of a lattice sum of orbital products. We shall discuss methods for evaluating Φ_{ij} and the quantities needed for kinetic energy integrals in the next section. We thus completely avoid the multicenter integral problem encountered in a direct-space approach.

Turning to the physical interpretation, we note that the use of the Fourier representation formalism has divided the Coulomb energy contributions into the lattice-structure term given in Eq. (54), plus nonuniformity terms as in Eqs. (45), (47), and (48). The exchange energy contribution, Eq. (56), can be similarly partioned. The term for $\vec{q}_\mu=0$ would be the only nonvanishing term if the orbitals were assumed to describe a uniform charge distribution, so that the terms for nonzero $\vec{q}_\mu$ can be thought of as nonuniformity exchange contributions.

Sorting the various energy contributions according to their anticipated size, we find three logical divisions. The contributions normally expected to be largest include the kinetic energy, the Coulomb lattice-structure term, and the uniform-distribution exchange energy. These are the terms which would be present in a crystal containing point nuclei immersed in a uniform electron gas. Next smaller in expected magnitude are the effects of nonuniformity upon the electron-nuclear attraction energy. These terms are smaller than those previously identified because they involve one factor Φ_{ij} with a non-zero argument, and $\Phi_{ij}(\vec{q})$ is a strongly decreasing function of q. Finally we have the terms normally expected to be smallest, namely the nonuniformity electron-repulsion contributions, both from the Coulomb and exchange integrals.

The above classification suggests several interesting observations. First, we note that the details of the electron-electron interactions only enter the energy expressions at the third level of magnitude, so that an independent-electron model of crystalline electronic structure might be expected to be rather good, and probably more satisfactory than for atoms and molecules. This observation may help us understand why one-electron models for solids have been so successful. Second, we note that the formula for the exchange energy, Eq. (56), becomes of the same general form as that for the Coulomb energy of electron repulsion, Eq. (45) in the limit $q_\mu \gg |\vec{k}-\vec{k}'|$. We therefore expect that as q_μ increases, $\langle\vec{k}\vec{k}'|r_{12}^{-1}|\vec{k}\vec{k}'\rangle$ and $\langle\vec{k}\vec{k}'|r_{12}^{-1}|\vec{k}'\vec{k}\rangle$ will contain increasingly similar nonuniformity contributions. It is thus clearly incorrect to conclude that exchange is unimportant in crystals; if the Hartree method reproduces some features of crystalline wavefunctions in a manner more qualitatively correct than the Hartree-Fock method, it must be because the neglect of exchange is to a great extent compensated by the neglect of electron correlation.

Evaluation of $\Phi_{ij}(\vec{q})$

It was shown in the preceding section that the integrals needed there for energy calculations could be reduced to forms related to Fourier transforms of lattice sums of atomic orbital products. In particular, it was found that all contributions to the potential energy could be expressed in terms of the quantity $\Phi_{ij}(\vec{q})$ defined in Eq. (37). This and a pair of related quantities were also found to occur in the evaluation of the kinetic-energy matrix elements. It is the purpose of the present section to show how these fundamental quantities can be evaluated in practical terms.

Two main approaches are available for evaluation of the Φ_{ij}. Either the lattice summation can be evaluated as it stands, or it can be manipulated using convolution and lattice orthogonality relations with the effective result of converting to a reciprocal-lattice summation. Direct evaluation would seem preferable when the orbitals do not overlap significantly, but has the drawback that the individual terms (which are transforms of orbital products) do not assume simple forms. On the other hand, conversion to a reciprocal-lattice summation yields summands which are simple but which may not show desirable convergence properties. For example, when Slater-type orbitals are used, the direct-space terms converge exponentially in the distance between the orbitals of a product, while the reciprocal-space summation converges only as a negative power of reciprocal-space distance. However, as we shall show, it seems possible to utilize the favorable properties of both the direct and reciprocal-space formulations by a technique similar to that introduced by Ewald for the classical Madelung problem.

The techniques which may usefully be employed in the evaluation of Φ_{ij} depend upon the atomic orbital basis which has been chosen. If Gaussian-type atomic orbitals are employed, the entire evaluation process becomes trivial owing to the fact that a product of Gaussians is essentially a single Gaussian centered at a point intermediate between the centers of the original Gaussians:

$$e^{-ar^2-b|\vec{r}-\vec{R}|^2} = e^{-\frac{abR^2}{a+b}} e^{-(a+b)|\vec{r}-\frac{b}{a+b}\vec{R}|^2} , \qquad (57)$$

$$\left[e^{-ar^2}\right]^T(\vec{q}) = \left(\frac{\pi}{a}\right)^{\frac{3}{2}} e^{-q^2/4a} . \qquad (58)$$

Unfortunately, the simplicity resulting from the use of Gaussian orbitals is compensated by their relatively poor representation they provide for actual electron distributions in crystals, and we have found it more advantageous to use a Slater-type orbital basis. It turns out that the crystal symmetry eliminates the most vexing aspects accompanying the use of Slater-type orbitals in polyatomic molecules. Accordingly, we restrict the remainder of this section to discussions based on the use of Slater-type orbitals (STO's).

Taking first the direct-space evaluation of Φ_{ij}, we see that it is necessary to have a tractable expression for the Fourier transform of a product of STO's whose centers are separated by a distance $\vec{R}_\mu+\vec{s}_u-\vec{s}_{u'}$. This Fourier transform has been discussed by several investigators (see, for example, Harris and Monkhorst, 1972), none of whom have presented its value in a convenient closed form. However, this transform can be evaluated by suitable analytical and numerical methods, and it is therefore possible to calculate Φ_{ij} by the direct summation indicated in Eq. (37). We have found the direct method cumbersome to program and rarely competitive in computer time, and have therefore abandoned it in favor of the approaches to be described below.

Turning next to methods based on reciprocal-space conversions, let us start by applying the Fourier convolution theorem to the transform appearing in Eq. (37). Writing $[\phi_i^*(\vec{r}-\vec{s}_u)\phi_j(\vec{r}+\vec{R}_\mu-\vec{s}_{u'})]^T(\vec{q})$ for $\langle\phi_i(\vec{r}-\vec{s}_u)|\exp(i\vec{q}\cdot\vec{r})|\phi_j(\vec{r}+\vec{R}_\mu-\vec{s}_{u'})\rangle$, the convolution theorem yields

$$[\phi_i^*(\vec{r}-\vec{s}_u)\phi_j(\vec{r}+\vec{R}_\mu-\vec{s}_{u'})]^T(\vec{q})$$

$$= \frac{1}{(2\pi)^3}\int d\vec{p}[\phi_i^*(\vec{r}-\vec{s}_u)]^T(\vec{p})[\phi_j(\vec{r}+\vec{R}_\mu-\vec{s}_{u'})]^T(\vec{q}-\vec{p})$$

$$= \frac{1}{(2\pi)^3}\int d\vec{p}\; e^{i\vec{p}\cdot\vec{s}_u+i(\vec{q}-\vec{p})\cdot(\vec{s}_{u'}-\vec{R}_\mu)}\phi_i^{*T}(\vec{p})\phi_j^T(\vec{q}-\vec{p}) \quad , \qquad (59)$$

where the last line was obtained using Eq. (42). Inserting Eq.(59) into Eq. (37), we have

$$\Phi_{ij}(\vec{q}) = \frac{1}{(2\pi)^3}\sum_\mu \sum_{u,u'=1}^{d} \int d\vec{p}$$

$$\times\; e^{i\vec{p}\cdot\vec{s}_u+i(\vec{q}-\vec{p})\cdot(\vec{s}_{u'}-\vec{R}_\mu)}\phi_i^{*T}(\vec{p})\phi_j^T(\vec{q}-\vec{p}) \quad . \qquad (60)$$

Invoking now the lattice orthogonality relation given in Eq. (17) and the definition of the structure factor given in Eq. (49), we may reduce Eq. (60) to the basic result

$$\Phi_{ij}(\vec{q}) = \frac{1}{v_o} \sum_{\mu} S(\vec{q}-\vec{p}_\mu)S(\vec{p}_\mu)$$

$$\times\ \phi_i^{*T}(\vec{q}-\vec{p}_\mu)\phi_j^{T}(\vec{p}_\mu) \quad , \tag{61}$$

where $\vec{p}_\mu$ are vectors of the lattice reciprocal to that of the $\vec{R}_\mu$. The obvious advantage of Eq. (61) is that it may be readily applied for any set of basis atomic orbitals whose individual transforms are known. We note that STO's have simple transforms; for example, the normalized 1s STO of screening parameter ζ has the transform

$$(1s)^T(q) = \frac{8\pi^{1/2}\,\zeta^{5/2}}{(q^2+\zeta^2)^2} \quad . \tag{62}$$

Equation (61) provides a simple and practical route to the Φ_{ij}, and is the formula we have used in many of the studies performed to date.. However, when very large numbers of Φ_{ij} are needed the efficiency of their evaluation becomes of paramount importance, and we have consequently maintained a program of study designed further to improve the efficiency of Φ_{ij} evaluation. Our efforts in this regard are best reported for a typical example, which we choose to be based on 1s STO's with screening parameters ζ_i and ζ_j, with Φ_{ij} evaluated for a simple cubic lattice with one atom per cell [for which $S(\vec{q})=1$]. Under these conditions Eqs. (37) and (61) yield

$$\Phi_{ij}(\vec{q}) = \frac{(\zeta_i\zeta_j)^{3/2}}{\pi} \sum_{\mu} \int d\vec{r}\ e^{-\zeta_i r-\zeta_j|\vec{r}+\vec{R}_\mu|+i\vec{q}\cdot\vec{r}} \tag{63}$$

$$= \frac{64\pi(\zeta_i\zeta_j)^{5/2}}{v_o} \sum_{\mu} \frac{1}{(|\vec{q}-\vec{p}_\mu|^2+\zeta_i^2)^2(p_\mu^2+\zeta_j^2)^2} \quad . \tag{64}$$

If we make changes to dimensionless variables such that in each of the above equations the summation is over a unit lattice of lattice vectors $\vec{\mu}$, and write $v_o=a^3$, $\delta_i=a\zeta_i/2\pi$, $\delta_j=a\zeta_j/2\pi$, we obtain

$$\Phi_{ij}\left(\frac{2\pi\vec{\nu}}{a}\right) = \frac{(\delta_i\delta_j)^{3/2}}{\pi} \sum_{\vec{\mu}} \int d\vec{r}\ e^{-\delta_i r-\delta_j|\vec{r}+2\pi\vec{\mu}|+i\vec{\nu}\cdot\vec{r}} \tag{65}$$

$$= \frac{8}{\pi^2} (\delta_i \delta_j)^{5/2} \sum_{\vec{\mu}} \frac{1}{(|\vec{\nu}-\vec{\mu}|^2+\delta_i^2)^2 (\mu^2+\delta_j^2)^2} \quad . \tag{66}$$

A more efficient approach to the evaluation of Φ_{ij} is provided by the introduction of integral transforms for $\exp(-\delta_i r)$ and $\exp(-\delta_j |\vec{r}-2\pi\vec{\mu}|)$ in Eq. (65). The transform which appears to be most useful is that suggested by Kikuchi (1954) and later used by Shavitt and Karplus (1965),

$$e^{-\delta r} = \frac{\delta}{\sqrt{\pi}} \int_0^\infty dx\, x^{-1/2}\, e^{-x\delta^2 - r^2/4x} \quad , \tag{67}$$

and leads to the formula

$$\Phi_{ij}\left(\frac{2\pi\vec{\nu}}{a}\right) = \frac{(\delta_i \delta_j)^{5/2}}{\pi^2} \sum_{\vec{\mu}} \int_0^\infty dx \int_0^\infty dy\, (xy)^{-1/2}\, e^{-\delta_i^2 x - \delta_j^2 y}$$

$$\times \int e^{-\frac{r^2}{4x} - |\vec{r}-2\pi\vec{\mu}|^2/4y + i\vec{\nu}\cdot\vec{r}}\, d\vec{r} \quad . \tag{68}$$

The $\vec{r}$ integration may be performed using Eqs. (57) and (58), so we have

$$\Phi_{ij}\left(\frac{2\pi\vec{\nu}}{a}\right) = \frac{8(\delta_i \delta_j)^{5/2}}{\sqrt{\pi}} \int_0^\infty dx \int_0^\infty dy\, \frac{xy}{(x+y)^{3/2}}$$

$$e^{-\delta_i^2 x - \delta_j^2 y - \nu^2 xy/(x+y)} \sum_{\vec{\mu}} e^{-\pi^2\mu^2/(x+y) + 2\pi i x\vec{\nu}\cdot\vec{\mu}/(x+y)} \quad . \tag{69}$$

Equation (69) is a starting point from which we may proceed as in the Ewald method, by dividing the (x,y) integration into regions, using the $\vec{\mu}$ summation as given in Eq. (69) for a small x+y, and the corresponding Poisson summation for large x+y:

$$\sum_{\vec{\mu}} e^{-\pi^2\mu^2/(x+y) + 2\pi i x\vec{\nu}\cdot\vec{\mu}/(x+y)}$$

$$= \left(\frac{x+y}{\pi}\right)^{3/2} \sum_{\vec{\mu}} e^{-(x+y)\mu^2 - 2x\vec{\mu}\cdot\vec{\nu} - x^2\nu^2/(x+y)} \quad . \tag{70}$$

For this purpose we transform from the variables (x,y) to new variables s-x+y, t=(x-y)/(x+y), in terms of which the overall integration region is $0 \leq s \leq \infty$, $-1 \leq t \leq 1$. We then divide the s integration into the two ranges (0,Z) and (Z,∞), using Eq. (70) in the latter range. It then becomes possible to integrate the (Z,∞) range in both s and t, and we have the result

$$\Phi_{ij}\left(\frac{2\pi\vec{\nu}}{a}\right) = \frac{8(\delta_i\delta_j)^{5/2}}{\pi^2} \sum_{\vec{\mu}} \left[\frac{e^{-\alpha Z}}{\alpha(\alpha-\beta)^2}\left(\frac{2}{\alpha-\beta} + \frac{1}{\alpha} + Z\right) + \frac{e^{-\beta Z}}{\beta(\alpha-\beta)^2}\left(\frac{-2}{\alpha-\beta} + \frac{1}{\beta} + Z\right)\right]$$

$$+ \frac{(\delta_i\delta_j)^{5/2}}{\pi^2} \sum_{\vec{\mu}} \int_0^Z ds \int_{-1}^{1} dt\ s^{3/2}(1-t^2)\ e^{-\gamma s-\pi^2\mu^2/s+\pi i(1+t)\vec{\mu}\cdot\vec{\nu}} , \tag{71}$$

with

$$\alpha = |\vec{\mu}-\vec{\nu}|^2 + \delta_i^2 , \tag{72}$$

$$\beta = \mu^2 + \delta_j^2 , \tag{73}$$

$$\gamma = \tfrac{1}{2}\delta_i^2(1+t) + \tfrac{1}{2}\delta_j^2(1-t) + \tfrac{1}{4}\nu^2(1-t^2) . \tag{74}$$

The value of Z used in Eq. (71) is arbitrary and should be chosen to obtain optimum convergence in both $\vec{\mu}$ summations. In the limit $Z\to 0$ the form given for Φ_{ij} reduces to that in Eq. (66); the limit $Z\to\infty$ yields integrals which can be manipulated to the form presented in Eq. (65). Intermediate values of Z enable advantage to be taken of the good features of both the limiting forms.

With the value Z=1 the first summation in Eq. (71) converges rapidly, while the second summation, whose summands are evaluated only with more difficulty, is also strongly convergent. In fact, seven significant figures in Φ_{ij} are apparently obtained with this choice of Z if the second summation only includes $\vec{\mu}=0$ and the (1,0,0) star, irrespective of the values of δ_i and δ_j. The terms of the second sum are readily evaluated by interchanging the order of the s and t integrations, performing the s integration analytically and then the t integration numerically. A Lobatto quadrature (Abramowitz and Stegun, 1964) is convenient for the numerical integration. For $\vec{\mu}=0$ the s integral is essentially an error function. For the (1,0,0) star an expansion of $e^{-\gamma s}$ in powers of s^{-1} leads for the s integral to the formula

$$\int_0^1 s^{3/2} e^{-\gamma s-\pi^2/s} ds = e^{-\gamma} \sum_{n=0}^{\infty} A_n \gamma^n/n! \quad , \tag{75}$$

where

$$A_n = \int_0^1 (1-s)^n s^{\frac{3}{2}-n} e^{-\pi^2/s} ds \quad . \tag{76}$$

Further details regarding the numerical procedures will appear elsewhere. Overall, the labor of evaluating each term of the second summation in Eq. (71) is somewhat smaller than that required for corresponding terms of Eq. (65).

The kinetic energy integrals can be reduced in a manner parallel to that used for the Φ_{ij}. However, far fewer kinetic energy integrals than the Φ_{ij} enter a calculation so that efficiency of evaluation is no longer of extreme importance and will not be discussed here.

ELECTRON CORRELATION

Most of the electron-correlation studies thus far carried out have been based on configuration-interaction (CI) methods. Much work on atomic and molecular systems indicates that the main drawback of the CI method for those applications is in the slow convergence of the configuration expansion. It appears to be possible to reach an adequate one-electron (orbital) basis without undue labor, but the number of many-electron functions constructible therefrom is so large that not all can be used. It turns out that, even using optimum-orbital methods (e.g., multi-configuration self-consistent field or iterative natural orbital), the convergence is disappointingly slow. With increasing numbers of electrons, the convergence problem becomes aggravated; in the limit of an extensive system the Hartree-Fock wavefunction and low-order orbital excitations therefrom become negligible components of the exact wavefunction. In addition, divergences may be introduced if Bloch states are used as the basis for a low-order CI, and a CI based on equivalent localized orbitals becomes inappropriate when the Hartree-Fock wavefunction cannot be so transformed (a circumstance which occurs in most three-dimensional problems; Monkhorst, 1972).

Some of the difficulties of the CI expansion may be avoided by use of a perturbation-theoretic formulation. Diagrammatic many-body perturbation theory (MBPT) provides a systematic way of organizing the various terms and enables the formal cancellation of those which do not scale correctly with system size. However, MBPT has the difficulty that the perturbation interaction is an inappropriate quantity in which to make an expansion, and some higher-order contributions are larger than many low-order terms. Included among the important higher-order terms are those which, in the MBPT formalism, describe repeated interactions of small numbers of electrons. The relative importance of various kinds of terms has been found to depend strongly upon the choice of zero-order states. Brueckner (1969) has demonstrated the inadequacy of plane-wave zero-order states for atomic problems, while Kelly (1969) has shown that Hartree-Fock states can be quite satisfactory. Monkhorst and Oddershede (1973) reached similar conclusions in their correlation study of the atomic hydrogen crystal.

Many extended systems also exhibit divergences in finite order MBPT contributions, typically due to excitations involving unoccupied zero-order states whose orbital energies join continuously to those of the occupied portion of an energy band. These divergences can usually be removed by judicious summation of various classes of contributions to infinite order, but the problem remains that there is no unambiguous significance ordering in the partially condensed MBPT expansion.

A possible solution to the problem of significance ordering is provided by the use of cluster-expansion methods. In the simplest methods of this type, the clusters are "uncoupled", i.e., the cluster wavefunctions are determined by a procedure which assumes only one cluster to be present. The "atomic Bethe-Goldstone method" introduced by Nesbet (1967) is of this type; it finds optimum cluster functions when all electrons not in the cluster are assumed to be in their Hartree-Fock distributions. The total energy is related to the sum of the individual cluster contributions, a procedure which can be motivated by considering a many-electron wavefunction which is the sum of all possible disjoint products of clusters, and for which interactions between clusters are ignored in evaluating the energy. The assumed independence of the clusters is the key drawback of the uncoupled cluster methods. When interactions between clusters of a certain size are important, their effect can included only by considering clusters of twice that size.

The main deficiency of the uncoupled cluster methods can be counteracted by including an inter-cluster coupling in the formalism. One then proceeds to determine cluster functions which in some sense are optimum in the presence of other clusters. Such formulations have the advantage of requiring clusters of lower order than for an uncoupled method of comparable accuracy, but have also the disadvantage that the resulting equations must be nonlinear. The coupled-cluster method which appears to be the most tractable was first introduced, for nuclear problem, by Kümmel (1962), and has been presented in diagrammatic form and applied to atomic and molecular problems by Cizek and Paldus (Cizek, 1966; Cizek and Paldus, 1971). The approach is nonvariational, and can be shown to enjoy a close relationship with partially summed perturbation-theoretic expansions. When the coupled-cluster formulation is extended through clusters of a given size, it implies the use of a many-electron wavefunction which is the sum of all possible disjoint products of clusters through that size, or equivalently, which includes high-order excitations from a reference state to the extent that they can be described as products of lower-order excitations. The formulation therefore implements in a quantitative and optimum fashion the ideas put forward by Sinanoglu (1962) in his "many-electron theory".

In this section we present the coupled-cluster equations, with particular attention to the reasons why the development is tractable for extended systems. An explicit relationship is identified between terms of these equations and partial diagram summations in MBPT. Some discussion of computational details is also included.

Coupled-Cluster Method

The cluster-expansion wavefunction Ψ is formed from an antisymmetrized-product reference wavefunction Φ by application of the operator $\exp(\hat{T})$, where $\hat{T}$ is an operator which generates all single clusters from Φ. In particular,

$$\Psi = e^{\hat{T}}\Phi \quad , \tag{77}$$

$$\hat{T} = \hat{T}_1 + \hat{T}_2 + \dots \quad , \tag{78}$$

$$\hat{T}_1 = \sum_{\alpha r} \hat{t}_\alpha^r \quad , \tag{79}$$

$$\hat{T}_2 = \frac{1}{4} \sum_{\alpha\beta rs} \hat{t}_{\alpha\beta}^{rs}, \quad \dots \quad , \tag{80}$$

where $\hat{T}_n$ is the operator generating all clusters of size n, the Greek indices $\alpha,\beta,\dots$ denote single-particle states occupied in Φ, and Latin indices $r,s,\dots$ denote unoccupied single-particle states. The operators $\hat{t}_\alpha^r$, $\hat{t}_{\alpha\beta}^{rs}$,... cause excitations of 1, 2, ... particles from the designated occupied states $(\alpha,\beta,\dots)$ to the designated unoccupied states (r,s,...), with coefficients appropriate to the amplitudes of their respective terms in the final wavefunction Ψ. Using a creation-annihilation operator formalism, we may write

$$\hat{t}_{\alpha}^{r} = t_{\alpha}^{r} a_{r}^{\dagger} a_{\alpha} \quad , \tag{81}$$

$$\hat{t}_{\alpha\beta}^{rs} = t_{\alpha\beta}^{rs} a_{r}^{\dagger} a_{s}^{\dagger} a_{\beta} a_{\alpha} \quad ,\ldots, \tag{82}$$

where t_{α}^{r}, $t_{\alpha\beta}^{rs}$,... are coefficients and $a_i^{\dagger}$ and a_i are respectively creation and annihilation operators for a particle in single-particle state i. For clarity and simplicity the foregoing equations have been written in a notation implying discrete indexing of single-particle states. When actual application is to be made to crystalline systems we must convert the discrete indices and their summations to continuous indices (e.g., $\vec{k}$) and their integrations.

The definitions of the preceding paragraph are such that the operator $\hat{T}$ converts Φ into a particular linear combination of determinants singly, doubly, ... excited therefrom. If the expansion for $\hat{T}$, Eq. (78), were extended to the total number of electrons in a system, then $\hat{T}\Phi$, with suitable values of the coefficients t_{α}^{r}, $t_{\alpha\beta}^{rs}$,... , could represent the exact many-electron wavefunction. The purpose of the cluster-expansion formalism is to provide satisfactory approximate wavefunctions when the expansion for $\hat{T}$ is truncated at clusters of modest size (e.g., after $\hat{T}_2$). Because Ψ is obtained from Φ by application of $\exp(\hat{T})$ rather than $\hat{T}$ itself, the final wavefunction will contain terms more highly excited than those present in $\hat{T}\Phi$.

A more complete understanding of the form of Ψ may be achieved by considering the expansion of $\exp(\hat{T})$ in Eq. (77). We have

$$\Psi = \Phi + \hat{T}\Phi + \frac{1}{2!}\hat{T}\hat{T}\Phi + \frac{1}{3!}\hat{T}\hat{T}\hat{T}\Phi + \ldots \quad . \tag{83}$$

If $\hat{T}$ is truncated at low order, higher-order excitations will be present in Ψ to the extent that they are expressed as products of excitations retained in $\hat{T}$. The excitations must be disjoint, as the properties of the annihilation and creation operators will cause a product $\hat{t}_{\alpha..}^{r..}\,\hat{t}_{\beta..}^{s..}$ to vanish if its factors have any index in common. Noting also that the terms of Eq. (83) have numerical coefficients appropriate to the numbers of permutations in which each particular $\hat{t}_{\alpha..}^{r..}$ product will occur, we see that Ψ consists of Φ, plus single-cluster excitations therefrom, plus all possible disjoint multiple-cluster excitations, with coefficients appropriate to independent excitation of the simultaneously occurring clusters. In the discussion to follow, we consider how the single-cluster operators $\hat{t}^{r\cdot\cdot}_{\alpha\cdot\cdot}$ may be determined so as to optimize in some sense the wavefunction Ψ which includes multiple-cluster terms. It will turn out that, because of the forms chosen for $\hat{T}$ and Ψ, there exists an optimization criterion whose formal application is satisfyingly simple.

To determine the coefficients in $\hat{T}$, we consider the Schrödinger equation, which if Ψ were exactly given by Eq. (77) would have the form

$$He^{\hat{T}}\Phi = Ee^{\hat{T}}\Phi \quad .$$

Multiplying both sides of this equation from the left by $\exp(-\hat{T})$, we have

$$e^{-\hat{T}}He^{\hat{T}}\Phi = E\Phi \quad . \tag{85}$$

Expanding the operator $\exp(-\hat{T})H\exp(\hat{T})$ according to a Baker-Campbell-Hausdorff formula (cf. Weiss and Maradudin, 1962), Eq. (85) may be brought to the form

$$\{H+(H,\hat{T})+\frac{1}{2!}((H,\hat{T}),\hat{T})+\frac{1}{3!}(((H,\hat{T}),\hat{T}),\hat{T})+\ \ldots\ \}\Phi = E\Phi \quad , \tag{86}$$

where we have introduced the standard commutator notation $(A,B) = AB-BA$.

If $\hat{T}$ has been truncated, Eq. (86) cannot be satisfied exactly. However, a value for E and an optimum set of coefficients for the truncated $\hat{T}$ can be obtained by requiring satisfaction of the projections of Eq. (86) on Φ and on a set of low-order excitations therefrom. If $\hat{T}$ has been truncated after $\hat{T}_n$, the truncated $\hat{T}$ will contain just enough coefficients to enable satisfaction of the projection of Eq. (86) on Φ and on all excitations therefrom through order n. The projection on Φ will give E in terms of the t coefficients, while the remaining projections will yield a set of equations from which these t coefficients may be determined. These equations will be nonlinear, as $\hat{T}$ occurs nonlinearly on the left hand side of Eq. (86).

Two important simplifications arise when the optimization procedure of the foregoing paragraph is carried out. First, because of the definition of $\hat{T}$, the commutator series in Eq. (86) rigorously terminates after the four-fold commutator, irrespective of the point at which $\hat{T}$ may have been truncated. And those portions of the series involving one-electron parts of the Hamiltonian terminate after the double commutator. For one-electron diagonal portions of H, only the single commutator survives. Second, because $\hat{T}$ is normally truncated at low order (typically after $\hat{T}_2$), many of the remaining terms of the now finite commutator series vanish after projection.

The above points and further details of the optimization procedure may be illustrated with an example. Consider a crystal possessing a single half-occupied band, and take Φ as the determinant of occupied Hartree-Fock crystal orbitals (with respective orbital energies $\varepsilon_\alpha, \varepsilon_\beta, \ldots$). We take the unoccupied Hartree-Fock states as the unoccupied single-particle states, with respective orbital energies $\varepsilon_r, \varepsilon_s, \ldots$. The Hamiltonian for the electronic energy of this system is of the form

$$H + H_o + V - V_{HF} \quad , \tag{87}$$

where H_o is the Hartree-Fock Hamiltonian, V is the electron-repulsion operator, and V_{HF} is the single-particle operator describing the interaction of an electron with all occupied Hartree-Fock distributions (included in H_o but, as indicated in Eq. (87), replaced in H by the exact operator V). Using creation-annihilation operator notation, we may write

$$H_o = \sum_i \varepsilon_i \, a_i^\dagger a_i \quad , \tag{88}$$

$$V = \frac{1}{4} \sum_{ijlm} \tilde{v}_{ij}^{lm} \, a_l^\dagger a_m^\dagger a_j a_i \quad , \tag{89}$$

$$V_{HF} = \sum_{ij\alpha} \tilde{v}_{\alpha i}^{\alpha j} \, a_j^\dagger a_i \quad , \tag{90}$$

where the indices i,j,l,m range over all single-particle states whether or not occupied in Φ. The quantities $\tilde{v}_{ij}^{lm}$ are antisymmetrized electron-repulsion matrix elements for normalized orbitals:

$$\tilde{v}_{ij}^{lm} = \langle lm|r_{12}^{-1}|ij\rangle - \langle lm|r_{12}^{-1}|ji\rangle \quad . \tag{91}$$

Note that Eq. (91) is such that

$$\tilde{v}_{ij}^{lm} = -\tilde{v}_{ji}^{lm} = -\tilde{v}_{ij}^{ml} = \tilde{v}_{ji}^{ml} \quad . \tag{92}$$

We consider here an approximation in which $\hat{T}$ is taken as $\hat{T}_2$ only. $\hat{T}_1$ has been neglected because almost all single-excitation determinants have translational symmetries preventing their mixing with the reference state Φ. We also require that the coefficients $t_{\alpha\beta}^{rs}$ be antisymmetrized in their index permutations:

$$t_{\alpha\beta}^{rs} = -t_{\alpha\beta}^{sr} = -t_{\beta\alpha}^{rs} = t_{\beta\alpha}^{sr} \quad . \tag{93}$$

Equation (93) does not constitute an actual restriction of $\hat{T}_2$, as the four operators $\hat{t}_{\alpha\beta}^{rs}$, $\hat{t}_{\alpha\beta}^{sr}$, $\hat{t}_{\beta\alpha}^{rs}$, and $\hat{t}_{\beta\alpha}^{sr}$ all convert Φ into the same determinant. The requirements embodied in Eq. (93) cause these four operators to become identical.

We are now ready to obtain an approximate solution to Eq. (86) for the problem at hand. Projecting first against Φ, we have

$$<\Phi|H+(H,\hat{T}_2)|\Phi> = E<\Phi|\Phi> \quad . \tag{94}$$

In Eq. (94), all the multiple commutators have been dropped because each occurrence of $\hat{T}_2$ must excite two particles from Φ, and H, being a two-electron operator, can at most de-excite two particles. This reasoning, together with the further observation that only the two-electron portion of H can perform the de-excitation, and only if it occurs to the left of $\hat{T}_2$, leads to

$$<\Phi|H+V\hat{T}_2|\Phi> = E<\Phi|\Phi> \quad . \tag{95}$$

Noting that

$$<\Phi|H|\Phi> = E_{HF}<\Phi|\Phi> \quad , \tag{96}$$

where E_{HF} is the Hartree-Fock energy, we can rearrange Eq. (95) to obtain

$$E = E_{HF} + \frac{<\Phi|V\hat{T}_2|\Phi>}{<\Phi|\Phi>} \quad . \tag{97}$$

We next observe that the only nonvanishing terms arising from $V\hat{T}_2$ will be those in which V and $\hat{T}_2$ involve the same indices, so that V de-excites the particles excited by $\hat{T}_2$. Examining all possible index permutations, utilizing the fact that fermion creation and annihilation operators satisfy the following anticommutation relations:

$$a_i a_j + a_j a_i = 0 \quad , \tag{98}$$

$$a_i^\dagger a_j^\dagger + a_j^\dagger a_i^\dagger = 0 \quad , \tag{99}$$

$$a_i a_j^\dagger + a_j^\dagger a_i = \delta_{ij} \quad , \tag{100}$$

and invoking the symmetry properties of the coefficients $\tilde{v}_{ij}^{lm}$ and $t_{\alpha\beta}^{rs}$, we obtain the final result

$$E + E_{HF} + \frac{1}{4} \sum_{\alpha\beta rs} \tilde{v}_{rs}^{\alpha\beta} \, t_{\alpha\beta}^{rs} \quad . \tag{101}$$

Equation (101) shows how the energy can be computed if the coefficients $t_{\alpha\beta}^{rs}$ are known.

Projection of Eq. (86) onto the determinants doubly-excited from Φ will provide a set of conditions from which we may determine the $t_{\alpha\beta}^{rs}$ coefficients. Defining

$$\Phi_{\alpha\beta}^{rs} = a_r^\dagger a_s^\dagger a_\beta a_\alpha \Phi \quad , \tag{102}$$

we obtain for each α, β, r, and s

$$\langle \Phi_{\alpha\beta}^{rs} | H + (H, \hat{T}_2) + \frac{1}{2}((H, \hat{T}_2), \hat{T}_2) | \Phi \rangle = 0 \quad . \tag{103}$$

Further multiple commutators do not enter Eq. (103) because they would necessarily generate too high a degree of particle excitation.

It is straightforward, albeit tedious, to insert into Eq. (103) the expressions for H and $\hat{T}_2$, removing all terms which individually vanish or which cancel against other terms. The procedure can be considerably simplified by noting that if a term of H and a term of $\hat{T}_2$ have no indices in common, their commutator vanishes and may be dropped from the evaluation. The surviving terms are therefore "linked" in the sense of having common indices. We shall see later that this definition of linking corresponds to connectedness in a diagrammatic formulation.

The various terms of Eq. (103) can be shown to have the following values:

$$\langle \Phi_{\alpha\beta}^{rs} | H | \Phi \rangle = \tilde{v}_{\alpha\beta}^{rs} \langle \Phi_{\alpha\beta}^{rs} | \Phi_{\alpha\beta}^{rs} \rangle \quad , \tag{104}$$

$$\langle\Phi_{\alpha\beta}^{rs}|(H_o,\hat{T}_2)|\Phi\rangle = (\varepsilon_r+\varepsilon_s-\varepsilon_\alpha-\varepsilon_\beta)t_{\alpha\beta}^{rs}\langle\Phi_{\alpha\beta}^{rs}|\Phi_{\alpha\beta}^{rs}\rangle \quad , \qquad (105)$$

$$\langle\Phi_{\alpha\beta}^{rs}|(V-V_{HF},\ \hat{T}_2)|\Phi\rangle = \left(\frac{1}{2}\sum_{\gamma\delta} \tilde{v}_{\alpha\beta}^{\gamma\delta}t_{\gamma\delta}^{rs} + \frac{1}{2}\sum_{uw} \tilde{v}_{uw}^{rs}t_{\alpha\beta}^{uw}\right.$$

$$\left. + \sum_{u\gamma}(\tilde{v}_{\alpha u}^{r\gamma}t_{\gamma\beta}^{us} + \tilde{v}_{\alpha u}^{s\gamma}t_{\gamma\beta}^{ru} + \tilde{v}_{u\beta}^{\gamma r}t_{\alpha\gamma}^{us} + \tilde{v}_{u\beta}^{\gamma s}t_{\alpha\gamma}^{ru})\right)\langle\Phi_{\alpha\beta}^{rs}|\Phi_{\alpha\beta}^{rs}\rangle \quad , (106)$$

$$\langle\Phi_{\alpha\beta}^{rs}|\frac{1}{2}((H,\hat{T}_2),\hat{T}_2)|\Phi\rangle = \langle\Phi_{\alpha\beta}^{rs}|\frac{1}{2}V\hat{T}_2\hat{T}_2 - \hat{T}_2V\hat{T}_2|\Phi\rangle$$

$$= \sum_{uw\gamma\delta} \tilde{v}_{uw}^{\gamma\delta}\left(\frac{1}{2}t_{\alpha\beta}^{ur}t_{\gamma\delta}^{sw} - \frac{1}{2}t_{\alpha\beta}^{us}t_{\gamma\delta}^{rw} + \frac{1}{2}t_{\alpha\gamma}^{rs}t_{\gamma\beta}^{uw} - \frac{1}{2}t_{\gamma\beta}^{rs}t_{\alpha\delta}^{uw}\right.$$

$$\left. + \frac{1}{4}t_{\alpha\beta}^{uw}t_{\gamma\delta}^{rs} + t_{\alpha\gamma}^{ru}t_{\beta\delta}^{sw} - t_{\alpha\gamma}^{su}t_{\beta\delta}^{rw}\right)\langle\Phi_{\alpha\beta}^{rs}|\Phi_{\alpha\beta}^{rs}\rangle \quad . \qquad (107)$$

There are no restrictions on the summations in Eqs. (106) and (107) other than as indicated by occupied and unoccupied-orbital indices. More specifically, the summed indices may duplicate those appearing in other factors multiplied therewith, even where a superficial study would suggest that such assignments correspond to exclusion principle violation. The apparently violating terms actually arise from operators whose sequential ordering is not the same as that of the coefficients in these equations. The relatively simple form of Eq. (106) is due to the choice of Hartree-Fock single-particle states.

With the aid of Eqs. (104)-(107), Eq. (103) can be manipulated into the form

$$-\varepsilon_{\alpha\beta}^{rs}t_{\alpha\beta}^{rs} = \tilde{v}_{\alpha\beta}^{rs} + \frac{1}{2}\sum_{\gamma\delta}\tilde{v}_{\alpha\beta}^{\gamma\delta}t_{\gamma\delta}^{rs} + \frac{1}{2}\sum_{uw}\tilde{v}_{uw}^{rs}t_{\alpha\beta}^{uw}$$

$$+ \sum_{u\gamma}\left(\tilde{v}_{\alpha u}^{r\gamma}t_{\gamma\beta}^{us} + \tilde{v}_{\alpha u}^{s\gamma}t_{\gamma\beta}^{ru} + \tilde{v}_{u\beta}^{\gamma r}t_{\alpha\gamma}^{us} + \tilde{v}_{u\beta}^{\gamma s}t_{\alpha\gamma}^{ru}\right)$$

$$+ \sum_{uw\gamma\delta}\tilde{v}_{uw}^{\gamma\delta}\left(\frac{1}{2}t_{\alpha\beta}^{ur}t_{\gamma\delta}^{sw} - \frac{1}{2}t_{\alpha\beta}^{us}t_{\gamma\delta}^{rw} + \frac{1}{2}t_{\alpha\gamma}^{rs}t_{\delta\beta}^{uw} - \frac{1}{2}t_{\gamma\beta}^{rs}t_{\alpha\delta}^{uw}\right.$$

$$+ \frac{1}{4} t^{uw}_{\alpha\beta} t^{rs}_{\gamma\delta} + t^{ru}_{\alpha\gamma} t^{sw}_{\beta\delta} - t^{su}_{\alpha\gamma} t^{rw}_{\beta\delta} \Bigg) \quad , \tag{108}$$

where

$$\varepsilon^{rs}_{\alpha\beta} = \varepsilon_r + \varepsilon_s - \varepsilon_\alpha - \varepsilon_\beta \quad . \tag{109}$$

The set of Eqs. (108) for all values of α, β, r, and s is now to be solved for the coefficients $t^{rs}_{\alpha\beta}$. As previously indicated, the equations are nonlinear, the nonlinearity describing the effect of inter-cluster interactions upon the coefficient determination.

To continue with the illustrative solid-state problem, we next convert Eq. (108) from discrete indices and summations to the corresponding continuous indices and integrations. In so doing it is natural to incorporate the effect of translational symmetry. Restricting consideration to the band of orbitals containing the occupied single-particle states, we replace $t^{rs}_{\alpha\beta}$ by $t(\vec{k},\vec{k}',\vec{q})$, where $\vec{k}$ and $\vec{k}'$ are Bloch wave vectors corresponding respectively to α and β, and $\vec{k}+\vec{q}$ and $\vec{k}'-\vec{q}$ play the roles of r and s. The index interchange symmetry yields the relations

$$t(\vec{k},\vec{k}',\vec{q}) = t(\vec{k}',\vec{k},-\vec{q}) = -t(\vec{k},\vec{k}',\vec{k}'-\vec{k}-\vec{q}) = -t(\vec{k}',\vec{k},\vec{k}-\vec{k}'+\vec{q}) \quad . \tag{110}$$

We adopt the convention that $t(\vec{k},\vec{k}',\vec{q})$ vanishes unless $\vec{k}$ and $\vec{k}'$ lie in the occupied portion of the band, and $\vec{k}+\vec{q}$ and $\vec{k}'-\vec{q}$ lie within its unoccupied portion. We also define

$$\varepsilon(\vec{k},\vec{k}',\vec{q}) = \varepsilon(\vec{k}+\vec{q}) + \varepsilon(\vec{k}'-\vec{q}) - \varepsilon(\vec{k}) - \varepsilon(\vec{k}') \quad , \tag{111}$$

$$\tilde{v}(\vec{k},\vec{k}',\vec{q}) = \frac{\langle \vec{k}+\vec{q},\vec{k}'-\vec{q}|r_{12}^{-1}|\vec{k}\vec{k}'\rangle - \langle \vec{k}+\vec{q},\vec{k}'-\vec{q}|r_{12}^{-1}|\vec{k}'\vec{k}\rangle}{[\langle\vec{k}|\vec{k}\rangle\langle\vec{k}'|\vec{k}'\rangle\langle\vec{k}+\vec{q}|\vec{k}+\vec{q}\rangle\langle\vec{k}'-\vec{q}|\vec{k}'-\vec{q}\rangle]^{1/2}} \quad , \tag{112}$$

the latter definition applying whenever $\vec{k}$, $\vec{k}'$, $\vec{k}+\vec{q}$, and $\vec{k}'-\vec{q}$ lie anywhere within the band. The quantity $\tilde{v}(\vec{k},\vec{k}',\vec{q})$ has an index interchange symmetry analogous to Eq. (110), and also, because it is real and Hermitian, satisfies the further relationship

$$\tilde{v}(\vec{k},\vec{k}',\vec{q}) = \tilde{v}(\vec{k}+\vec{q},\vec{k}'-\vec{q},-\vec{q}) \quad . \tag{113}$$

In replacing the index sums by integrals, we note that translational symmetry restrictions remove one of the index sums, and we make use of the fact that there are $Nv_0/8\pi^3$ states per unit volume in reciprocal space. Using Eqs. (110)-(113), the expression for the total energy, Eq. (101), becomes

$$E + E_{HF} + \frac{1}{4}\left(\frac{Nv_o}{8\pi}\right)^3 \int d\vec{k} \int d\vec{k}' \int d\vec{q}\ \tilde{v}(\vec{k},\vec{k}',\vec{q})\,t(\vec{k},\vec{k}',\vec{q}) \quad . \quad (114)$$

Equation (114) describes an energy which scales correctly with the number of particles, as both $\tilde{v}(\vec{k},\vec{k}',\vec{q})$ and $t(\vec{k},\vec{k}',\vec{q})$ are proportional to N^{-1}. The integrals in Eq. (114), and in equations to follow, are over the entire range such that the integrand is defined (i.e., such that the various Bloch wave vectors lie within the band or appropriate portions thereof). The equation for the t coefficients, Eq. (108), becomes

$$\begin{aligned}
-\varepsilon(\vec{k},\vec{k}',\vec{q})t(\vec{k},\vec{k}',\vec{q}) &= \tilde{v}(\vec{k},\vec{k}',\vec{q}) + \frac{Nv_o}{8\pi^3}\int d\vec{p}\left(\frac{1}{2}\tilde{v}(\vec{k},\vec{k}',\vec{p})t(\vec{k}+\vec{p},\vec{k}'-\vec{p},\vec{q}-\vec{p})\right.\\
&+ \frac{1}{2}\tilde{v}(\vec{k}+\vec{p},\vec{k}'-\vec{p},\vec{q}-\vec{p})t(\vec{k},\vec{k}',\vec{p}) + \tilde{v}(\vec{k},\vec{p}+\vec{q},\vec{q})t(\vec{p},\vec{k}',\vec{q})\\
&+ \tilde{v}(\vec{p},\vec{k}',\vec{q})t(\vec{k},\vec{p}+\vec{q},\vec{q}) + \tilde{v}(\vec{k},\vec{k}'+\vec{p}-\vec{q},\vec{p})t(\vec{k}+\vec{p},\vec{k}',\vec{k}'-\vec{k}-\vec{q})\\
&\left.+ \tilde{v}(\vec{k}+\vec{p},\vec{k}',\vec{k}'-\vec{k}-\vec{q})t(\vec{k},\vec{k}'+\vec{p}-\vec{q},\vec{p})\right)\\
&+ \left(\frac{Nv_o}{8\pi^3}\right)^2\int d\vec{p}\int d\vec{p}'\left[t(\vec{k},\vec{k}',\vec{q})\left(\frac{1}{2}\tilde{v}(\vec{p},\vec{p}',\vec{k}'-\vec{p}-\vec{q})t(\vec{p},\vec{p}',\vec{k}'-\vec{p}-\vec{q})\right.\right.\\
&- \frac{1}{2}\tilde{v}(\vec{p},\vec{p}',\vec{k}-\vec{p}+\vec{q})t(\vec{p},\vec{p}',\vec{k}-\vec{p}+\vec{q}) + \frac{1}{2}\tilde{v}(\vec{p},\vec{k}',\vec{p}')t(\vec{p},\vec{k}',\vec{p}')\\
&\left.- \frac{1}{2}\tilde{v}(\vec{k},\vec{p},\vec{p}')t(\vec{k},\vec{p},\vec{p}')\right) + \frac{1}{4}\tilde{v}(\vec{k}+\vec{p},\vec{k}'-\vec{p},\vec{p}'-\vec{p})t(\vec{k},\vec{k}'\vec{p})t(\vec{k}+\vec{p}',\vec{k}'-\vec{p}',\vec{q}-\vec{p})\\
&+ \tilde{v}(\vec{p},\vec{p}',\vec{q})t(\vec{k},\vec{p}',\vec{q})t(\vec{p},\vec{k}',\vec{q})\\
&\left.- \tilde{v}(\vec{p},\vec{p}',\vec{k}'-\vec{k}-\vec{q})t(\vec{k},\vec{p}',\vec{k}'-\vec{k}-\vec{q})t(\vec{p},\vec{k}',\vec{k}'-\vec{k}-\vec{q})\right] \quad . \quad (115)
\end{aligned}$$

Equation (115) may be solved by iterative techniques appropriate to systems of nonlinear equations. Zivkovic (1976) has shown that under conditions applying to most problems these equations will have real solutions, and studies presently underway in the author's laboratory (Freeman, 1976) confirm the utility of straightforward iterative approaches.

Relation to Perturbation Theory

Although Eq. (115) is tractable, it is sufficiently complicated that the identification and neglect of its less significant terms may be well worth consideration. Moreover, additional insight can come from an improved understanding of the contents of that equation and of the energy expression, Eq. (114). It turns out to be possible to relate the coupled-cluster equations, at any level of approximation, with terms arising in many-body perturbation theory. As we shall see, even relatively preliminary coupled-cluster approximations correspond to extensive partial summations in MBPT.

The relationships we seek are conveniently discussed in a diagrammatic formulation. We introduce Goldstone-type diagram elements (cf. Mattuck, 1967) for $t^{rs}_{\alpha\beta}$ and for the two terms of $\tilde{v}^{kl}_{ij}$ as shown in Figure 4. Note that the element for $t^{rs}_{\alpha\beta}$ is actually defined as standing for $-\varepsilon^{rs}_{\alpha\beta}t^{rs}_{\alpha\beta}$. Complete diagrams constructed from these elements are to be interpreted conventionally: energy denominators are to be associated with the vertical regions between horizontal interaction lines (of either t or $\tilde{v}$), intermediate-state indices aré to be summed ignoring the exclusion principle, the usual sign rules apply, and a factor 1/2 is to be associated with closed diagrams possessing complete left-right symmetry. One additional rule is needed because of the antisymmetrization condition on $t^{rs}_{\alpha\beta}$, namely that *a factor of 1/2 must be associated with each t element from which two particle (or two hole) lines are connected to the same $\tilde{v}$ element.* Such pairs of lines are termed equivalent. (If a t element has both a pair of equivalent particle lines and a pair of equivalent hole lines, only one factor of 1/2 is applied; this is so because interchanging the particle-line connections simultaneously with the hole-line connections leaves a diagram unchanged.)

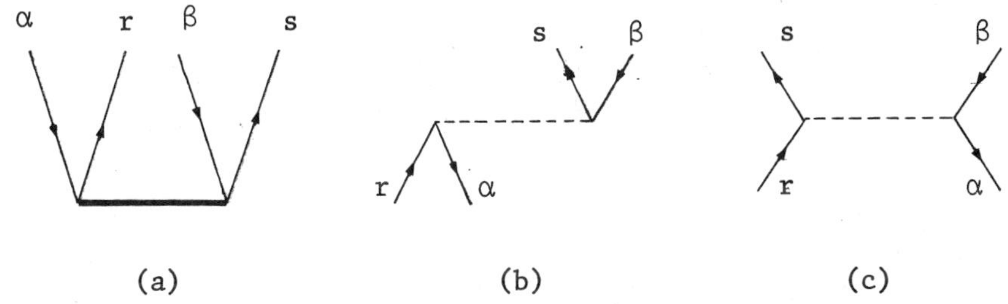

Figure 4. Diagram elements: (a) $-\varepsilon^{rs}_{\alpha\beta}t^{rs}_{\alpha\beta}$, (b) $\langle \alpha s|r_{12}^{-1}|r\beta\rangle$, (c) $\langle s\alpha|r_{12}^{-1}|r\beta\rangle$. Elements (b) and (c) together stand for $\tilde{v}^{\alpha s}_{r\beta}$.

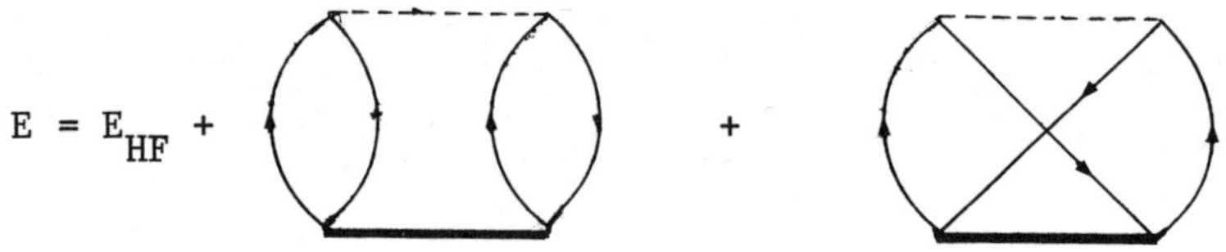

Figure 5. Diagrammatic equation for the correlation energy.

The correlation energy $E-E_{HF}$ is then given by the diagrammatic equivalent of Eq. (114), namely by the set of two diagrams shown in Figure 5. We note that each of the diagrams in Figure 5 has a factor 1/2 due to symmetry and a factor 1/2 due to equivalent lines. The resultant factor, 1/4, is seen to be consistent with Eq. (114).

The diagrammatic equivalent of Eq. (115), giving the t element in terms of more complex diagrammatic structures, is presented in Figure 7. The right hand side of this diagrammatic equation consists of all topologically distinct diagrams which contain exactly one $\tilde{v}$ element and which have the same open lines as does the t element, with the following exceptions: (i) only one diagram is to be retained from a set of diagrams which can be converted into one another by interchanging the connections of inequivalent lines to a t element (a consequence of t antisymmetrization), and (ii) no diagrams may include the "bubble" or "open-oyster" structures shown in Figure 6 (a consequence of the use of Hartree-Fock reference states). To verify the equation in Figure 7 it is necessary to combine pairs of diagrams which differ only in the relative vertical placement of two t elements. Each such pair of diagrams can then be identified in Eq. (115).

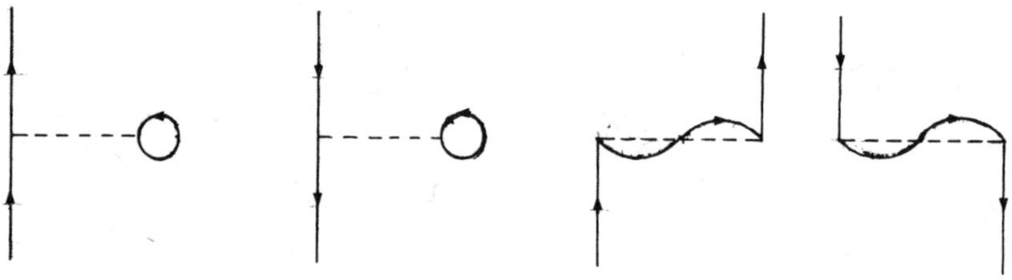

Figure 6. "Bubble" and "open-oyster" structures which do not occur when Hartree-Fock reference states are used.

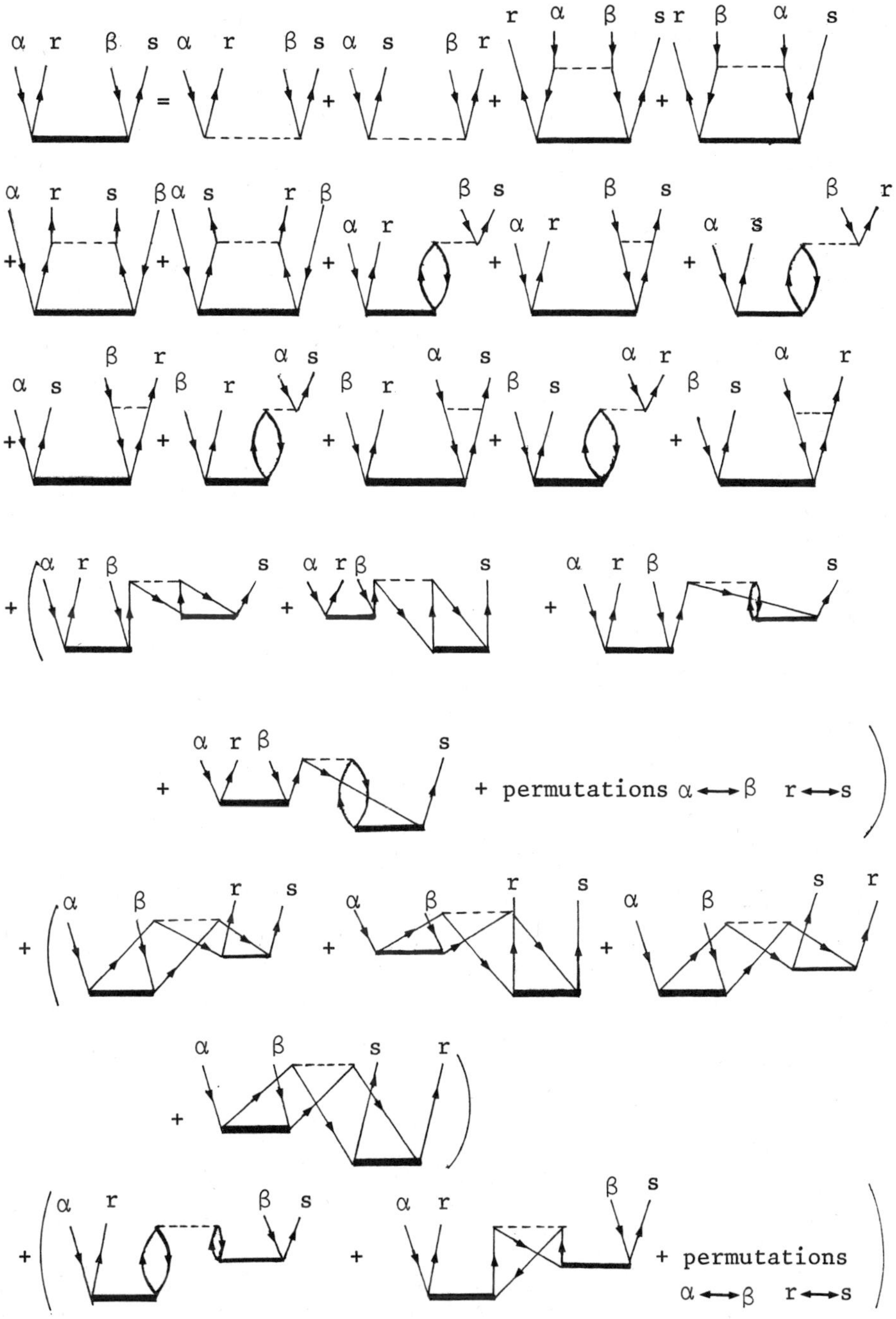

Figure 7. Diagrammatic equation for $t_{\alpha\beta}^{rs}$.

We may obtain insight into the content of the coupled-cluster expansion by substituting the diagrammatic expansion for the t element (Figure 7) where this element occurs in the diagrammatic energy expression (Figure 5). In making the substitution, each term on the right hand side of the equation in Figure 7 must be connected in a way consistent with the index labellings. For example, the hole line connecting into the t element on the left (corresponding to "α" on the left hand side in Figure 7) must connect with the hole lines marked "α" on the right hand side. After the substitution is complete, all duplicate diagrams must be deleted. The remaining diagrams may then be interpreted according to the rules already set forth. The substitution process may be continued at the occurrences of t resultant after the first substitution, and iteratively thereafter *ad infinitum*.

Individual diagrams in the equation for $t_{\alpha\beta}^{rs}$ will, upon iteration, generate series of energy diagrams. For example, iteration of the diagram in Figure 8(a) will yield the particle ladders illustrated in Figure 8(b); iteration of the diagram in Figure 8(c) will yield the rings shown in Figure 8(d). These observations mean that solution for $t_{\alpha\beta}^{rs}$ in a given approximation will be equivalent to the MBPT partial summations thereby indicated. If no terms are dropped from the equation for $t_{\alpha\beta}^{rs}$, it is clear that the solution will correspond to far more than has ever been included in an actual MBPT calculation. It is also clear that more drastic approximations will still compare favorably with contemporary MBPT efforts. For example, a suitable choice of terms of $t_{\alpha\beta}^{rs}$ can be equivalent to a MBPT partial summation containing both ring and ladder contributions in a consistent and justifiable fashion.

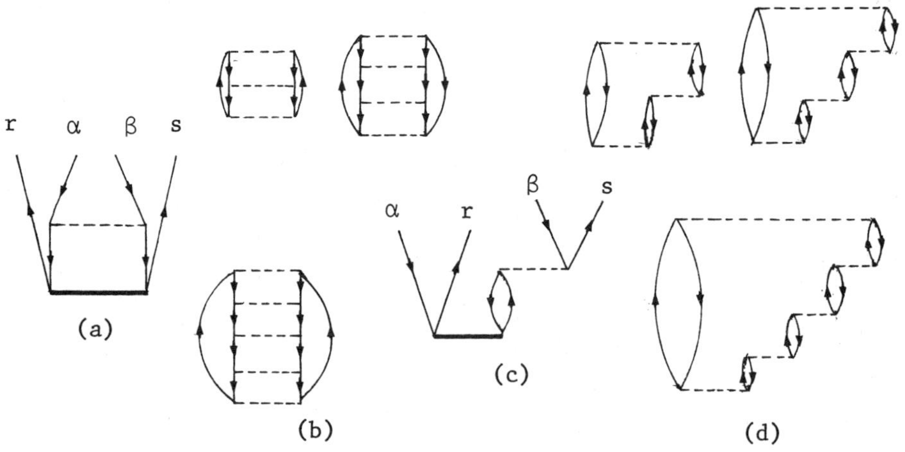

Figure 8. Diagrams of $t_{\alpha\beta}^{rs}$ and the classes of energy diagrams they generate: (a) and (b), particle ladders; (c) and (d), rings.

ACKNOWLEDGEMENTS

The work described here has received generous past support from the U.S. Air Force Office of Scientific Research, and is presently supported in part by the U.S. National Science Foundation, Grant No. CHE75-01284 A01. The author gratefully acknowledges his colleague Professor Hendrik Monkhorst, who has helped shape our solid-state theory efforts since their inception, and the valuable contributions reported herein of Dr. David Freeman, Dr. Lalit Kumar, Dr. Jens Oddershede, Dr. David Ramaker, and Dr. Tomislav Zivkovic.

REFERENCES

Abramowitz, M. and Stegun, I.A. (eds.) 1964. *Handbook of Mathematical Functions* (U.S. National Bureau of Standards, Washington), p. 888.

Bertaut, E.F. 1952. J. Phys. Radium 13, 499.

Bonham, R.A., Peacher, J.L. and Cox, H.L., Jr. 1964. J. Chem. Phys. 40, 3083.

Brener, N.E. 1975a. Phys. Rev. B 11, 929.

Brener, N.E. 1975b. Phys. Rev. B 11, 1600.

Brueckner, K.A. 1969. Adv. Chem. Phys. 14, 215.

Calais, J.L. and Sperber, G. 1972. J. Phys. (Paris) 33, 205.

Calais, J.L. and Sperber, G. 1973. Int. J. Quantum Chem. 7, 501; 7, 521; 7, 537.

Cizek, J. 1966. J. Chem. Phys. 45, 4256.

Cizek, J. and Paldus, J. 1971. Int. J. Quantum Chem. 5, 359.

Emersleben, O. 1923. Physik. Z. 24, 97.

Emersleben, O. 1950. Z. Angew. Math. Mech. 30, 252.

Euwema, R.N. and Surratt, G.T. 1975. J. Phys. Chem. Solids 36, 67.

Euwema, R.N. Wilhite, D.L. and Surratt, G.T. 1973. Phys. Rev. B 7, 818.

Euwema, R.N., Webfer, G.G., Surratt, G.T. and Wilhite, D.L. 1974. Phys. Rev. B 9, 5249.

Evjen, H.M. 1932. Phys. Rev. 39, 675.

Ewald, P.P. 1921. Ann. Physik 64, 253.

Freeman, D.L. 1976. Private communication.

Gaspar, R. 1954. Acta Phys. Hung. 3, 263.

Harris, F.E. 1975. *Theoretical Chemistry*, H. Eyring and D. Henderson, eds. (Academic Press, New York), Vol. 1, pp. 147-218.

Harris, F.E., Kumar, L. and Monkhorst, H. 1972. J. Phys. (Paris) 33, 99.

Harris, F.E., Kumar, L. and Monkhorst, H. 1973. Phys. Rev. B 7 2850.

Harris, F.E. and Monkhorst, H. 1969. Phys. Rev. Lett. 23, 1026.

Harris, F.E. and Monkhorst, H. 1971. *Computational Methods in Band Theory*, P.M. Marcus, J.F. Janak and A.R. Williams, eds. (Plenum Press, New York), pp. 517-541.

Harris, F.E. and Monkhorst, H. 1972. Int. J. Quantum Chem. 6, 601.

Johnson, K.H. and Smith, F.C., Jr. 1971. Int. J. Quantum Chem. 5S, 429.

Kelly, H.P. 1969. Adv. Chem. Phys. 14, 129.

Kikuchi, R. 1954. J. Chem. Phys. 22, 148.

Kohn, W. and Sham, L.J. 1965. Phys. Rev. 140, A1133.

Kumar, L. and Monkhorst, H. 1974. J. Phys. F 4, 1135.

Kümmel, H. 1962. *Lectures on the Many-Body Problem*, E.R. Caianiello, ed. (Academic Press, New York), p. 265.

Mattuck, R.D. 1967. *A Guide to Feynman Diagrams in the Many-Body Problem* (McGraw-Hill, New York).

Monkhorst, H. 1972. Chem. Phys. Lett. 17, 461.

Monkhorst, H. and Oddershede, J. 1973. Phys. Rev. Lett. 30, 797.

Nesbet, R.K. 1967. Phys. Rev. 155, 51.

Ramaker, D.E., Kumar, L. and Harris, F.E. 1975. Phys. Rev. Lett. 34, 812.

Shavitt, I. and Karplus, M. 1965. J. Chem. Phys. 43, 398.

Sinanoglu, O. 1962. J. Chem. Phys. 36, 706.

Slater, J.C. 1951. Phys. Rev. 81, 385.

Slater, J.C. 1970. Int. J. Quantum Chem. 4S, 3.

Slater, J.C. 1971. Int. J. Quantum Chem. 5S, 403.

Tosi, M.P. 1964. *Solid State Physics*, F. Seitz and D. Turnbull, eds. (Academic Press, New York), pp. 1-120.

Weiss, G.H. and Maradudin, A.A. 1962. J. Math. Phys. 3, 771.

Zivkovic, T. 1976. Private communication.

METHODS OF CALCULATION OF ENERGY BANDS IN SOLIDS*

Joseph Callaway

Dept. of Physics & Astronomy, Louisiana State University

Baton Rouge, Louisiana 70803 USA

ABSTRACT

These lectures present a review of the basic principles and methods of the calculation of energy bands. The procedures in current use are surveyed briefly. Recent developments are noted. The problems of the construction of a crystal potential are examined in regard to the representation of the exchange interaction.

* Supported in part by the U.S. National Science Foundation.

CONTENTS

I. THE FOUNDATIONS OF BAND THEORY

There are two basic and rather obvious questions associated with the calculation of energy levels of electrons in solids: (1) What equation should be used? (2) How should it be solved? The first question is the more difficult, since it involves the problem of the relation between a many body theory which gives exact results in principle but is often rather impractical; and the single particle approach which is perhaps a rather crude approximation, but can be applied to a wide variety of systems. For this reason, we will discuss the simpler second question first:

Assume that we have to solve an effective one-electron Schrodinger equation of the form

$$[-\nabla^2+V(\vec{r})]\psi_n(\vec{k},\vec{r}) = E_n(\vec{k})\psi_n(\vec{k},\vec{r}) \ . \qquad (1)$$

Atomic units, with energies in Rydbergs will be used throughout. We will consider only infinite, perfect (no impurities or defects) crystals, as time limitations imposed on these lectures do not permit discussion of the interesting problems associated with surfaces, imperfections, alloys, or disordered materials.

The potential function V(r) in an infinite, perfect crystal is periodic. This means that there exist a set of vectors $\vec{R}_\mu$ (direct lattice vectors) such that

$$V(\vec{r}+\vec{R}_\mu) = V(\vec{r}) \ . \qquad (2)$$

Translations by the $\vec{R}_\mu$ form a group (the translation group). The $\vec{R}_\mu$ are combinations of three independent (but not necessarily orthogonal) primitive lattice vectors $\vec{a}_\nu$ with <u>integer</u> coefficients

$$\vec{R}_\mu = \sum_{\nu=1}^{3} n_{\mu\nu}\vec{a}_\nu \ . \qquad (3)$$

We will not discuss the possible crystal structures. A good source is Slater's book.[1]

The reciprocal lattice is constructed from vectors $\vec{b}_t$ which are related to the a_ν by

$$\vec{b}_t\cdot\vec{a}_\nu = 2\pi\delta_{t\mu} \ . \qquad (4)$$

Explicitly, we have

$$\vec{b}_1 = 2\pi\frac{\vec{a}_2 x\vec{a}_3}{\vec{a}_1\cdot(\vec{a}_2 x\vec{a}_3)} \ . \qquad (5)$$

The other $\vec{b}$ are obtained by cyclic permutation. A general reciprocal lattice vector is

$$\vec{K}_s = \sum_{t=1}^{3} m_{st} \vec{b}_t \tag{6}$$

in which the m_{st} are integers. Evidently we have

$$\vec{K}_s \cdot \vec{R}_\mu = 2\pi \times (\text{integer}) \tag{7}$$

for all $\vec{K}_s$ and $\vec{R}_\mu$.

We will now consider Bloch's theorem: The essential point is that since the crystal potential is periodic, and the potential is constructed from the charge density; the electron charge density must be periodic. This means that, if $\psi(\vec{r})$ is the wave function for an electron, we must have

$$|\psi(\vec{r}+\vec{R}_\mu)|^2 = |\psi(r)|^2 , \tag{8}$$

and this implies

$$\psi(\vec{r}+\vec{R}_\mu) = e^{i\theta_\mu} \psi(\vec{r}) , \tag{9}$$

where θ_μ is a real number which in general depends on μ. Suppose we consider as well a translation through $\vec{R}_\nu$, and let $\vec{R}_\mu+\vec{R}_\nu=\vec{R}_\rho$

$$\psi(\vec{r}+\vec{R}_\mu+\vec{R}_\nu) = e^{i\theta_\rho} \psi(\vec{r}) = e^{i\theta_\nu} \psi(\vec{r}+\vec{R}_\mu) = e^{i\theta_\mu+\theta_\nu} \psi(\vec{r}) .$$

The phases are additive

$$\theta_\rho = \theta_\nu + \theta_\mu .$$

This result will be satisfied for all μ, ν, etc. if

$$\theta_\mu = \vec{k}\cdot\vec{R}_\mu$$

where $\vec{k}$ is a constant vector. Thus we have

$$\psi(\vec{r}+\vec{R}_\mu) = e^{i\vec{k}\cdot\vec{R}_\mu} \psi(\vec{r}) , \tag{10}$$

Eq. (10) expresses Bloch's theorem.[2]

The vector $\vec{k}$ in effect describes the behavior of the wave function ψ under lattice translations. We include $\vec{k}$ in the specification of the wave function. The energy of a state will also depend on $\vec{k}$, since (10) may also be interpreted as a boundary condition on the solution of (1). Of course, there may be different solutions ψ

of (1) for a given $\vec{k}$: these will be denoted by an index n. Usually we simply number the states for fixed $\vec{k}$ in order of increasing energy. Thus, we write the energies as $E_n(\vec{k})$, (the energy band function), and the wave functions as $\psi_n(\vec{k},\vec{r})$. We shall limit the range of $\vec{k}$ to the first Brillouin zone: this comes about as follows: If we consider two vectors $\vec{k}$ and $\vec{k}'$ which are related by

$$\vec{k}' = \vec{k} + \vec{K}_s \ , \qquad (11)$$

where $\vec{K}_s$ is some reciprocal lattice vector, it follows from Eq. (7) that the eigenvalues of $\psi_n(\vec{k},\vec{r})$ and $\psi_n(\vec{k}',\vec{r})$ under lattice translations through R_μ are the same for all R_μ

$$e^{i\vec{k}\cdot\vec{R}_\mu} = e^{i\vec{k}'\cdot\vec{R}_\mu} \ .$$

Therefore these functions can be said to describe the same state, and $\vec{k}$ and $\vec{k}'$ are said to be equivalent. It is only necessary to consider vectors in (or on the surface) of a region such that (11) is never satisfied for $\vec{k}$, $\vec{k}'$ inside the region. This region is the first Brillouin zone.

We state without proof that the Bloch functions are orthogonal

$$\int \psi_n^*(\vec{k},\vec{r})\psi_\ell(\vec{q},\vec{r})d^3r = \delta_{n\ell}\ \delta(\vec{k}-\vec{q}) \ , \qquad (12)$$

and complete

$$\sum_n \int \psi_n^*(\vec{k},\vec{r}')\psi_n(\vec{k},\vec{r})d^3k = \delta(\vec{r}-\vec{r}') \ . \qquad (13)$$

In (12), the integral includes all space, while in (13) the integral runs over the first Brillouin zone.

II. METHODS OF SOLUTION OF THE ONE ELECTRON EQUATION

Our attention now focuses on problems of calculating the Bloch function $\psi_n(\vec{k},\vec{r})$ for a known crystal potential $V(\vec{r})$. There are many methods of doing this. There are several reasonably comprehensive surveys.[1,3-9] We shall not consider group theory and symmetry properties here. A thorough exposition has been given by Bradley and Cracknell.[10]

Three independent methods of energy band calculation were introduced in the early 1930's: (1) the cellular method, (2) the plane wave expansion, and (3) the tight binding approximation. Limitations of calculating equipment available at the time meant that none could yield really quantitative results, except occasionally for specific states of extremely high symmetry. For example, one could calculate with the cellular method the energy of the Γ_1 state at the bottom of the lowest valence electron band in alkali metals,[11] and slightly higher states could be obtained by $\vec{k}\cdot\vec{p}$ perturbation theory; but states in higher bands could not be determined reliably. The complexity of the actual boundary condition on the surface of the cellular polyhedron requires a fairly complicated expansion in spherical harmonics. In the case of the plane wave expansion, the problem of boundary conditions is not present, but the number of terms required to give an accurate representation of the wave function in the region where the potential is strong becomes extremely large. The tight binding method runs into the problem of the evaluation of large numbers of multicenter integrals between overlapping atomic wave functions. The obvious elementary approximations involving neglect of three center integrals of two center integrals beyond first neighbors or worse are seldom quantitatively valid.

Later on (late '30's, early '40's) hybrid methods began to be developed: the augmented plane wave method as a combination of the cellular and plane wave methods; the orthogonalized plane wave method combined plane waves for valence electrons with a tight binding treatment of core states. In the 50's the formal developments continued: the Green's function method as a more powerful combination of plane waves with cellular functions ; the pseudopotential grew from the OPW method, but the most important advance was the introduction of electronic computers. This at last enabled computations of a reasonable standard of numerical accuracy and freed the theory from the burden of unjustified approximations of a computational nature. Formal and technical developments have continued at a rapid pace.

All methods of calculating energy bands are expansion methods: the unknown $\psi_n(\vec{k},\vec{r})$ is expanded in some set of functions with useful properties. The problem is to solve the effective Schrodinger equation (1) subject to the boundary conditions imposed by Bloch's theorem. As the system does not have a simple symmetry which permits separation of variables the problem of boundary conditions is potentially quite complicated. The two basic choices are: (1) invent a set of basis functions which satisfy the boundary conditions automatically, but do not satisfy the Schrödinger equation, or (2) combine a set of solutions of the Schrödinger equation for the same energy but different angular momenta with the objective of determining coefficients so as to satisfy the boundary conditions. A third possibility develops later: partition the unit cell so that in the outer region, functions of type (1) can be used; in an inner spherical region functions of type (2) are employed. Then one faces a different type of boundary problem: to join the two sets of functions smoothly across the inner surface.

A. Methods Based on Plane Waves

The simplest set of functions which incorporate the exact boundary conditions are plane waves: Let $\vec{K}_s$ be any reciprocal lattice vector. Then consider the function

$$e^{i(\vec{k}+\vec{K}_s)\cdot\vec{r}} .$$

Let $\vec{r}$ be displaced to $\vec{r}+\vec{R}_\mu$. In view of (7) we have

$$e^{i(\vec{k}+\vec{K}_s)\cdot(\vec{r}+\vec{R}_\mu)} = e^{i\vec{k}\cdot\vec{R}_\mu} e^{i(\vec{k}+\vec{K}_s)\cdot\vec{r}} .$$

Evidently the function satisfies Bloch's theorem. Hence we can introduce the expansion

$$\psi_n(\vec{k},\vec{r}) = (N\Omega)^{-1/2} \sum_s b_n(\vec{k}+\vec{K}_s) e^{i(\vec{k}+\vec{K}_s)\cdot\vec{r}} . \tag{14}$$

The sum runs over all reciprocal lattice vectors. The coefficients $b_n(\vec{k}+\vec{K}_s)$ have to be calculated by diagonalizing the Hamiltonian on the basis chosen.

It is assumed that the system contains N unit cells of volume Ω. The factor $(N\Omega)^{-1/2}$ insures that individual plane waves are normalized. The matrix elements of the Hamiltonian are given by the very simple expression

$$\begin{aligned} H_{st}(\vec{k}) &= \int e^{-i(\vec{k}+\vec{K}_s)\cdot\vec{r}} [-\nabla^2+V(\vec{r})] e^{i(\vec{k}+\vec{K}_t)\cdot\vec{r}} d^3r \\ &= (\vec{k}+\vec{K}_t)^2 \delta_{st} + V(\vec{K}_t-\vec{K}_s) , \end{aligned} \tag{15}$$

in which

$$V(\vec{K}_t-\vec{K}_s) = \frac{1}{\Omega} \int_\Omega e^{i(\vec{K}_t-\vec{K}_s)\cdot\vec{r}} V(\vec{r}) d^3r . \tag{16}$$

The integral in (16) is restricted to a single unit cell. $V(\vec{K}_t-\vec{K}_s)$ will be referred to as a Fourier coefficient of potential.

Unfortunately, the wave functions for most interesting states in crystals are rapidly varying functions of position near any nucleus. This means that plane waves of large wave vectors are required to describe the wave function accurately. In short, the plane wave expansion converges so slowly except for weak pseudopotentials as to be useless for practical first principles computations. The principal value of the plane wave expansion is as a tool to establish general theorems.

Herring[12] observed that the convergence of the plane wave expansion could be improved if the plane waves were made orthogonal to the wave functions of the tightly bound core electrons. This is accomplished by the Schmidt process.

The calculational method based on these functions is the orthogonalized plane wave (OPW) method. It has been reviewed by Woodruff[13] and by Herman.[14] The first step is to construct linear combinations of the localized core wave functions which satisfy Bloch's theorem. Let $u_j(\vec{r}-\vec{R}_\mu)$ denote a wave function for state j centered at lattice site $\vec{R}_\mu$. We construct a new function $\phi_j(\vec{k},\vec{r})$ as follows

$$\phi_j(\vec{k},\vec{r}) = \frac{1}{\sqrt{N}} \sum_\mu e^{i\vec{k}\cdot\vec{R}_\mu} u_j(\vec{r}-\vec{R}_\mu) \ . \tag{17}$$

It is easy to show that ϕ_j satisfies

$$\phi_j(\vec{k},\vec{r}+\vec{R}_\nu) = e^{i\vec{k}\cdot\vec{R}_\nu} \phi_j(\vec{k},\vec{r}) \tag{18}$$

for all direct lattice vectors $\vec{R}_\nu$. The function ϕ_j will be a reasonable approximation to the wave functions of a core electron in the solid if the overlap between core wave functions centered on different atoms is small. In fact, the OPW method becomes quite cumbersome if there is appreciable overlap of core functions, and we will therefore assume this does NOT occur. An OPW can now be defined

$$\chi(\vec{k},\vec{r}) = (N\Omega)^{-1/2} e^{i\vec{k}\cdot\vec{r}} - \sum_j \mu_j(\vec{k})\phi_j(\vec{k},\vec{r}) . \tag{18}$$

The quantity $\mu_j(\vec{k})$ is called an "orthogonality coefficient". It is determined by the requirement that $\chi(\vec{k},\vec{r})$ be orthogonal to all the core functions $\phi_c(\vec{k},\vec{r})$ of the same $\vec{k}$.

$$\int \phi_c^*(\vec{k},\vec{r})\chi(\vec{k},\vec{r})d^3r = 0 \ . \tag{19}$$

χ is orthogonal by reasons of symmetry to core functions of different $\vec{k}$. Equation (19) leads to

$$\mu_c(\vec{k}) = \frac{1}{\Omega^{1/2}} \int e^{i\vec{k}\cdot\vec{r}} u_c^*(\vec{r})d^3r \ . \tag{20}$$

The orthogonalization of the plane wave to core wave functions builds in the rapidly varying nature of core functions near a nucleus. An expansion of the crystal Bloch functions in terms of orthogonalized plane waves should converge much more rapidly than if ordinary plane waves are used. There is a price to pay for this improvement: the expression for the Hamiltonian matrix elements is slightly more complicated than (15), and since orthogonalized plane waves are not orthogonal to each other, there is an overlap matrix to be considered.

$$<\chi_{\vec{k}+\vec{K}_s}|H|\chi_{\vec{k}+\vec{K}_t}> = (\vec{k}+\vec{K}_s)^2\delta_{st} + V(\vec{K}_t-\vec{K}_s)$$

$$- \sum_j \mu_j^*(\vec{k}+\vec{K}_s)\mu_j(\vec{k}+\vec{K}_t)E_j \ , \qquad (21)$$

$$<\chi_{\vec{k}+\vec{K}_s}|\chi_{\vec{k}+\vec{K}_t}> = \delta_{st} - \sum_j \mu_j^*(\vec{k}+\vec{K}_s)\mu_j(\vec{k}+\vec{K}_t) \ , \qquad (22)$$

in which E_j is the energy of the core state j.

The second and third terms in (21) tend to cancel. This occurs for the following reasons: The potential V is attractive, and the Fourier coefficients $V(\vec{K}_t-\vec{K}_s)$ are generally negative. On the other hand, the orthogonality coefficients are slowly varying real functions, so that the product $\mu_j^*(\vec{k}+\vec{K}_s)\mu_j(\vec{k}+\vec{K}_t)$ is positive, unless possibly $\vec{K}_s$ and $\vec{K}_t$ differ widely. The energies of the core states are negative. Thus, the orthogonality term tends to cancel the potential energy term. Numerical calculations have shown for a wide variety of materials that this cancellation can be quite effective. A few orthogonalized plane waves may give a much better estimate of the energy of a state than many plane waves. The OPW method has been successfully employed in the calculation of energy bands in many substances; the best known applications concern Group IV and III-V semiconductors.[15,16]

However, there are two remaining problems which have tended to limit the applicability of the OPW method[17]: (1) The core functions must be accurately known, and they must be eigenfunctions in the same potential used in the band calculation. The potential functions used in band computations are usually somewhat different from those employed in atomic structure calculations because the electron density distribution changes in going from an isolated atom to a solid composed of many atoms. This means that the core functions must be specifically computed in the potential used in the solid state calculation. They cannot be taken from atomic self-consistent fields; and in the case of an iterative calculation, they must be recomputed at each stage. Errors resulting from use of incorrect core functions can be large, and correction formulas are complicated.

(2) A major difficulty arises if the states of interest are orthogonal, by reasons of symmetry, to all the core states. This occurs in important cases: for example, the p states in diamond are orthogonal to the 1s-core; the 3d states in third row transition elements are orthogonal to all the core states. The result is that the states which are orthogonal by reasons of symmetry are

in fact expanded in ordinary plane waves. However, usually such states have sufficiently rapid variation near a nucleus, even though there are no radial nodes so that the convergence of the plane wave expansion is quite slow.

These difficulties have diminished the popularity of the OPW method in its original form; and have spurred the development of so-called mixed basis methods.[18-22] In fact, both difficulties can be cured by essentially the same procedure. We can construct linear combinations of arbitrary localized functions which have a definite wave vector $\vec{k}$ according to (17). It is useful to consider two sorts of localized functions: (1) approximations (not necessarily exact) to the core functions, and/or (2) approximations to the radial wave function of states orthogonal by reasons of symmetry to the core states (3d). It may be convenient to choose these functions so that the overlap of such functions on different atoms is negligible. This can be done in the case of 3d functions for example simply by applying some smooth cut off so that the function goes to zero on the boundary of an atomic cell. The basis set for the expansion of the Bloch state then consists of the combinations $\phi_j(\vec{k},\vec{r})$ (j here denotes any of the functions added to the basis) and plane waves of wave vector $\vec{k}+\vec{K}_s$. It is now unnecessary to orthogonalize the plane waves to the ϕ_j. To be specific, we can write

$$\psi_n(\vec{k},\vec{r}) = \frac{1}{\sqrt{N\Omega}} \sum_s a_n(\vec{k}+\vec{K}_s)e^{i(\vec{k}+\vec{K}_s)\cdot\vec{r}} + \sum_\ell c_{n\ell}(\vec{k})\phi_\ell(\vec{k},\vec{r}) \ . \quad (23)$$

The Hamiltonian and overlap matrices are partitioned as follows

$$\begin{pmatrix} \ell-\ell' & \ell-K \\ K-\ell & K-K' \end{pmatrix} . \quad (24)$$

(where we use ℓ to denote the Bloch sums of localized functions and K to denote plane waves. Formal expressions are readily obtained for the elements of the Hamiltonian and overlap matrices. It is explicitly assumed here that the u's are strictly confined to a single unit cell.

$$\langle \ell\vec{k}|H|\ell'\vec{k}\rangle = \int u_\ell^*(\vec{r})H\, u_{\ell'}(\vec{r})d^3r \ , \quad (25)$$

$$\langle \ell\vec{k}|H|\vec{k}+\vec{K}_s\rangle = (\vec{k}+\vec{K}_s)^2\mu_\ell(\vec{k}+\vec{K}_s) + \frac{1}{\Omega^{\frac{1}{2}}}\int u_\ell^*(\vec{r})V(\vec{r})e^{i(\vec{k}+\vec{K}_s)\cdot\vec{r}}\, d^3r \ , \quad (26)$$

$$\langle\vec{k}+\vec{K}_s|V|\vec{k}+\vec{K}_t\rangle = (\vec{k}+\vec{K}_t)^2\delta_{st} + V(\vec{K}_t-\vec{K}_s) \quad , \tag{27}$$

$$\langle\ell\vec{k}|\vec{k}+\vec{K}_s\rangle = \mu_\ell(\vec{k}+\vec{K}_s) \quad , \tag{28}$$

and μ_ℓ is given by (20), with the appropriate selection of functions u and $V(K_t-K_s)$ is given by (16).

The physical idea underlying the mixed basis method is, given some reasonable but not necessarily exact functions describing the core states, or the behavior of the function of interest in the core region, a small number of plane waves can describe both the overlapping of extended wave functions and necessary corrections in the core. The former is more plausible than the latter, and the practical success of mixed basis methods depends critically on obtaining an adequate description of functions in the atomic core. Otherwise the convergence may be slow, and it appears that the importance of this problem has been underestimated in some calculations.

B. Pseudopotentials

Perhaps the most important outgrowth of the OPW method is the pseudopotential. There is now an enormous literature on pseudopotentials (some reviews are listed in the references[23-26]), and here we shall have space and time enough to review only the most basic ideas.[27] We shall not discuss the extensive empirical use of pseudopotentials: this has been reviewed in detail by Cohen and Heine.[25]

It was observed above that the orthogonalization terms in the OPW matrix elements tend to cancel the potential energy. Let us try to formulate a procedure in which this cancellation is explicit[27]: We express the wave function for a band state as

$$\psi_n(\vec{k},\vec{r}) = f_n(\vec{k},\vec{r}) - \sum_j a_{nj}(\vec{k})\phi_j(\vec{k},\vec{r}) \quad , \tag{29}$$

in which f_n is a "smooth" function, extending throughout the crystal and the ϕ_j are still given by (17). The coefficients a_{nj} are determined by the requirement that ψ_n should be orthogonal to the core states.

$$\int \phi_c^*(\vec{k},\vec{r})\psi_n(\vec{k},\vec{r})d^3r = 0 \quad . \tag{30}$$

One finds, with the use of Bloch's theorem

$$a_{nc}(\vec{k}) = N^{\frac{1}{2}} \int u_c^*(\vec{r}) f_n(\vec{k},\vec{r})d^3r \ . \tag{31}$$

Note that, if f_n were a single plane wave $e^{ik\cdot r}/(N\Omega)^{\frac{1}{2}}$, a_{nc} would reduce to the orthogonality coefficient $\mu_c(k)$ given by (20).

We substitute (29) into (1) in order to determine an equation satisfied by f_n. It is explicitly assumed here that the ϕ_j are eigenfunctions of the crystal Hamiltonian (energies E_j). One obtains

$$Hf_n(\vec{k},\vec{r}) - \sum_j a_{nj}(\vec{k})[E_j-E_n(\vec{k})]\phi_j(\vec{k},\vec{r}) = E_n(\vec{k})f_n(\vec{k},\vec{r}) \ . \tag{32}$$

Eq. (32) is actually an integro differential equation for the function f_n (recall that a_n involves f_n). To make this clear, let us express H as

$$H = T + V_c(\vec{r}) \ ,$$

where T is the kinetic energy and V_c is the real crystal potential. We also introduce a non local potential $V_R(\vec{r},\vec{r}')$

$$V_R(\vec{r},\vec{r}') = \sum_\nu \sum_j u_j^*(\vec{r}-\vec{R}_\nu)[E_n(\vec{k})-E_j]u_j(\vec{r}'-\vec{R}_\nu) \ , \tag{33}$$

and form the pseudopotential $V_p(\vec{r},\vec{r}')$ as a combination of V_c and V_R

$$V_p(\vec{r},\vec{r}') = V_c(\vec{r})\delta(\vec{r}-\vec{r}') + V_R(\vec{r},\vec{r}') \ . \tag{34}$$

These formulas enable us to rewrite (32) as

$$Tf_n(\vec{k},\vec{r}) + \int V_p(\vec{r},\vec{r}')f_n(\vec{k},\vec{r}')d^3r' = E_n(\vec{k})f_n(\vec{k},\vec{r}') \ . \tag{35}$$

The "smooth part" of the Bloch function is seen to satisfy an effective Schrodinger equation in which a non local, energy dependent pseudopotential appears. It is possible to solve (35) directly by an iterative procedure, however it is customary instead to approximate the "first principles" pseudopotential by a local, energy independent potential. This approximation will be a reasonable one since the u_j are tightly bound core states, so that if r and r' differ substantially, V_R will vanish. Further, the large negative core energy E_j dominates E_n. Since the core function product is positive for r=r', we see that V_R is also positive (repulsive) and tends to cancel the negative (attractive) V_c.

It is important to note that the expressions for the pseudopotential, Eqs. (33) and (34) are not unique. Austin et al.[28] showed that any expression for V_R in (33) which has the form

$$V_R(\vec{r},\vec{r}') = \sum_j F_j(r')\phi_j(\vec{k},\vec{r}) \ , \tag{36}$$

in which the $F_j(r')$ are entirely arbitrary functions will yield a solution for f_n from the equation

$$[T+V_c(\vec{r})]f_n(\vec{k},\vec{r}) + \int V_R(\vec{r},\vec{r}')f_n(\vec{k},\vec{r}')d^3r' = E_n(\vec{k})f_n(\vec{k},\vec{r}) \ , \tag{37}$$

which has the same energy as the correct Bloch function. This arbitrariness indicates that there should also be many different empirical local pseudopotentials which yield energy bands of good quality.

C. Tight Binding

At this point, we will change directions and consider the tight binding method and its generalizations. The elementary notion on which the method is based is that, if the electron states of interest in a solid do not overlap "very much" from site to site, a properly phased linear combination of atomic wave functions of the form (17) should be a reasonable approximation to the solid state wave function. We have used this idea for the core functions in the OPW method, but now extend the procedure to include cases where there is at least some overlap. Originally it was intended that (17) should be used in a calculation by first order perturbation theory (or degenerate perturbation theory if the atomic states considered have non-zero angular momentum). Unfortunately, the method as originally formulated became bogged down in complicated multicenter integrals which were too numerous and difficult to calculate, but often too important to ignore. Recently Lin and collaborators[29,30] have reformulated the method in a manner which makes accurate calculation possible. We shall discuss their procedure. An alternative successful approach has been developed by Ellis and Painter.[31]

Given functions $\phi_i(\vec{k},\vec{r})$ formed by (17) from some set of localized functions u_i; we expand an arbitrary Bloch function as

$$\psi_n(\vec{k},\vec{r}) = \sum_i c_{ni}(\vec{k})\phi_i(\vec{k},\vec{r}) \ . \tag{38}$$

The coefficients $c_{ni}(\vec{k})$ are determined by the simultaneous diagonalization of the Hamiltonian and overlap matrices which are constructed on the basis of the ϕ_i. The elements of the Hamiltonian are

$$H_{ij}(\vec{k}) = \int \phi_i^*(\vec{k},\vec{r})H\phi_j(\vec{k},\vec{r})d^3r = \sum_\sigma e^{i\vec{k}\cdot\vec{R}_\sigma} E_{ij}(R_\sigma) \ , \tag{39}$$

in which

$$E_{ij}(\vec{R}_\sigma) = \int u_i^*(\vec{r}) H\, u_j(\vec{r}-\vec{R}_\sigma) d^3r \ . \tag{40}$$

Use has been made of the translation invariance of the Hamiltonian: $H(\vec{r}+\vec{R}_\mu) = H(\vec{r})$. A similar expression is obtained for elements of the overlap matrix

$$S_{ij}(\vec{k}) = \int \phi_i^*(\vec{k}) \phi_j(\vec{k}) d^3r = \sum_\sigma e^{i\vec{k}\cdot\vec{R}_\sigma} S_{ij}(\vec{R}_\sigma) \ , \tag{41}$$

with

$$S_{ij}(\vec{R}_\sigma) = \int u_i^*(\vec{r}) u_j(\vec{r}-\vec{R}_\sigma) d^3r \ . \tag{42}$$

The localized functions u_i will normally be products of radial functions times spherical harmonics:

$$u_i(\vec{r}) = R_n(r) Y_{\ell m}(\theta,\phi) \ , \tag{43}$$

(the index i then represents the triple numbers n, ℓ, m). Thus the $\vec{u}_i$ transform under a rotation of coordinates β according to

$$\beta u_i(\vec{r}) = u_i(\beta^{-1}\vec{r}) = \Sigma\, D_{ji}^{(\ell)}(\beta^{-1}) u_j(\vec{r}) \ , \tag{44}$$

in which the $D_{ji}^{(\ell)}$ are representation matrices for the rotation which are appropriate for angular momentum ℓ. Eq. (44) enables us to deduce transformation properties of the E_{ij} and S_{ij}. For example, we have

$$E_{ij}(\beta\vec{R}_\sigma) = \sum_{i',j'} D_{i'i}(\beta^{-1}) D_{j'j}(\beta^{-1}) E_{i'j'}(\vec{R}_\sigma) \ . \tag{45}$$

Relations of the type (45) enable us to reduce the numbers of independent parameters necessary to specify a matrix element. An additional relation can be derived from the Hermitean nature of the Hamiltonian

$$E_{ij}(\vec{R}_\sigma) = E_{ji}^*(-\vec{R}_\sigma) \ . \tag{46}$$

The central problem in the tight binding method is the accurate evaluation of the E_{ij} and S_{ij}. We separate kinetic and potential energy parts

$$E_{ij}(\vec{R}_\sigma) = T_{ij}(\vec{R}_\sigma) + V_{ij}(\vec{R}_\sigma) \ , \tag{47}$$

in which

$$T_{ij}(\vec{R}_\sigma) = \int u_i(\vec{r}-\vec{R}_\sigma)(-\nabla^2)u_j(\vec{r})\ d^3r \quad . \tag{48}$$

The matrix elements of the kinetic energy can be evaluated by differentiation of the overlap matrix elements S_{ij}. The potential matrix elements V_{ij} can be evaluated using a Fourier representation of the crystal potential

$$V(\vec{r}) = \sum_s V(\vec{K}_s)e^{i\vec{K}_s\cdot\vec{r}} \quad , \tag{49}$$

in which the $V(K_s)$ are given by (16). Thus

$$V_{ij}(\vec{R}_\sigma) = \sum_s V(\vec{K}_s) \int u_i^*(\vec{r}-\vec{R}_\sigma)e^{i\vec{K}\cdot\vec{r}}\ u_j(\vec{r})\ d^3r \ . \tag{50}$$

Use of the Fourier representation has the advantage that the troublesome three center integrals of earlier approaches are avoided. On the other hand, a sum over reciprocal lattice vectors has to be performed, and this sum may be slowly convergent. Hence it is desirable that the basic integrals, involving the matrix elements between orbitals of

$$e^{i\vec{K}_s\cdot\vec{r}}$$

should be done rapidly. This requirement strongly motivates use of Gaussian type orbitals for the radial functions in (43)

$$R \sim r^\ell\ e^{-\alpha r^2}$$

since the required integrals can be performed analytically. The summation over reciprocal lattice vectors can be facilitated by Ewald type procedures.[30] However, the use of analytic orbitals, particularly Gaussians, becomes quite cumbersome for elements with large atomic numbers since the number of orbitals required to describe a wave function with many nodes becomes large. This is a major limitation of present LCAO methods.

For the purpose of making self-consistent band calculations, it is necessary to have an interative process in which the wave functions and energies from one stage of the calculation are used to calculate a revised crystal potential for use in the next stage. This process seems to work rather well within the framework of the tight binding method as discussed here, since it is not necessary to construct the potential in "real" space directly. Rather, one can work directly with the Fourier coefficients.[32]

Poisson's equation (in atomic units) relates the potential energy to the electron charge distribution.

$$\nabla^2 V = 8\pi\rho_e \tag{51}$$

In terms of the Fourier coefficients (using (49)), we have

$$V(\vec{K}_s) = \frac{-8\pi}{K_s^2}\rho(\vec{K}_s) , \tag{52}$$

in which $\rho(\vec{K}_s)$ is a Fourier coefficient of the electron charge density

$$\rho(\vec{K}_s) = \frac{1}{N\Omega}\int \rho(\vec{r})e^{i\vec{K}_s\cdot\vec{r}}\,d^3r , \tag{53}$$

(here the integral includes the entire crystal). The charge density is expressed as (using (38))

$$\rho(\vec{r}) = \sum_{nk}^{occ}\sum_{ij} c_{ni}^*(\vec{k})c_{nj}(\vec{k})\phi_i^*(\vec{k},\vec{r})\phi_j(\vec{k},\vec{r}) . \tag{54}$$

Eq. (54) is substituted into (53), and (17) is used

$$\rho(\vec{K}_s) = \frac{1}{N\Omega}\sum_{nk}\sum_{ij} c_{ni}^*(\vec{k})c_{nj}(\vec{k})S_{ij}(\vec{k},\vec{K}_s) . \tag{55}$$

The $S_{ij}(\vec{k},\vec{K}_s)$ are generalized overlap integrals

$$S_{ij}(\vec{k},\vec{K}) = \sum_\sigma e^{i\vec{k}\cdot\vec{R}_\sigma}\int u_i^*(\vec{r})e^{i\vec{K}_s\cdot\vec{r}}u_j(\vec{r}-\vec{R}_\sigma)d^3r . \tag{56}$$

Eq. (52) appears to fail for $K_s=0$, however a value for this case may be obtained by a limiting process.

The quantities $S_{ij}(\vec{k},\vec{K})$ can be computed once at the beginning of a calculation; the computation of the change in the potential is then straightforward. A somewhat similar procedure is available for an exchange potential which is a function of the local charge density. The $S_{ij}(\vec{k},\vec{K}_s)$ are quite numerous. For example, if one considers approximately 50 orbitals 90 $\vec{k}$ points in the Brillouin zone, and 40 stars of reciprocal lattice vectors, there will be of the order of 10^8 integrals. It is evidently quite important to have rapid methods of generating these.

The tight binding method has also found important a quite successful application in semi-empirical interpolation schemes, both alone[33,34] and in combination with plane waves in mixed basis methods.[35-39] We will not discuss this work in detail here.

D. The APW and Green's Function Methods

We now turn to the consideration of a pair of related methods: the augmented plane wave method (APW)[40-45] and the Green's function[46-50] (or KKR) method. These are two approaches to the problem of combining a plane wave expansion (which satisfies the correct boundary conditions on the atomic cell) with exact solutions of the radial Schrödinger equation.

To begin, suppose we had a spherically symmetric potential inside of a unit cell. Then we could separate variables in the wave equation in spherical coordinates. For any angular momentum we would have

$$\frac{1}{r^2}\frac{d}{dr}\left(r^2\frac{dR_\ell}{dr}\right) + [E-V(r) - \frac{\ell(\ell+1)}{r^2}]R_\ell = 0 . \tag{57}$$

If the boundary of the cell were spherically symmetric, we could stop. However, let us allow the cell its actual polyhedral shape while keeping the potential spherical. Then we must consider a combination of solutions for different ℓ; thus

$$\psi_n(\vec{k},\vec{r}) = \sum_{\ell m} i^\ell a_{\ell m}(\vec{k})R_\ell(E,r)Y_{\ell m}(\theta,\phi) . \tag{58}$$

The coefficients $a_{\ell m}$ must be chosen so that the boundary conditions are satisfied. This process determines the allowed $E(\vec{k})$. The factor i^ℓ is included for convenience in subsequent work. Unfortunately, the convergent of (58) is not rapid.

The essential idea of the APW method is the following: Define a sphere of radius r_i which lies entirely within a single unit cell. Usually this is taken to be the largest sphere which can be inscribed within the cell. Inside this sphere, we will write the wave function in a form resembling (58); outside we will use plane waves. Specifically let $\varepsilon(x)$ be a unit step function: $\varepsilon(x)=1$ if $x>0$, $\varepsilon(x)=0$ if $x<0$. Then an augmented plane wave (APW) is defined as

$$\Phi(\vec{k},\vec{r}) = \varepsilon(r-r_i)e^{i\vec{k}\cdot\vec{r}} + \sum_L i^\ell \beta_L(\vec{k})\varepsilon(r_i-r)R_\ell(E,r)Y_L(\theta,\phi) . \tag{59}$$

The symbol L is used for simplicity to represent the pair of numbers (ℓ,m). The coefficients β_L are determined by the requirement that Φ should be continuous across the inscribed sphere.

$$\beta_L = 4\pi \; j_\ell(kr_i) Y_L^*(\hat{k})/R_\ell(E,r_i) \; . \tag{60}$$

The function $\Phi(\vec{k},\vec{r})$ has a discontinuous derivative on the inscribed sphere even when an infinite number of spherical waves are used in (59). We know from scattering theory that to join smoothly plane waves with spherical waves a phase shift is required. However, we can write the Bloch wave function $\psi_n(\vec{k},\vec{r})$ as a linear combination of augmented plane waves.

$$\psi_n(\vec{k},\vec{r}) = \sum_s C_n(\vec{k}+\vec{K}_s)\Phi(\vec{k}+\vec{K}_s,r) \; . \tag{61}$$

In the limit that an exact solution of the effective Schrödinger equation is obtained, the combination (61) has a continuous first derivative on the inscribed sphere.

The calculation of matrix elements of the Hamiltonian between APW's is rather complicated, since allowance must be made for the discontinuous derivative of the individual Φ on the inscribed sphere. It is useful to consider a stationary expression whose variation yields the Schrödinger equation

$$I = \int \psi^* [-\nabla^2 + V(r) - E]\psi d^3r$$
$$- \int \psi_i^* \frac{\partial}{\partial r}(\psi_o - \psi_i) ds \; . \tag{62}$$

In the second term we have a surface integral to be performed over the inscribed sphere in which the wave function in the inner (ψ_i) and outer region (ψ_o) have been distinguished. It has been assumed that $\psi_o(\vec{r}_i) = \psi_i(\vec{r}_i)$; however, the second term includes allowance for the discontinuous derivative on the sphere.

A major simplification is usually introduced. It is assumed that the crystal potential is constant (and the constant is taken to be zero) in the region between the inscribed sphere and the boundary of the atomic cell. This is the "famous" "muffin tin" approximation. Eq. (61) is substituted into (62). The volume integration is simple since the interior function is an exact solution of the Schrödinger equation. The result is

$$\int \Phi^*(\vec{k}+\vec{K}_s,\vec{r})[-\nabla^2+V-E]\Phi(\vec{k}+\vec{K}_t,\vec{r})d^3r$$
$$= [(\vec{k}+\vec{K}_s)^2-E] \int_o e^{i(\vec{K}_t-\vec{K}_s)\cdot\vec{r}} d^3r$$
$$= [(\vec{k}+\vec{K}_s)^2-E][\Omega\delta_{st} - 4\pi r_i^2 \frac{j_1(|\vec{K}_t-\vec{K}_s|r_i)}{|\vec{K}_t-\vec{K}_s|}] \; . \tag{63}$$

In the second line of (63), the subscript o on the integral indicates that only the region between the sphere and the cell is included. As before Ω is the volume of the cell. The matrix elements coming from the surface term in (62) can be written as

$$\int \Phi_o(\vec{k}+\vec{K}_s,\vec{r})\ \{\partial/\partial r\ [\Phi_o(\vec{k}+\vec{K}_t,\vec{r}) - \Phi_i(\vec{k}+\vec{K}_t,\vec{r})]\}_{r_i}\ ds =$$

$$-4\pi r_i^2 \sum_\ell (2\ell+1)P_\ell(\cos\theta_{st})j_\ell(|\vec{k}+\vec{K}_s|r_i)j_\ell(|\vec{k}+\vec{K}_t|r_i)$$

$$\times\left(\frac{1}{j_\ell(|\vec{k}+\vec{K}_t|r)}\ \frac{dj_\ell(|\vec{k}+\vec{K}_t|r)}{dr} - \frac{1}{R_\ell(E,r)}\ \frac{d\,R_\ell(E,r)}{dr}\right)_{r_i}. \tag{64}$$

This term can be regarded as a kind of pseudopotential. The combination of (63) and (64) leads to

$$I = \sum_{st} C_n^*(\vec{k}+\vec{K}_s)I_{st}\ C_n(\vec{k}+\vec{K}_t)\ , \tag{65}$$

in which

$$I_{st} = [(\vec{k}+\vec{K}_t)^2-E]\Omega\delta_{st}-4\pi r_i^2 \Big\{[(\vec{k}+\vec{K}_s)\cdot(\vec{k}+\vec{K}_t)-E]$$

$$\times\frac{j_1(|\vec{K}_s-\vec{K}_t|r_i)}{|\vec{K}_s-\vec{K}_t|} - \sum_\ell (2\ell+1)P_\ell(\cos\theta_{st})j_\ell(|\vec{k}+\vec{K}_s|r_i)$$

$$\times\frac{j_\ell(|\vec{k}+\vec{K}_t|r_i)}{R_\ell(E,r_i)}\left(\frac{dR_\ell}{dr}(E,r)\right)_{r_i}\Big\}\ . \tag{66}$$

An apparent peculiarity of (66) is that matrix elements of the crystal potential do not appear explicitly. The potential enters implicitly through the logarithmic derivative. This can be expressed in terms of the scattering phase shift through a formula which is valid for a muffin tin potential

$$\left(\frac{1}{R_\ell}\ \frac{dR_\ell}{dr}\right)_{r_i} = \frac{j_\ell'(kr_i)-\tan\delta_\ell\ n_\ell'(kr_i)}{j_\ell(kr_i)-\tan\delta_\ell\ n_\ell(kr_i)}\ , \tag{67}$$

in which n_ℓ is a spherical Neumann function and the prime indicates the radial derivative.

The energy levels are obtained from

$$\det[I_{st}] = 0 \ . \tag{68}$$

The elements, I_{st}, are complicated functions of energy (through the logarithmic derivatives). Therefore roots of (68) have to be found individually: it is not possible to diagonalize the effective Hamiltonian to find all energies for a given $\vec{k}$. Instead, the determinant (68) must be evaluated at different energies to locate zeros. The size of the determinant required for reasonable convergence is not unduly large: approximately 50 APW basis functions are required for each atom in the unit cell in order to determine d band levels which are converged within 0.001 Ry. One requires of the order of 10-12 angular momenta ℓ in Eq. (66) to obtain accurate values of I_{st}.[44]

The APW method has become one of the standard methods of energy band calculations. There have been important extensions of the basic framework described here. Relativistic effects can be included in a straightforward way by allowing the wave functions involved to be solutions of the Dirac equation rather than the Schrodinger equation.[43] Relativistic corrections can be significant for the band structures of heavy elements. Recent work has emphasized the development of correction procedures to remove the limitations of the muffin tin potential and to allow for non spherical components of the potential in the interior region.[51-53] The APW method has been quite extensively applied to study band structures of metals for which the atomic number is too large to permit application of other methods, and for which relativistic effects may be important.[54-55] A recent development which may increase the speed of APW calculation is the introduction of energy independent augmented plane waves by Andersen.[56]

The final method that we shall discuss is the Green's function method (often called the KKR -- for the Korringa,[46] Kohn, and Rostoker[47] method, and by Ziman,[7] the "Greenian"). We shall keep to the formulation of the Kohn-Rostoker paper, but note the existence of an elegant alternative procedure involving "muffin-tin orbitals".[59] We begin with the integral form of the Schrödinger equation:

$$\psi_n(\vec{k},\vec{r}) = \int G(\vec{k},\vec{r}-\vec{r}\,')\ V(\vec{r}\,')\psi_n(\vec{k},\vec{r}\,')d^3r' \ , \tag{69}$$

in which the Green's function $G(\vec{k},\vec{r}-\vec{r}\,')$ is a solution of the equation

$$(\nabla^2+E)G(\vec{k},\vec{r}-\vec{r}\,') = \delta(\vec{r}-\vec{r}\,') \quad . \tag{70}$$

The utility of the method arises from the possibility of incorporating the boundary conditions from Bloch's theorem into G from the start (hence the inclusion of the wave vector $\vec{k}$ into the arguments of G). Specifically G is required to satisfy

$$G(\vec{k},\vec{r}+\vec{R}_\mu) = e^{i\vec{k}\cdot\vec{R}_\mu} G(\vec{k},\vec{r}) \quad . \tag{71}$$

The homogeneous integral Eq. (69) possesses solutions only for specific energies: the allowed energies for the given $\vec{k}$. A formal expression for G which satisfies (70) and (71) can easily be written:

$$G(\vec{k},\vec{r}-\vec{r}\,') = \frac{1}{\Omega} \sum_s \frac{e^{i(\vec{k}+\vec{K}_s)\cdot(\vec{r}-\vec{r}\,')}}{E-(\vec{k}+\vec{K}_s)^2} \quad . \tag{72}$$

It is evident immediately that (71) is obeyed; satisfaction of (70) is guaranteed by the identity

$$\frac{1}{\Omega} \sum_s e^{i(\vec{k}+\vec{K}_s)\cdot(\vec{r}-\vec{r}\,')} = \delta(\vec{r}-\vec{r}\,') \quad . \tag{73}$$

Instead of working directly with (69), it is useful to set up a functional $\Lambda(\vec{k})$ from which (69) is derived by a variational method. A suitable functional is

$$\Lambda = \int \psi^*(\vec{r}) \; V(\vec{r})\psi(\vec{r})d^3r - \iint \psi^*(\vec{r})V(\vec{r})G(\vec{k},\vec{r}-\vec{r}\,') \times V(\vec{r}\,')\psi(\vec{r}\,')d^3r \, d^3r' \quad . \tag{74}$$

The condition

$$\delta\Lambda = 0 \tag{75}$$

leads immediately to (69). The variational calculation is carried out by expanding ψ in a set of known functions, which are conventionally (but not necessarily) spherical waves, according to Eq. (58). Eq. (58) is substituted into (74). The result is put in the form

$$\Lambda = \sum a_L^*(\vec{k})\Lambda_{LL'}(\vec{k})a_i(\vec{k}) \tag{76}$$

in which the index L is used to denote the pair (ℓ,m) and $\Lambda_{LL'}$ is a matrix element of the operator $V-VGV$.

The heart of the computations is the determination of the matrix elements of Λ. This appears to be -- and is -- a very complicated matter. Only an outline of the procedure will be given here. For additional details, the papers of Kohn and

Rostoker,[47] and Ham and Segall[48] should be consulted. Considerable simplifications result if the potential in each cell is assumed to be of the muffin tin form, as in the APW method. We will adopt this assumption. Then it becomes possible to express Λ in a form which does not involve the crystal potential.

To see how this is done we observe that since $\psi(\vec{k},\vec{r})$ in (69) satisfies the Schrödinger equation

$$\int G(\vec{k},\vec{r}-\vec{r}')V(\vec{r}')\psi_n(\vec{k},\vec{r}')d^3r' = \int G(\vec{k},\vec{r}-\vec{r}')(\nabla'^2+E)\psi_n(\vec{k},\vec{r}')d^3r' \tag{77}$$

(in which ∇' refers to coordinates $\vec{r}'$, Green's theorem can now be applied to convert the right side of (77) into

$$\int(\nabla'^2+E)G(\vec{k},\vec{r}-\vec{r}')\psi_n(\vec{k},\vec{r}')d^3r' + \int[G(\vec{k},\vec{r}-\vec{r}')\nabla'\psi_n(\vec{k},\vec{r}') - \psi_n(\vec{k},\vec{r}')\nabla G(\vec{r}-\vec{r}')]\cdot d\vec{s}' \quad . \tag{78}$$

We can now apply (70). It is convenient to choose the surface of integration in (78) to be the inscribed sphere (more properly, a sphere whose radius is infinitesimally smaller than the inscribed sphere). Then (69) reduces to

$$\int[G(k,\vec{r}-\vec{r}')\frac{\partial\psi_n(\vec{k},\vec{r}')}{\partial r'} - \psi_n(\vec{k},\vec{r}')\frac{\partial}{\partial r'}G(\vec{k},\vec{r}-\vec{r}')]ds' = 0 \quad . \tag{79}$$

This procedure enables us to reduce the expression for Λ, Eq. (74) to the following form

$$\Lambda = \int\psi^*(\vec{k},\vec{r})V(\vec{r})[\psi(\vec{k},\vec{r}) - \int G(\vec{k},\vec{r}-\vec{r}')V(\vec{r}')\psi(\vec{k},\vec{r}')d^3r']d^3r$$
$$= \int ds\int ds'\left[\frac{\partial\psi^*(\vec{k},\vec{r})}{\partial r} - \psi^*(\vec{k},\vec{r}')\frac{\partial}{\partial r}\right]\left[\psi(\vec{k},\vec{r}')\frac{\partial}{\partial r'}G(\vec{k},\vec{r}-\vec{r}') - G(\vec{k},\vec{r}-\vec{r}')\frac{\partial}{\partial r'}\psi(\vec{k},\vec{r}')\right] \quad . \tag{80}$$

The coordinate r' in (80) is infinitesimally smaller than the radius of the inscribed sphere, while the coordinate r is infinitesimally smaller than r'. Ultimately r, r' are allowed to go on to the sphere.

It is now necessary to consider the Green's function in detail. Use of standard expressions for the expansion of a plane wave leads to

$$G(\vec{k},\vec{r}-\vec{r}') = \frac{(4\pi)^2}{\Omega} \sum_{L,L'} \sum_s (i)^{\ell-\ell'} \frac{j_\ell(Kr) j_\ell'(Kr')}{E-(K_s+k)^2} Y_L(\hat{r}) Y_{L'}(\hat{r}')$$

$$\times\; Y_L^*(\hat{K}) Y_{L'}(\hat{K}) \;, \tag{81}$$

in which K is an abbreviation for $|\vec{k}+\vec{K}_s|$, $\hat{K}$ indicates the angles specifying the orientation of $\vec{k}+\vec{K}_s$ relative to fixed axes. Eq. (81) can be manipulated into a different form. The one which is most useful for the present purposes is

$$G(\vec{k},\vec{r}-\vec{r}') = \sum_{L,L'} [A_{L,L'}(\vec{k},E) j_\ell(\kappa r) j_\ell(\kappa r')$$

$$+ \kappa\delta_{LL'}\, j_\ell(\kappa r) n_\ell(\kappa r')] Y_L(\hat{r}) Y_L^*(\hat{r}') \;, \tag{82}$$

where n_ℓ is a spherical Neumann function,

$$\kappa^2 = E \;,$$

and the $A_{LL'}$ are structure constants to be determined. Eqs. (82) and (58) are inserted into (80). The spherical harmonics are eliminated by the angular integration. The matrix elements, $\Lambda_{LL'}$ are given by

$$\Lambda_{LL'} = -(L_\ell j_\ell - j_\ell')(L_{\ell'} j_{\ell'} - j_{\ell'}') \left[A_{LL'} + \kappa\delta_{LL'} \frac{n_\ell' - n_\ell L_\ell}{j_\ell' - j_\ell L_\ell}\right] \tag{83}$$

in which L_ℓ is the logarithmic derivative of the radial function R_ℓ, and j_ℓ' is the derivative of j_ℓ

$$L_\ell = \left(\frac{1}{R_\ell}\frac{dR_\ell}{dr}\right)_{r_i} \quad ; \quad j_\ell' = \left(\frac{dj_\ell}{dr}\right)_{r_i} \;.$$

The normalization $R_\ell(r_i) = 1$ was used in deriving (83). The condition that Λ should be stationary means that we can differentiate (76) with respect to the coefficients a_L^*. An infinite set of homogeneous linear equations results. The condition for a non-trivial solution is that the determinant of the coefficients should vanish:

$$\det(\Lambda_{LL'}) = 0 \;. \tag{84}$$

This equation determines the relation between the energy and the wave vector: in other words, the energy bands. Eq. (84) can be simplified by dividing out the factors $L_\ell j_\ell - j_\ell'$ etc. and also by introducing the scattering phase shifts from Eq. (67). Thus, Eq. (84) becomes

$$\det(A_{LL'} + \kappa\delta_{LL'} \operatorname{ctn} \delta_\ell) = 0 \quad . \tag{85}$$

Eq. (85) indicates that the K.K.R. determinant is a function of the "structure constants, $A_{LL'}$" and the scattering phase shifts δ_L. The computation of the structure constants is a fairly lengthy and complicated process, involving either direct or reciprocal lattice summations. We shall not discuss the evaluation of the $A_{LL'}$ here. It should be noted that they are functions of $\vec{k}$ and E, dependent on the crystal lattice, but are independent of the potential. Hence, the expenditure of computational effort to determine the $A_{LL'}$ for one lattice structure may be useful in many different band structure calculations. Experience has shown that only relatively small angular momenta (through ℓ=2 or ℓ=3) are required to obtain convergence. A relativistic formulation has also been developed.[58-59]

There is evidently a close relation between the APW and Green's function methods. This has been discussed in some detail by Ziman[60] and Slater.[61] Although the KKR matrix from (85) can be transformed to a plane wave basis, and written in a form resembling (66), the results are not identical. The fundamental reason for the inequivalence of the APW and Green's function methods is that the wave functions in the interstitial region between the inscribed sphere and the cell boundary are not the same. In fact, the plane wave expansion of a continuous wave function in the entire unit cell is unique; however, the set of all plane waves which are truncated so as to vanish inside the muffin tin sphere is overcomplete in the interstitial region.

Relations between the KKR method and the general pseudopotential have been discussed by several authors.[7,61,62,63] The phase shifts which appear in both the KKR and APW secular Eqs. [(85) and (66) with (67)] can be regarded as parameters to be determined empirically so as to obtain a band structure or Fermi surface in agreement with experiment.[64,65]

III. EXCHANGE AND THE CRYSTAL POTENTIAL

Let us now turn to an altogether different topic: how do we determine the potential $V(\vec{r})$ to use in the calculations. We shall not be concerned with the mechanics of constructing muffin tin potentials or of carrying out direct or reciprocal lattice sums, but rather will address the question "What equation should we be trying to solve?".

The concept of energy bands is quite clear within the framework of a one-electron approximation, but becomes fuzzy when many body theory is considered, except very close to the Fermi surface. In the single particle approach, two answers to our question have been given. The most obvious is the Hartree-Fock equations: In this case, the Bloch function satisfies

$$\left[-\nabla^2 - 2\sum_{\mu} \frac{Z_\mu}{|\vec{r}-\vec{R}_\mu|} + 2\sum_{\ell\vec{q}} \int \frac{|\psi_\ell(\vec{q},\vec{r}\,')|^2}{|\vec{r}-\vec{r}\,'|} d^3r'\right]\psi_n(\vec{k},\vec{r})$$

$$- 2\sum_{\ell q} \int \frac{\psi_\ell^*(\vec{q},\vec{r}\,')\psi_n(\vec{k},\vec{r}\,')}{|\vec{r}-\vec{r}\,'|} d^3r' \; \psi_\ell(\vec{q},\vec{r}) = E_n(\vec{k})\psi_n(\vec{k},\vec{r}) \; . \qquad (86)$$

For simplicity in writing, we (here) include spin components in the wave function. The summation over bands ℓ and wave vectors $\vec{q}$ includes only occupied states.

These equations are obviously quite complicated and have to be solved by an iterative procedure which leads to self-consistency. However, there has been considerable success in the application of the Hartree-Fock equation to fairly large molecules, and recently, there have been several Hartree-Fock energy band calculations. A representative sample is listed in the references.[66-70] What success can be expected in a solid? The results of Hartree-Fock calculations of the total energy and geometric properties of molecules, while not quantitatively correct, are seldom really bad in a qualitative sense at actual interatomic separations. It is therefore surprising that the Hartree-Fock approximation can lead to both serious quantitative and qualitative errors in solid state problems.

The most significant of these for the study of energy bands is that the density of states, in the Hartree-Fock approximation, must vanish at the Fermi energy. This is not a difficulty in a semiconductor, but is significant in a metal, as it leads to consqeuences such as a $T/\ell n\ T$ dependence of the electronic specific heat which are in direct conflict with experiment. The origin of this disaster is the long range of the Coulomb interaction which leads to an infinite value for $d\varepsilon_x/dk$ (ε_x is the exchange energy) as k crosses the Fermi surface. The original argument (see, for instance Seitz's book[71]) was developed for the free electron gas, but can be extended in a straightforward fashion to any metal. Of course, numerical band calculations may miss this feature, either because of numerical inaccuracies or because of some implicit effective truncation of the Coulomb interaction. A failure to find the required singularity should not be taken as evidence of a meaningful calculation.

The Fermi energy of a pure semiconductor is in the band gap, so the problem with the density of states does not arise. However, the apparent excessively large band and gap widths which have been reported in some calculations may be a related phenomena.[67] Average ground state properties which do not depend strongly on the detailed band structure may be given more accurately.

Two advantages associated with the Hartree Fock approximation are (1) the Koopmans' theorem interpretation of energy level differences as transition energies, and (2) the availability of many body perturbation theory for the calculations of correction due to the electron interaction. However, corrections to Koopmans' theorem may be important[72] and the reliable calculation of correlation corrections in real metals would seem to be quite difficult at the present time for both theoretical and computational reasons.

On what, then, should band calculations be based? This author advocates the density functional formulation[73,74] at least for metals. It seems to yield band structures which are free of obvious diseases and yields Fermi surfaces in reasonably good agreement with experiment. Instead of (86) one solves the equation

$$[-\nabla^2 - 2\Sigma \frac{Z_\mu}{|\vec{r}-\vec{R}_\mu|} + 2\int \frac{\rho(\vec{r}')}{|\vec{r}-\vec{r}'|} d^3r' - 6\alpha \left(\frac{3\rho_\sigma}{4\pi}\right)^{1/3}] \psi_{n\sigma}(\vec{k},\vec{r})$$

$$= E_n(k)\psi_n(k,r) \ , \qquad (87)$$

in which σ denotes the spin of the electron state considered, ρ_σ is the charge density of electrons of spin σ, and α is a parameter (the α of the "Xα" method). Our experience has shown that $\alpha = 2/3$ (the "Kohn-Sham[74]-Gaspar[75] value) or close to 2/3 yields good results when (87) is solved self-consistently. (On the other hand, α=1 seems to give good results in non self-consistent "one shot" calculations). Since the exchange interaction is represented by a local potential, there is no pathology associated with the density of states.

This equation is derived from the theory of the inhomogeneous electron gas as developed by Hohenberg and Kohn.[73] These authors showed that the energy of the ground state of any system of interacting electrons is uniquely determined in principle if the charge density is specified. Formally, the ground state energy is a functional of the charge density ρ

$$E_G = E[\rho] \ . \qquad (88a)$$

If the derivation of Hohenberg and Kohn is extended by being based on relativistic quantum theory, a slight generalization is obtained[76]

$$E_G = E[\rho,\vec{j}] \ , \qquad (88b)$$

where $\vec{j}$ is the current density (including the spin current). A variational principle applies: the energy attains its minimum value for the correct charge density function: The variational principle can be shown to lead to a set of single particle Schrödinger-like equations for an N particle system involving functions $\psi_n(r)$ which are related to the charge density in the usual way

$$\rho = \sum_{n}^{N} |\psi_n(\vec{r})|^2 . \tag{89}$$

These equations have a form similar to (87) except that they contain a general "exchange-correlation potential"

$$V_{xc}(\vec{r}) = \frac{\delta}{\delta\rho} E_{xc}[\rho] , \tag{90}$$

which is the functional derivative of the exchange plus correlation contribution to $E[\rho]$.

The most obvious difficulty in applying the preceeding remarks is that the general functional $E[\rho]$ is not known. If the charge density ρ were a slowly varying function of position, it would be permissible to use expressions obtained for the energy of a free electron gas, where the density $\rho=\rho_{fe}$ is constant; but with the replacement of the constant density ρ_{fe} by the actual position dependent density $\rho(r)$. In spite of the fact that the charge density tends to be a rapidly varying function near atomic nuclei, this procedure is adopted precisely because we do not have a better one. If the Hartree-Fock expression for the exchange energy of a free electron gas is used (correlation being ignored) we find, from (90), with

$$E_x = -3 \left(\frac{3}{8\pi}\right)^{1/3} \rho^{4/3} , \tag{91}$$

so

$$V_x = -4 \left(\frac{3\rho}{8\pi}\right)^{1/3} . \tag{92}$$

This corresponds to the use of $\alpha=2/3$ in (87). Correlation corrections are easily found IF the free electron gas correlation energy is given as a function of ρ. In addition, corrections involving derivatives of the density may be applied.[77] From the present point of view we simply regard α in (87) as an adjustable parameter which allows a very simple modification of (92) to correct for omitted correlation and gradient effects.

A further problem arises concerning the interpretation of energies and wave functions which are solutions of (87). Since the Hohenberg-Kohn-Sham theory works directly with the charge density of the ground state, there is no readily available procedure for constructing a many body wave function from the single particle solution of (87), and no immediate relation necessarily exists between the energy levels and the energies of transitions. Slater et al.[78] have shown that these eigenvalues do give a good approximation to the derivative of the energy with respect to the occupation number of the orbitals. This is what is required in an energy band calculation to determine the Fermi surface.

Considerable calculational experience has shown that rather good quality Fermi surfaces result from (87). Ground state charge, spin, and momentum densities seem to be in good agreement with experiment. Even optical properties in metals, which involve transitions to excited states, are fairly good, though discrepancies arise as the transition energy increases.[79] The defect which is most obvious to this author at the present time is that in a spin polarized situation (a metallic ferromagnet for example), the exchange splitting between majority and minority spin such bands appears to be overestimated.[80] Efforts are now in progress to use more sophisticated spin dependent exchange-correlation potentials[81,82] to allow, at least in part, for correlation effects. The situation is less satisfactory in a material like LiF where the local exchange approximation does not give good results,[83] although perhaps not as wildly bad as the Hartree-Fock equations.

The reason for the success of the local exchange approximation appears clearly in a paper by Duff and Overhauser.[84] They define a "correlation" operator, then calculate the matrix elements of the exchange and correlation operators for a model system. The local exchange approach turns out to be a poor approximation to exchange alone, but a not unreasonable first approximation to the sum of exchange and correlation. The exchange only, (Hartree-Fock) approximation would appear to be a much poorer starting point, since large correlation corrections then appear which cancel pathologies of the exchange operator.

IV. CONCLUDING REMARKS

At last, we venture some evaluations.

1. The APW and KKR methods remain powerful tools in common use for calculations which are now rather routine. This author would prefer the KKR method over all others for computations in which only energy levels were desired. A drawback here is that most existing KKR programs of which I am aware refer only to cubic lattices, while the APW method has been applied to a wide variety of structures. The recent development of linear muffin tin orbital methods shows great promise for wide applicability.

2. Since the appearance of Ziman's review,[7] the most important development in the technology of band structure calculation has been the emergence of a modernized form of the LCAO method as a competitive technique for accurate calculations. Comparison of numerical results between self-calculations for lithium using the LCAO[85] and APW[86] method shows agreement to about four significant figures for the lowest band. The major advantage of LCAO is that wave functions are obtained in a form which facilitates calculations of matrix elements involved in optical property[80,87] and dielectric function calculations.[88]

3. At the present time, it is quite difficult to combine realistic band computations with accurate calculations of many body effects. Thus, one wants a single particle potential which gives bands and related properties in reasonable agreement with experiment. Perfection is not to be expected. From this view, the local exchange, or Xα, potential is the best we have. It will be highly worthwhile to investigate other exchange-correlation potentials to see whether improvements can be made in a systematic and convincing manner.

V. REFERENCES

1. J. C. Slater "Quantum Theory of Molecules and Solids, Vol. 2, Symmetry and Energy Bands in Crystals", McGraw Hill, New York, 1965.

2. F. Bloch, Z. Phys. 52, 555 (1928).

3. J. Callaway "Quantum Theory of the Solid State", Academic Press, New York, 1974, Ch. 4.

4. G. C. Fletcher, "The Electron Band Theory of Solids", North Holland, Amsterdam, 1971.

5. P. M. Marcus, J. F. Janak, and A. R. Williams, eds. "Computational Methods in Band Theory", Plenum Press, New York, 1971.

6. B. Alder, S. Fernbach, and M. Rotenberg, eds., "Energy Bands in Solids", Vol. 8 of "Methods in Computational Physics", Academic Press, New York, 1968.

7. J. M. Ziman, Solid State Physics 26, 1 (1971).

8. J. Callaway "Energy Band Theory", Academic Press, New York, 1964.

9. H. Jones "The Theory of Brillouin Zones and Electronic States in Crystals", North-Holland, Amsterdam, 1960.

10. C. J. Bradley and A. P. Cracknell "The Mathematical Theory of Symmetry in Solids", Oxford University Press, 1972.

11. E. Wigner and F. Seitz, Phys. Rev. 43, 804 (1933).

12. C. Herring, Phys. Rev. 57, 1169 (1940).

13. T. O. Woodruff, Solid State Phys. 4, 367 (1967).

14. F. Herman, Rev. Mod. Phys. 30, 102 (1958).

15. F. Herman, R. L. Kortum, C. D. Kuglin, and R. A. Short in "Quantum Theory of Atoms, Molecules, and the Solid State", edited by P. O. Löwdin, Academic Press, New York, 1966, p. 381.

16. D. J. Stuckel, R. N. Euwema, T. C. Collins, F. Herman, and R. L. Kortum, Phys. Rev. 179, 740 (1969); T. C. Collins, D. J. Stuckel, and R. N. Euwema, Phys. Rev. B 1, 724 (1970); D. J. Stuckel and R. N. Euwema, Phys. Rev. B 1, 1635 (1970).

17. J. Callaway, Phys. Rev. 97, 933 (1955).

18. E. Brown and J. A. Krumhansl, Phys. Rev. 109, 30 (1958); F. A. Butler, F. K. Bloom, and E. Brown, Phys. Rev. 180, 744 (1969); D. M. Gray and R. M. Karpien in Marcus et al., Ref. 5, p. 144.

19. A. B. Kunz, Phys. Rev. 180, 934 (1969).

20. R. A. Deegan and W. D. Twose, Phys. Rev. 164, 993 (1967).

21. R. N. Euwema, Phys. Rev. B 4, 4332 (1971).

22. L. Kleinman and E. Carruthers, Phys. Rev. B 10, 3213 (1974).

23. W. A. Harrison "Pseudopotentials in the Theory of Metals", W. A. Benjamin, New York, 1966.

24. V. Heine, Solid State Physics 24, 1 (1970).

25. M. L. Cohen and V. Heine, Solid State Physics 24, 37 (1970).

26. V. Heine and D. Weaire, Solid State Phys. 24, 249 (1970).

27. J. C. Phillips and L. Kleinman, Phys. Rev. 116, 287 (1959).

28. B. J. Austin, V. Heine, and L. J. Sham, Phys. Rev. 127, 276 (1962).

29. E. Lafon and C. C. Lin, Phys. Rev. 152, 579 (1966).

30. R. C. Chaney, T. K. Tung, C. C. Lin, and E. E. Lafon, J. Chem. Phys. 52, 361 (1970).

31. D. E. Ellis and G. S. Painter, Phys. Rev. B 2, 2887 (1970).

32. J. Callaway and J. L. Fry in Marcus et al., Ref. 5, p. 512.

33. J. C. Slater and G. F. Koster, Phys. Rev. 94, 1498 (1954).

34. G. Dresselhaus and M. S. Dresselhaus, Phys. Rev. 160, 649 (1967).

35. L. Hodges, H. Ehrenreich, and W. D. Lang, Phys. Rev. 152, 505 (1966).

36. F. M. Mueller, Phys. Rev. 153, 659 (1967).

37. V. Heine, Phys. Rev. 153, 673 (1967).

38. H. Ehrenreich and L. Hodges, Methods. Comp. Phys. 8, 149 (1968).

39. E. I. Zornberg, Phys. Rev. B 1, 244 (1970).

40. J. C. Slater, Phys. Rev. 51, 846 (1937).

41. M. M. Saffren and J. C. Slater, Phys. Rev. 92, 1126 (1953).

42. H. Schlosser and P. M. Marcus, Phys. Rev. 131, 2529 (1963).

43. T. L. Loucks "Augmented Plane Wave Method: A Guide to Performing Electronic Structure Calculations", W. A. Benjamin, New York, 1967.

44. L. F. Mattheiss, J. H. Wood, and A. C. Switendick, Methods Computat. Phys. 8, 64 (1968).

45. J. O. Dimmock, Solid State Physics 26, 103 (1971).

46. J. Korringa, Physica 13, 392 (1947).

47. W. Kohn and N. Rostoker, Phys. Rev. 94, 1111 (1954).

48. F. S. Ham and B. Segall, Phys. Rev. 124, 1786 (1961).

49. A. R. Williams, J. F. Janak, and V. L. Moruzzi, Phys. Rev. B 6, 4509 (1972).

50. G. S. Painter, Phys. Rev. B 7, 3521 (1973).

51. L. Kleinman and R. Shurtleff, Phys. Rev. 188, 1111 (1969).

52. D. D. Koelling, A. J. Freeman, and F. M. Mueller, Phys. Rev. B 1, 1318 (1970).

53. E. O. Kane, Phys. Rev. B 4, 1917 (1971).

54. S. C. Keeton and T. L. Loucks, Phys. Rev. 168, 672 (1968).

55. D. D. Koelling and A. J. Freeman, Phys. Rev. B 7, 4454 (1973).

56. O. K. Andersen, Phys. Rev. B 12, 3060 (1975).

57. O. K. Andersen in Marcus et al., Ref. 5, p. 178; O. K. Andersen and R. V. Kasowski, Phys. Rev. B 4, 1064 (1971).

58. Y. Onodera and M. Okazaki, J. Phys. Soc. Japan 21, 1273 (1966).

59. S. Takada, Prog. Theor. Phys. Suppl. 36, 224 (1966).

60. J. M. Ziman, Proc. Phys. Soc. (London) 86, 337 (1965).

61. J. C. Slater, Phys. Rev. 145, 599 (1966).

62. J. Hubbard, Proc. Phys. Soc. (London) 92, 921 (1967).

63. D. G. Pettifor, J. Phys. C 5, 97 (1972).

64. M. J. G. Lee, Phys. Rev. 178, 953 (1969).

65. J. C. Shaw, J. B. Ketterson, and L. R. Windmiller, Phys. Rev. B 5, 3894 (1972).

66. R. N. Euwema, D. L. Wilhite, and G. T. Surratt, Phys. Rev. B 7, 818 (1972); R. N. Euwema, G. G. Wepfer, G. T. Surratt and D. L. Wilhite, Phys. Rev. B 9, 5249 (1974).

67. A. B. Kunz, D. J. Mickish, and T. C. Collins, Phys. Rev. Letts. 31, 756 (1973).

68. A. B. Kunz, Phys. Rev. B 8, 1690 (1973), D. J. Mickish, A. B. Kunz, and S. K. Pantelides, Phys. Rev. B 10, 1369 (1974).

69. L. Kumar, H. J. Monkhorst, and F. E. Harris, Phys. Rev. B 9, 4084 (1974).

70. F. E. Harris, L. Kumar and H. J. Monkhorst, Phys. Rev. B 7, 2850 (1973).

71. F. Seitz "The Modern Theory of Solids", McGraw Hill, New York, 1940, p. 421. J. Bardeen, Phys. Rev. 50, 1098 (1936).

72. S. Doniach in Marcus et al., Ref. 5, p. 500.

73. P. Hohenberg and W. Kohn, Phys. Rev. 136, B864 (1964).

74. W. Kohn and L. J. Sham, Phys. Rev. 140, A1133 (1965).

75. R. Gaspar, Acta Phys. Acad. Sci. Hung. 3, 263 (1954).

76. A. K. Rajagopal and J. Callaway, Phys. Rev. B 7, 1912 (1973).

77. F. Herman, J. P. van Dyke, and I. B. Ortenburger, Phys. Rev. Letts. 22, 807 (1969).

78. J. C. Slater, J. B. Mann, T. M. Wilson, and J. H. Wood, Phys. Rev. 184, 672 (1969).

79. J. F.Janak, A. R. Williams, and V. L. Moruzzi, Phys. Rev. B 11, 1522 (1975).

80. C. S. Wang and J. Callaway, Phys. Rev. B 9, 4897 (1974).

81. U. von Barth and L. Hedin, J. Phys. C 5, 1629 (1972).

82. O. Gunnarson and B. Lundqvist, Phys. Rev. B 13, 4274 (1976).

83. W. P. Menzel, C. C. Lin, D. F. Fouquet, E. E. Lafon, and R. C. Chaney, Phys. Rev. Letts. 30, 813 (1973).

84. K. J. Duff and A. W. Overhauser, Phys. Rev. B 5, 2799 (1972).

85. W. Y. Ching and J. Callaway, Phys. Rev. B 9, 5115 (1974).

86. L. Dagens and F. Perot, Phys. Rev. B 8, 1281 (1973).

87. W. Y. Ching and J. Callaway, Phys. Rev. Letts. 30, 441 (1973), Phys. Rev. B 11, 1324 (1975).

88. S. P. Singhal, Phys. Rev. B 12, 564 (1975).

COHESIVE PROPERTIES OF SOLIDS

JEAN-LOUIS CALAIS

Quantum Chemistry Group, University of Uppsala
Box 518
S-751 20 Uppsala, Sweden

Introduction

The adjective "cohesive" refers to those properties which depend the way the constituents of a solid are held together. The traditional division of solids in ionic, valence, molecular crystals and metals is based mainly on qualitative aspects of the cohesive properties. It is of interest and indeed also possible to make quantitative calculations of a number of cohesive properties such as cohesive energies, lattice constants, compressibilities, elastic constants etc. It is also possible in certain cases to predict theoretically which one of several possible crystal structures has the lowest energy and therefore is the one found. The study of cohesive properties is of essential importance for connecting lattice dynamics to the theory of the electronic structure.
The purpose of these lectures is to review certain fundamental aspects of the theory of cohesive properties and to give examples of different types of calculations.

This set of notes is of course not meant to give an exhaustive survey of all that has been done in this area. The intention is only to sum up certain aspects which will be covered in the lectures, to give a few more details in the mathematical derivations than time will permit during the lectures, and to give literature references which can serve as a starting point for further work. We will try to indicate clearly how one goes from one step to the next, but we will not give too many details. Everybody who works with these notes can easily make up a set of good exercises simply by carrying out all the derivations in detail.

The primary aim of these lectures is to show something of "the state of the art": what type of calculations are actually carried out; for what type of crystals; which approximations are involved; possibilities of improvements. We will limit ourselves to three types of calculations: LCAO work on ionic crystals, pseudopotential calculations for sp bonded metals and certain semiconductors and $X\alpha$ -type calculations. Before proceeding to a more detailed description of these three methods we will discuss certain general aspects of the calculation of cohesive properties of solids.

Despite the fact that cohesion is one of the most fundamental properties of a solid, one finds relatively little about this topic in most text books on solid state physics. Very much research has certainly been done on the practical, mechanical side, but there remains an enormous gap to be filled between that area and the microscopic theory of cohesive properties. The aim of the latter is to predict how much energy is gained (or lost) when a set of atoms, ions or molecules are brought together in a particular crystal lattice, whether there is an equilibrium position and for which distances it occurs, in what way a solid responds to stress both under normal conditions (elastic constants) and under high pressure. The theory should also explain the phase transitions that occur and predict such which have not yet been found because of difficult experimental situations. All these properties depend on temperature, but here we will restrict ourselves to 0 K, i.e. the study of the electronic structure as a function of different fixed nuclear frameworks.

The most detailed survey of LCAO (Linear Combination of Atomic Orbitals) calculations of cohesive properties of solids is Löwdin's paper from 1956[1]. This also contains a large amount of general fundamental theory of cohesion and is despite its age very much to be recommended for anybody who is interested in this subject. References to more recent work using the LCAO method will be given in the section in question. Cohesion from the point of view of pseudopotentials is treated in detail in the article by Heine and Weaire[2] in volume 24 of "Solid State Physics". The first two articles in that volume discuss other aspects of pseudopotential theory. The $X\alpha$ -method including applications to cohesion is thoroughly described in volume IV of Slater's series "Quantum Theory of Molecules and Solids"[3].

Generalities

More about most of the topics treated in this section can be found in ref. 1, where also further references are given.
We start by assuming that the Born-Oppenheimer approximation is

valid, which means that we study the electronic structure of the solid for fixed nuclear configurations. The total "electronic" energy then provides (in principle) a potential for setting up a Schrödinger equation for the motion of the nuclei. We are primarily interested in this total "electronic" energy in the neighbourhood of the equilibrium position for the state under consideration. The distance in energy between the minimum and the energy of the free constituents of the crystal is the cohesive energy (neglecting zero-point vibrations). The position of the minimum determines the lattice constants and the various second derivatives with respect to the components of the strain tensor determine the elastic constants.

Calculations of these quantities in a certain model followed by comparison with experimental data are certainly of interest per se, not least because they provide information about the validity of the model. But as will be seen in the next section, careful so-called ab initio calculations are also valuable from another point of view since they can help us to form new concepts of importance for understanding the bonding in crystals. Semi-empirical calculations are usually much simpler and they can therefore relatively easily furnish information about many different solids and many different properties. It is important though that both semi-empirical and ab initio methods can be expressed in a common language, so that they can be compared and their relative positions can be determined.

The cohesive energy of a solid is thus defined by

$$E_{coh} = E_{cr} - E_{fr} , \tag{1}$$

where E_{cr} is the total energy of the solid and E_{fr} that of its free constituents. Those constituents can be atoms, ions or molecules. This means that E_{coh} is a negative number (provided the solid exists). Properly speaking the cohesive energy is the value of E_{coh} at equilibrium, i.e. when $-E_{coh}$ has a maximum. We must further specify in which states the crystal and the constituents are; normally this should be the ground state in both cases. The total energies E_{fr} and E_{cr} are of the same order of magnitude and much larger than their difference. This implies that great care has to exercised to get a reasonable accuracy for the cohesive energy. One way to handle this problem is to combine terms in E_{cr} and E_{fr} so as to achieve maximal cancellation. Another way is to avoid calculating E_{coh} directly via the total energies and instead try to find methods to calculate it directly.

In practically all cases both E_{cr} and E_{fr} have to be obtained from approximate calculations. Not least for making comparisons possible it is however desirable to have a framework which is in principle exact, and in which one can also situate the various approximations.

One such framework is supplied by the reduced density matrices, of which we only need the first two, since the basic Hamiltonian only contains one and two electron operators. If a quantum mechanical state is described by an (exact or approximate) wave function Ψ , we can express all properties in terms of the first, γ , and second, Γ , order reduced density matrices

$$\gamma(x_1|x_1') = N\int\Psi(x_1,x_2,\ldots x_N)\Psi^*(x_1',x_2,\ldots x_N)dx_2\ldots dx_N;$$
$$\Gamma(x_1x_2|x_1'x_2') = \binom{N}{2}\int\Psi(x_1,x_2,x_3\ldots x_N)\Psi^*(x_1',x_2',x_3,\ldots x_N)dx_3\ldots dx_N, \qquad (2)$$

instead of Ψ , which means a great simplification from several points of view. (It should be noticed that this notation differs slightly from that used in ref. 1.) The first order density matrix can be obtained from the second order one, but not vice versa. The diagonal elements have the following physical interpretation

$\gamma(x|x)dv$: the number of electrons times the probability of finding an electron at $\vec{r}$ around dv with a certain spin σ , irrespective of the other (N-1) electrons;

$\Gamma(x_1x_2|x_1x_2)dv_1dv_2$: the number of pairs of electrons times the probability of finding one at $\vec{r}_1$, around dv_1 with spin σ_1 and another one at $\vec{r}_2$ around dv_2 with spin σ_2 , irrespective of the other (N - 2) electrons.

We are using the convention that x_k denotes the combined spin space - ordinary space coordinate

$$x_k = (\vec{r}_k;\sigma_k). \qquad (3)$$

We further assume that the wave function Ψ is normalized so that (2) implies

$$\int\gamma(x|x)dx = N;$$
$$\int\Gamma(x_1x_2|x_1x_2)dx_1dx_2 = \binom{N}{2}. \qquad (4)$$

The electron density is essentially

$$\gamma(\vec{r}) = \int\gamma(x|x)d\sigma \qquad (5)$$

Although for heavier atoms relativistic effects have to be taken into account, we will here work only with the basic Hamiltonian

$$H = H_o + \sum_{i=1}^{N} h_i + \frac{1}{2} \sum_{i,j}{}' h_{ij}, \tag{6}$$

where

$$H_o = \frac{1}{2} \sum_{g,h}{}' \frac{Z_g Z_h}{R_{gh}} ; \tag{7a}$$

$$h_i = - \frac{1}{2} \Delta_i - \sum_g \frac{Z_g}{r_{ig}} ; \tag{7b}$$

$$h_{ij} = \frac{1}{r_{ij}} \tag{7c}$$

In most of this paper we use atomic units derived from setting $e = m = \hbar = 1$, and with 1 Hartree = 2 Rydberg $\approx$ 27.2 eV as the energy unit. The (within the Born-Oppenheimer approximation) constant term H_o represents the repulsion between the nuclei, the one electron operator h_i the kinetic energy of electron i and its attraction to the nuclei, and the two electron operator h_{ij} the repulsion between electrons i and j. Z_g is the charge of nucleus g .

The expectation value of H with respect to the wave function Ψ can be written as

$$\langle H \rangle = \frac{\langle \Psi | H | \Psi \rangle}{\langle \Psi | \Psi \rangle} = E = H_o + \int h_1 \gamma(x_1 | x_1') dx_1 + \tag{8}$$

$$+ \int h_{12} \Gamma(x_1 x_2 | x_1 x_2) dx_1 dx_2$$

The prime in the second argument of the first order density matrix indicates that the operator h_1 should work only on the first argument. After that operation the prime is taken away and the integration is carried out. In the last term of (8) no prime is needed, since h_{12} is a purely multiplicative operator.

The expression (8) for the total energy is valid as such for any system of nuclei and electrons - atoms, molecules, solids. For a solid we are interested not so much in the total energy for the whole solid, but in the total energy per atom, per electron, per ion pair or per some other constituent. In order to be able to make meaningful statements about the total energy we must therefore have a quantity which is proportional to N (the number of electrons). Such is the case for E as defined in (8). We might, however, also want to discuss various parts of the total energy, and then the situation is less simple. One meaningful (in this sense) way of dividing the energy in "convergent" parts (proportional to N) is the following.

$$E = E_1 + E_2 + E_3; \tag{9a}$$

$$E_1 = -\frac{1}{2}\int \Delta_1 \gamma(x_1|x_1')dx_1; \tag{9b}$$

$$E_2 = \frac{1}{2}\sum_{g,h}{}' \frac{Z_g Z_h}{R_{gh}} - \sum_g \int \frac{\gamma(x_1|x_1)}{r_{1g}} dx_1 + \frac{1}{2}\int \frac{\gamma(x_1|x_1)\gamma(x_2|x_2)}{r_{12}} dx_1 dx_2; \tag{9c}$$

$$E_3 = \int \frac{\Gamma(x_1x_2|x_1x_2) - \frac{1}{2}\gamma(x_1|x_1)\gamma(x_2|x_2)}{r_{12}} dx_1 dx_2 \tag{9c}$$

E_1 is the total kinetic energy, E_2 the total electrostatic energy of a neutral system, and E_3 in a certain sense represents the contribution from what is usually called exchange and correlation. It must be stressed though that these terms should not be used without further specification.

It is often desirable to distribute in some way the electron density $\gamma(x) = \gamma(x|x)$ (for a particular spin) among the atoms (nuclei) in the crystal:

$$\gamma(x) = \sum_g \gamma_g(x) \tag{10}$$

There is no objection to such a procedure, but it is very important to remember that it is not unique - there are many different ways of alloting parts of the density to the nuclei. In the next section we will see one example of such a procedure.

Given such a procedure we can calculate the charge associated with nucleus g

$$n_g = \int \gamma_g(x)dx, \tag{11}$$

and the potential $V_g(1)$ due to the charge distribution γ_g:

$$V_g(1) = \int \frac{\gamma_g(x_2)}{r_{12}} dx_2 = \frac{n_g}{r_{1g}} - \omega_g(1). \tag{12}$$

We can rewrite (9c) in terms of these quantities

$$\begin{aligned} E_2 &= \frac{1}{2} \sum_{g,h}{}' \frac{(Z_g-n_g)(Z_h-n_h)}{R_{gh}} + \\ &+ \frac{1}{2} \sum_{g,h}{}' \left[(2Z_g - n_g)\omega_h(g) - \int \gamma_g(x_1)\omega_h(x_1)dx_1\right] + \\ &+ \frac{1}{2} \sum_g \left[-2Z_g V_g(g) + \int \gamma_g(x_1)V_g(x_1)dx_1\right] . \end{aligned} \tag{13}$$

The first term in (13) represents the electrostatic energy of a set of point charges; it is usually called the Madelung energy. The remaining terms in (13) represent corrections for the fact that the constituents of the crystal are not point charges, but have extended charge distributions. It is a good exercise to analyse the meaning of the various terms in (13).

An alternative starting point for the calculation of total energies and differences is to express the potential energy [$E_2 + E_3$ in (9)] in terms of a generalized polarizability. We follow here Lindner and Goscinski[4], in which also further references can be found.
The density can be thought of as an operator

$$\gamma(\bar{r}) = \rho(\bar{r}) = \sum_{i=1}^{N} \delta(\bar{r} - \bar{r}_i), \tag{14}$$

since

$$\langle\Psi|\sum_i \delta(\bar{r}-\bar{r}_i)|\Psi\rangle = \int \delta(\bar{r}-\bar{r}_1)\gamma(x_1|x_1')dx_1 = \gamma(\bar{r}) \tag{15}$$

The Fourier components of the density are therefore

$$\rho_{\bar{q}} = \int e^{-i\bar{q}\cdot\bar{r}} \rho(\bar{r})dv = \sum_{i=1}^{N} e^{-i\bar{q}\cdot\bar{r}_i} \tag{16}$$

The Fourier components of a Coulomb potential can be writtén

$$V_{\bar{k}} = \frac{1}{\Omega}\int \frac{e^{-i\bar{k}\cdot\bar{r}}}{r}\, dv = \frac{4\pi}{\Omega k^2} . \tag{17}$$

Ω is the Born-Kármán volume (the piece of the crystal considered). The Coulomb potential itself can therefore be written

$$\frac{1}{r} = \frac{\Omega}{8\pi^3}\int V_{\bar{k}}\, e^{i\bar{k}\cdot\bar{r}}\, d\bar{k} = \sum_{\bar{k}} V_{\bar{k}}\, e^{i\bar{k}\cdot\bar{r}}. \tag{18}$$

In (18) we have used the fact that the "k-points" form a quasicontinuum, so that sums can be replaced by integrals and vice versa, provided we keep track of the density in k-space.

The potential energy parts can now be written in terms of the Fourier components as for e.g. the attraction term:

$$-\sum_{j,g} \frac{Z_g}{r_{jg}} = -\sum_{j,g} Z_g \sum_{\bar{k}} V_{\bar{k}}\, e^{i\bar{k}\cdot(\bar{r}_g - \bar{r}_j)} =$$

$$= \sum_{\bar{k}} V_{\bar{k}}\left(-\sum_g Z_g\, e^{i\bar{k}\cdot\bar{r}_g}\right)\left(e^{-i\bar{k}\cdot\bar{r}_j}\right) =$$

$$= \sum_{\bar{k}} V_{\bar{k}}\, S_{-\bar{k}}\, \rho_{\bar{k}} , \tag{19}$$

where we have introduced the notation

$$S_{\bar{k}} = -\sum_g Z_g\, e^{i\bar{k}\cdot\bar{r}_g} . \tag{20}$$

The total potential part of the Hamiltonian can therefore be written as follows

$$\frac{1}{2}\sum_{k} V_{\bar{k}} \left[(\rho_{\bar{k}} + S_{\bar{k}})(\rho_{-\bar{k}} + S_{-\bar{k}}) - N - \sum_g Z_g^{\,2}\right]. \tag{21}$$

(An important misprint in eq. (1) of ref. 4 and in eq. (14) of ref. 19 should be noticed) Inserting the resolution of the identity

$$1 = |0\rangle\langle 0| + \sum_{s\neq 0} |s\rangle\langle s| = |0\rangle\langle 0| + P, \tag{22}$$

with respect to the eigenstates of H , between the parentheses in (21), and taking the expectation value of (21) with respect to the ground state of the system, we get

$$E_2 + E_3 = U = \frac{1}{2}\sum_{k} V_{\bar{k}} \left[\langle 0|\rho_{\bar{k}} + S_{\bar{k}}|0\rangle\langle 0|\rho_{-\bar{k}} + S_{-\bar{k}}|0\rangle - \sum_{g} Z_g{}^2\right] +$$

$$+ \frac{1}{2}\sum_{\bar{k}} V_{\bar{k}} \left[\langle 0|(\rho_{\bar{k}} + S_{\bar{k}})P(\rho_{-\bar{k}} + S_{-\bar{k}})|0\rangle - N\right]. \tag{23}$$

Within the Born-Oppenheimer approximation the quantities $S_{\pm\bar{k}}$ are constants, so that

$$U = \frac{1}{2}\sum_{k} \left[(\langle 0|\rho_{\bar{k}}|0\rangle + S_{\bar{k}})(\langle 0|\rho_{-\bar{k}}|0\rangle + S_{-\bar{k}}) - \sum_{g} Z_g{}^2\right] +$$

$$+ \frac{1}{2}\sum_{\bar{k}} V_{\bar{k}} \left[\langle 0|\rho_{\bar{k}}\, P\, \rho_{-\bar{k}}|0\rangle - N\right]. \tag{24}$$

One can show (good exercise) that the first line of (24) is equal to E_2 defined by (9c) and that the second line of (24) is therefore equal to E_3 (9d). Using the identity (check!)

$$P = \frac{2}{\pi}\int_0^\infty \left[H - E_o\right]\left[(H - E_o)^2 + \omega^2\right]^{-1} P\, d\omega, \tag{25}$$

where E_o is the ground state energy of the system, we can then write E_3 in terms of the generalized polarizability for imaginary frequencies,

$$\alpha(\bar{k},\bar{k};i\omega) = \sum_{s\neq 0}\left[\frac{\langle 0|\rho_{\bar{k}}|s\rangle\langle s|\rho_{-\bar{k}}|0\rangle}{E_s - E_o - i\omega} + \frac{\langle 0|\rho_{-\bar{k}}|s\rangle\langle s|\rho_{\bar{k}}|0\rangle}{E_s - E_o + i\omega}\right], \tag{26}$$

as

$$E_3 = \frac{1}{2}\sum_{\bar{k}} V_{\bar{k}}\left[\frac{1}{\pi}\int_0^\infty \alpha(\bar{k},\bar{k};i\omega)d\omega - N\right]. \tag{27}$$

It is also a good idea to identify the various parts of the first line of (24) with the various terms of E_2.

The formulae given so far are exact within the Born-Oppenheimer approximation of the non-relativistic Schrödinger equation. To go on we have to make other approximations. To begin with some type of independent particle model is nearly always invoked, which leads to effective one electron operators. The perhaps most fundamental one of this type of approximations - although not the only one as we will see in the Xα-section - is the Hartree-Fock approximation. In that model the total wave function is approximated by a single determinant, which amounts to approximating a series by its first term. One can then show that the first order density matrix reduces to the so-called Fock-Dirac matrix,

$$\gamma(x|x') = \rho(x,x') = \sum_{k=1}^{N} \Psi_k(x)\Psi_k^*(x'). \tag{28}$$

Here the N spin orbitals (one electron functions) $\Psi_k(x)$ are those functions which make up the determinant. The form (28) presupposes that these spin orbitals are orthogonal-something which can always be achieved within one determinant without changing it. Another characteristic aspect of the Hartree-Fock approximation is the fact that here the second order density matrix can be expressed in terms of the first order matrix:

$$\Gamma(x_1x_2|x_1'x_2') = \frac{1}{2}\left[\rho(x_1,x_1')\rho(x_2,x_2') - \rho(x_1,x_2')\rho(x_2,x_1')\right] \tag{29}$$

Thus the Hartree-Fock level of formulae (5), (8)-(13) are obtained from the special forms (28) and (29).

The error in the Hartree-Fock approximation is usually called correlation. The general formulae given in this section provide a framework also for treating correlation problems. The total energy of any system can be written as a sum of a Hartree-Fock contribution and a correlation energy

$$E = E_{HF} + E_{corr}. \tag{30}$$

For the cohesive energy this implies

$$E_{coh} = E_{HF}^{cr} - E_{HF}^{fr} + E_{corr}^{cr} - E_{corr}^{fr} \tag{31}$$

If the correlation error for the crystal is of the same order of magnitude as that of the free constituents, the Hartree-Fock level should provide a good approximation of the cohesive energy. In many cases this is, however, not the case. Then great care has to be exercised since the correlation error can be of the same order of magnitude as the cohesive energy itself.

The types of calculations of cohesive properties to be discussed in these lectures fall between an empirical treatment like e.g. the Born-Mayer model of ionic crystals and a hopefully asymptotically reachable exact treatment. The fact that comparatively little has been done, in particular on the ab initio side, is connected with the many, not least numerical, difficulties involved. More specifically it is probably due to the fact that most work so far has been put into the solution of the one electron equations (band theory), and it is left to the future to use those solutions, among other things, to the calculation of cohesive properties. It is tempting to make an analogy. The basic principles of band theory were set out at the end of the twenties and during the thirties, and a few pilot calculations were made. But it was not until around 1960 and later that "full scale" band calculations really got started. Similarly the basic principles of the calculation of cohesive properties have been known for a long time and a number of calculations have been carried out, but we are still far from the situation band structure calculations are in now.

LCAO Calculations for Ionic Crystals

The first type of calculations to be discussed in the lectures will be L(inear) C(ombination) A(tomic) O(rbitals) type calculations of the cohesive properties of ionic crystals. This method is not confined to such systems, but it is particularly suitable to systems with closed shells. As discussed in detail in ref. 1, the names Hylleraas[5)], Landshoff[6)], Löwdin[7)] mark important steps in the earlier calculations. The recent work in this area has been summed up by the author[8)].

We neglect in this case the correlation problem. The calculations made so far seem to confirm the expectations that in this case the Hartree-Fock level should give quite satisfactory results. Our first job will therefore be to specialize the general formulae given in the previous section to the Hartree-Fock level. We use the notation ρ for the Fock-Dirac matrix of the free constituents (i.e. for all of them), and $\bar{\rho}$ for that of the crystal. Then (8), (28), (29) give

$$E^{cr}_{HF} = \frac{1}{2}\sum_{g,h}{}' \frac{Z_g Z_h}{R_{gh}} - \frac{1}{2}\int \Delta_1 \bar{\rho}(x_1,x_1')dx_1 - \sum_g Z_g \int \frac{\bar{\rho}(x_1,x_1)}{r_{1g}}dx_1 + \tag{32}$$

$$+ \frac{1}{2}\int \frac{\bar{\rho}(x_1,x_1)\bar{\rho}(x_2,x_2) - \bar{\rho}(x_1,x_2)\bar{\rho}(x_2,x_1)}{r_{12}} dx_1 dx_2 .$$

The total Fock-Dirac matrix for the non-interacting constituents can be written as

$$\rho = \sum_g \rho_g , \tag{33}$$

and the corresponding total energy in the Hartree-Fock approximation is

$$E_{HF}^{fr} = \sum_g \{- \frac{1}{2}\int \Delta_1 \rho_g(x_1,x_1')dx_1 - Z_g \int \frac{\rho_g(x_1,x_1)}{r_{1g}} dx_1 + \\ + \frac{1}{2}\int \frac{\rho_g(x_1,x_1)\rho_g(x_2,x_2) - \rho_g(x_1,x_2)\rho_g(x_2,x_1)}{r_{12}} dx_1 dx_2\}. \tag{34}$$

The effective, one electron Hartree-Fock operator for ion g (i.e. for the set of electrons associated with ion g) can be written as

$$H_{eff,g}(1) = - \frac{1}{2}\Delta_1 - \frac{Z_g}{r_{1g}} + \int \frac{(1-P_{12})\rho_g(x_2,x_2')}{r_{12}} dx_2. \tag{35}$$

Here the operator P_{12} permutes x_1 and x_2 , when the effective operator works on a function of x_1. We assume in principle that ρ_g is built up of the "eigenfunctions" of $H_{eff,g}$:

$$H_{eff,g}(1)\phi_\mu(x_1) = \varepsilon_\mu \phi_\mu(x_1); \tag{36}$$

$$\rho_g(x_1,x_1') = \sum_{\mu=1}^{N_g} \phi_\mu(x_1)\phi_\mu^*(x_1'). \tag{37}$$

We also assume that the same situation holds for $\bar{\rho}$ and the effective Hartree-Fock operator for the crystal. Thus we need the solution of the band structure problem to start our cohesive energy calculations.

We proceed to a division of the crystal Fock-Dirac matrix (cf.(10))

$$\bar{\rho} = \sum_g \bar{\rho}_g \tag{38}$$

It is therefore preferable to express $\bar{\rho}$ in terms of the "localised" Wannier functions rather than the delocalised Bloch functions. Since we are working closed shell systems it is immaterial which one of these two representations that is used.

We are now ready to subtract (34) from (32) to get the cohesive energy at the Hartree-Fock level. Instead of $\bar{\rho}$ we will use the difference

$$\Delta\rho = \bar{\rho} - \rho = \sum_g \Delta\rho_g. \tag{39}$$

After some tedious but perfectly straight-forward algebra (which is very much recommended to be checked - this gives some feeling for the various terms) we get the following expression

$$E_{HF}^{coh} = E_{elstat} + E_{exch} + E_S \tag{40a}$$

$$E_{elstat} = \frac{1}{2} \sum_{g,h}{}' \frac{Z_g Z_h}{R_{gh}} - \sum_{g,h}{}' Z_g \int \frac{\rho_h(x_1,x_1)}{r_{1g}} dx_1 + \\ + \frac{1}{2} \sum_{g,h}{}' \int \frac{\rho_g(x_1,x_1)\rho_h(x_2,x_2)}{r_{12}} dx_1 dx_2 \; ; \tag{40b}$$

$$E_{exch} = -\frac{1}{2} \sum_{g,h}{}' \int \frac{\rho_g(x_1,x_2)\rho_h(x_2,x_1)}{r_{12}} dx_1 dx_2 \; ; \tag{40c}$$

$$E_S = - \sum_{g,h}{}' Z_g \int \frac{\Delta\rho_h(x_1,x_1)}{r_{1g}} dx_1 + \\ + \sum_{g,h}{}' \int \frac{\Delta\rho_g(x_1,x_1)\rho_h(x_2,x_2) - \Delta\rho_g(x_1,x_2)\rho_h(x_2,x_1)}{r_{12}} dx_1 dx_2 + \\ + \frac{1}{2} \int \frac{\Delta\rho(x_1,x_1)\Delta\rho(x_2,x_2) - \Delta\rho(x_1,x_2)\Delta\rho(x_2,x_1)}{r_{12}} dx_1 dx_2 + \\ + \sum_g \{\int H_{eff,g}(1)\Delta\rho_g(x_1,x_1')dx_1\} \tag{40d}$$

It should be noticed that (40b) and (40c) only contain "non-diagonal" terms with $g \neq h$ - the large "diagonal" terms have been cancelled out. E_{elstat} is the classical electrostatic interaction between the ions. As sketched in (13) one can show that E_{elstat} is the sum of the Madelung energy and corrections for the fact that the ions are extended charge distributions. E_{exch} is the exchange part of the interactions between the ions. In E_S finally (the S will be explained shortly) we have collected all terms that contain the difference $\Delta\rho$ or parts thereof, and E_S also contains both "electrostatic" and "exchange parts".

So far nothing has been said about LCAO. Formula (40) is general and can be used for any set of exact or approximate Wannier functions for the filled bands of the crystal and any set of Hartree-Fock solutions for the free ions. The particular form (40) was derived by Löwdin[7] and has so far been used mainly in an LCAO version, as has been summarized in ref.8. The main difficulty at this stage is connected with the fact that one of the assumptions made in the derivation of (40) has only recently tended to be reasonably realistic. We refer to the lack of Hartree-Fock functions for the crystal. Thanks mainly to the work in the groups at Urbana, Dayton and Salt Lake City we are now beginning to get good Hartree-Fock functions for a number of crystals.

The remaining raw material needed, namely the Hartree-Fock functions for the free ions, presents no problem. Several sets of such functions have been available for many years, both in numerical[9] and analytical[10] (expressed as expansions of basis functions) form.

Even though we now begin to get Hartree-Fock functions also for the crystal, there is still a problem. We need such functions not only for one lattice constant but as functions of the lattice constant, in order to be able to calculate the potential energy curve in the neighbourhood of the equilibrium. The procedure which has been used in most of the calculations so far satisfies that requirement, although it only represents an approximation to the exact Wannier functions. One way to describe this procedure is to say that we approximate the exact Wannier functions by O(rthogonalized) A(tomic) O(rbitals). We collect the free ion orbitals (solutions of (36)) in a row

$$\boldsymbol{\Phi} = |\boldsymbol{\Phi}\rangle = [\ldots \phi_{\mu g} \ldots] = [\boldsymbol{\Phi}_1, \boldsymbol{\Phi}_2, \ldots \boldsymbol{\Phi}_g \ldots] \,. \tag{41}$$

The quantity defined in (41) can also be thought of as a matrix whose row index is a continuous variable. A(tomic) O(rbitals) situated at different lattice sites are in general not orthogonal, which can be expressed as follows

$$\boldsymbol{\Phi}^+ \boldsymbol{\Phi} = \langle \boldsymbol{\Phi} | \boldsymbol{\Phi} \rangle = \boldsymbol{\Delta} \neq \mathbf{1} \tag{42}$$

Such a set of functions can, however, be orthogonalized[11] (see also ref.1) in different ways, of which one particularly suited for solid state problems is the symmetric orthogonalisation which gives the following set of OAO's ϕ

$$\boldsymbol{\varphi} = \boldsymbol{\Phi}\boldsymbol{\Delta}^{-\frac{1}{2}} \,. \tag{43}$$

The overlap matrix $\mathbf{\Delta}$ and $\mathbf{\Delta}^{-\frac{1}{2}}$ have elements labelled by both the site of the AO's and the type of orbital. The OAO's ϕ are therefore labelled in the same way. It is important to realize though the OAO's are linear combinations of in principle all AO's on all centres.

The exact Wannier functions are thus approximated by the OAO's (43). The Fock-Dirac matrix for the crystal is then approximated (cf.(28)) by

$$\bar{\boldsymbol{\rho}} = \boldsymbol{\varphi}\boldsymbol{\varphi}^{\dagger} = |\boldsymbol{\varphi}\rangle\langle\boldsymbol{\varphi}| = \boldsymbol{\Phi}\boldsymbol{\Delta}^{-1}\boldsymbol{\Phi}^{\dagger} = |\boldsymbol{\Phi}\rangle\,\boldsymbol{\Delta}^{-1}\langle\boldsymbol{\Phi}| \;, \tag{44}$$

and the difference (39) by

$$\Delta\boldsymbol{\rho} = \bar{\boldsymbol{\rho}} - \boldsymbol{\rho} = |\boldsymbol{\Phi}\rangle\,[\boldsymbol{\Delta}^{-1} - \mathbf{1}]\langle\boldsymbol{\Phi}|. \tag{45}$$

The elements of this matrix are thus

$$\Delta\rho(x,x') = -\sum_{\mu}^{\text{all}}\sum_{\nu}^{\text{all}} \phi_{\mu}(x)\, P_{\mu\nu}\phi_{\nu}^{*}(x'), \tag{46}$$

where we have introduced the notation

$$\mathbf{P} = \mathbf{1} - \boldsymbol{\Delta}^{-1} . \tag{47}$$

The summations in (46) are carried out over the spin orbitals on all the ions. A division of $\Delta\rho$ of the type (10) or (38), which is natural in the LCAO method is

$$\Delta\rho_g(x,x') = -\sum_{\mu}^{\text{all}}\sum_{\nu}^{g} \phi_{\mu}(x)\, P_{\mu\nu}\phi_{\nu}^{*}(x'). \tag{48}$$

Here the sum over ν is carried out only over those spin orbitals, which are centered at ion g .

The fundamental expression (40) for the cohesive energy can now be evaluated by means of (37) and (48). This involves as usual in the LCAO method calculations of a large number of integrals, which is, however, quite feasible with the present day computers. As a matter of fact Löwdin's first calculations from 1947 and 1948 were carried out by means of electric desk computers!

When the crystal density matrix has been expressed in terms of the free ion orbitals as has been done here (cf.(44)-(48)) the last term in (40d) vanishes which can be seen as follows

$$\sum_g \left[\int H_{eff,g}(1)\Delta\rho_g(x_1,x_1')dx_1 \right] =$$

$$= -\sum_g \sum_\mu^{all} \sum_\nu^{g} \int H_{eff,g}(1)\phi_\mu(x_1)P_{\mu\nu}\phi_\nu^*(x_1')dx_1 =$$

$$= -\sum_\mu^{all} \sum_\nu^{all} \varepsilon_\mu P_{\mu\nu} \int \phi_\nu^*(x_1)\phi_\mu(x_1)dx_1 =$$

$$= -\sum_\mu^{all} \sum_\nu^{all} \varepsilon_\mu (\mathbf{1} - \mathbf{\Delta}^{-1})_{\mu\nu} \Delta_{\nu\mu} =$$

$$= -\sum_\mu^{all} \varepsilon_\mu (\mathbf{\Delta} - \mathbf{1})_{\mu\mu} = 0. \tag{49}$$

Here we have used (36), (47), (48) and we have further assumed that the AO's ϕ_μ are normalized. Landshoff derived (49) and Lundqvist (cf. ref.1) pointed out that the AO's actually used are not exact solutions of (36). The relation (49) should therefore be tested on the raw material used in the calculation.

A few words should be said about $\mathbf{\Delta}^{-1}$ which is needed to calculate the cohesive energy if we approximate the Wannier functions by OAO's. The elements of the overlap matrix $\mathbf{\Delta}$ are sufficiently small to allow a power series expansion of $\mathbf{\Delta}^{-1}$:

$$\mathbf{\Delta}^{-1} = (\mathbf{1} + \mathbf{S})^{-1} = \mathbf{1} - \mathbf{S} + \mathbf{S}^2 - \mathbf{S}^3 \ldots \tag{50}$$

Here we have introduced the notation $\mathbf{S}$ for the non-diagonal part of $\mathbf{\Delta}$.- The fact that the $S_{\mu\nu}$ are small may not be used as a reason to neglect them - there are very many of them! Actually the overlap integrals are absolutely essential quantities in the LCAO formulation, and their inclusion gives so to speak the theory a new dimension. It is this S which appears in (40d).
In the calculations of this type done up to now (cf. ref.8) the so-called S^2-approximation has been used. This means for $\mathbf{\Delta}^{-1}$ that the expansion (50) is truncated after the first two terms. In the expression (40) $\mathbf{\Delta}^{-1}$, however, does not appear alone, but multiplied by various integrals, and a consistent S^2-approximation requires that all terms of order S^2 are neglected. As pointed out in ref.8 some interesting improvements in the procedure for calculating the inverse of the overlap matrix have recently been introduced.

The fundamental expression (40) was first used by Löwdin[7] to evaluate the cohesive energy for many of the alkali halides as a function of the lattice parameter. Both cohesive energies, equilibrium distances and compressibilities agreed quite well with experimental data. In these calculations the sums over g and h were carried out only over nearest neighbour ion pairs, except for the Madelung energy and corresponding lattice sums needed for the calculation of $\Delta\rho$. Later calculations (see ref.8) have shown that the agreement with experiment gets worse when we also include sums of next nearest neighbour pairs in (40). Particularly important are the large negative ions with relatively large overlap also with the next nearest neighbour distances.

This situation is connected with the fact that we use free ion orbitals in the crystal instead of the exact Hartree-Fock solutions. Various attempts have been made (cf. ref.8) to adapt the free ion orbitals to the crystalline surrounding by relatively simple means, but unfortunately without too much success. What is needed is a better approximation of the exact Wannier functions than can be provided by the OAO's (cf. ref.8).

The good agreement between Löwdin's calculations and experimental data is certainly rewarding. By means of these calculations it was possible to "explain" the repulsive term in the Born-Mayer model both qualitatively and quantitatively, thus giving a good example of one reason for making ab initio calculations. But this quantum mechanical calculation goes far beyond the Born-Mayer model, which is perhaps best seen in the treatment of elastic constants.

The (second order) elastic constants of a solid are defined by (see further ref.1)

$$c_{ij} = \frac{1}{V_o}\left(\frac{\partial^2 F}{\partial x_i \partial x_j}\right)_o;\ ij = 1,2,3,\ldots 6. \tag{51}$$

Here V_o is the (unstrained) volume of the piece of crystal considered and x_i one of the six components of the strain tensor. F is the free energy $F = E_{coh} - ST$, which at 0 K reduces to the cohesive energy.

The Cauchy relations for the elastic constants are obtained under the assumption that the effective forces between the constituents of the crystal (ions in this case) are central, two-body forces. In general there are 36 elastic constants, but since they are symmetric (cf. (51)) the number is reduced to 21. There are at most 6 Cauchy relations.

For cubic crystals (e.g. alkali and silver halides) there are only three independent elastic constants c_{11}, c_{12}, and c_{44}. There is one Cauchy relation, namely

$$c_{12} = c_{44}. \tag{52}$$

Experimentally the Cauchy relations are not satisfied for the alkali halides, c_{44} always being somewhat larger than c_{12}. For the silver halides the discrepancy is much larger.

These experimental deviations from the Cauchy relations have naturally for a long time constituted a challenge to the theoreticians. It was therefore an important step forward when Löwdin[7] was able to show that the quantum mechanical calculation based on (40) gave different values for c_{12} and c_{44}, which furthermore also went in the right direction. The difference between these two elastic constants could be traced to certain terms in E_S, thus depending on the inclusion of the overlap integrals. What was particularly interesting was that they could be interpreted as due to effective many ion forces. An example is provided by the term

$$\sum_{\mu}^{\text{all}} \sum_{\nu}^{g} (\mathbf{\Delta}^{-1} - \mathbf{1})_{\nu\mu} \int \phi_\mu^*(x_1) \left[\sum_{h \neq g} \frac{Z_h - n_h}{r_{1h}} \right] \phi_\nu(x_1) dx_1, \tag{53}$$

in which three ions are involved if ϕ_μ is centered on an ion different from both g and h. The appearance of these effective many ion forces is thus a result of a careful ab initio calculation, and therefore shows an excellent reason for carrying out such calculations. The many ion forces have also been thoroughly discussed in connection with rare gas crystals (see ref.8).

As pointed out in ref.8 there are now good reasons for taking up such calculations again as have been sketched in this section. The numerical difficulties are not insurmountable and the conceptual framework makes the LCAO model very attractive for physical interpretations.

Pseudopotential Theory of Cohesive Properties

The LCAO model emphasizes the fact that solids consist of atoms (ions, molecules). An alternative and in a way opposite view is provided by the pseudopotential approach, in which the solid is described as an electron gas perturbed by atomic pseudopotentials. The starting point in the pseudopotential model is an exact transformation of the effective one electron equation for the solid, but in actual applications it is used as a semi-empirical method. A number of "recipes" lead to relatively simple expressions, which can be applied to a large number of solids (mainly sp-bonded metals and semiconductors) for describing various cohesive properties.

The use of the pseudopotential model for cohesive properties is described in detail in ref.2. The first contribution to that volume of "Solid State Physics" discusses various aspects of the pseudopotential concept and the second the fitting of pseudopotentials to experimental data. In the present lectures we will only try to indicate briefly the main ideas in this connection and we will also make a few remarks concerning the comparison with other methods. Such comparisons are not made in order to show that one method is better or worse in some respect than another one, but to make it easier to see what is actually being done in the two methods. The LCAO version of the Hartree-Fock method as described in the previous section is applicable to ionic crystals. The pseudopotential model in certain respects goes beyond the Hartree-Fock level, but on the other hand neglects a number of contributions to the cohesive energy included in the LCAO model, and a direct comparison is therefore not possible.

The pseudopotential method represents one way of solving one of the fundamental problems in solid state theory - the need to describe both the tightly bound core levels and the more extended valence and conduction states. It does this so to speak by "putting the core levels under the carpet". Although the concept of a pseudopotential can be used in connection with several methods, it is probably easiest to describe it by reference to the OPW method, in which the solutions of the effective one electron equation are expanded in plane waves orthogonalized to the Bloch functions for the core levels,

$$\psi(\bar{k},\bar{r}) = \sum_{\bar{K}} \phi(\bar{k}+\bar{K},\bar{r})\alpha(\bar{k}+\bar{K}). \tag{54}$$

Here $\bar{K}$ is a reciprocal lattice vector and the coefficients are determined by the variational principle. The basis functions are constructed from plane waves

$$\eta(\bar{k},\bar{r}) = e^{i\bar{k}\cdot\bar{r}} , \tag{55}$$

and Bloch functions $\phi_c(\bar{k},\bar{r})$ for the core states

$$\phi(\bar{k}+\bar{K},\bar{r}) = \eta(\bar{k}+\bar{K},\bar{r}) - \sum_c \phi_c(\bar{k},\bar{r})\int\phi_c^*(\bar{k},\bar{r}_1)\eta(\bar{k}+\bar{K},\bar{r}_1)dv_1. \tag{56}$$

The orthogonalization in (56) can be thought of as a projection of the plane wave into a space orthogonal to the core states:

$$\begin{aligned} |\phi\rangle &= |\eta\rangle - \sum_c |\phi_c\rangle\langle\phi_c|\eta\rangle = \\ &= \left[1 - \sum_c |\phi_c\rangle\langle\phi_c|\right]|\eta\rangle = P|\eta\rangle \end{aligned} \tag{57}$$

The pseudo wave function is an expansion in plane waves with the same coefficients as in the OPW expansion of the "real" function (54):

$$\chi(\bar{k},\bar{r}) = \sum_{\bar{K}} \eta(\bar{k}+\bar{K},\bar{r})\alpha(\bar{k}+\bar{K}). \tag{58}$$

An effective one electron equation

$$H_{eff}\psi(\bar{k},\bar{r}) = \varepsilon(\bar{k})\psi(\bar{k},\bar{r}) \ , \tag{59}$$

where H_{eff} is not necessarily the Hartree-Fock operator, can be rewritten as an equation for the pseudo wave function

$$\left[H_{eff} + \sum_c(\varepsilon - \varepsilon_c)|\phi_c\rangle\langle\phi_c|\right] \chi = \varepsilon\chi, \tag{60}$$

with the same eigenvalues as (59). The potential part of H_{eff} together with the sum over the core states form the pseudopotential. What has been gained by this procedure is an equation, the solutions of which can be expanded in a (under certain conditions) relatively rapidly convergent series of plane waves. The pseudo wave function has this attractive property, since it has no nodes in the core region; outside the core region it is equal to the "real" function. It is important though to realize that the requirement $\chi = \psi$ outside the core region does not determine χ uniquely. - There are several other important aspects of the pseudopotential concepts. In particular the description in terms of scattering theory is very illuminating.

The equation (60) for the pseudo wave function is not solved as it stands. The pseudopotential is expressed in terms of a few parameters which are determined by fitting to experimental data. Since the solution of (60) is expanded in plane waves one needs the matrix elements of the pseudopotential with respect to the plane waves, i.e. the Fourier transform of the pseudopotential.

The total energy of the solid per atom with respect to separated ion cores and electrons is written as

$$U = U_o + U_E + U_{bs} \ . \tag{61}$$

If the neutral atoms are regarded as the constituents of the solid, the cohesive energy as defined in (1) is

$$E_{coh} = I - U, \tag{62}$$

where I is the total ionization energy for removal of all the valence electrons. In (61) U_o is structure independent and depends only on the electron density. U_E is the Madelung (or Ewald) energy of the ion cores, and U_{bs} is a correction for the fact that the band structure of the solid is not identical to that of a free electron gas.

The density of a solid is often expressed in terms of a radius r_s, such that a sphere with that radius on the average contains one electron. The contribution to the total energy per atom, which is due to the electron gas and its interaction with the ion cores, can be approximately written as (in Hartree)

$$U_o = z(1.105/r_s^2) - z(0.458/r_s) + zU_c(r_s) - \tag{63}$$
$$- (0.9z^2/R_a) + (3z^2/2R_a)(R_M/R_a)^2 - A_o z(R_M/R_a)^2.$$

Here z is the number of free electrons per atom, i.e. z is also the charge of the ion cores. R_a is the radius of an atomic sphere, i.e. a sphere containing z electrons, so that

$$R_a = z^{\frac{1}{3}} r_s \tag{64}$$

R_M and A_o are pseudopotential parametres. The pseudopotential of the ion core is written as

$$v^{ion}(r) = \begin{cases} -A_o \text{ for } r < R_M; \\ -\frac{z}{r} \text{ for } r \geqslant R_M, \end{cases} \tag{65}$$

in this simple case. The first three terms of (63) represent the kinetic, exchange and correlation energy of the electron gas. There are several functions U_c available for the correlation energy, the main point being that this quantity can be calculated quite accurately. The last three terms originate from the interaction between the electron gas and the ion cores. - The expression (63) can be minimized with respect to r_s (or R_a) to find the equilibrium density. For several elements the results are quite satisfactory, but there are also important systematic deviations, which is not surprising since we are so far only working with a free electron model.

The Madelung energy U_E is the electrostatic energy of the ion cores considered as point charges immersed in a uniform compensating background of charge. One obtains a conditionally convergent series which must be summed with great care. A number of procedures for doing this exist. In the most common one (see ref.2) the summation is carried out partially in real and partially in reciprocal space. An interesting alternative method is used by Harris and collaborators[12]. The net result of all such calculations is that the Madelung energy can be written as

$$U_E = \alpha_{EW}(z^{*2}/2R_a). \tag{66}$$

Here α_{EW} is the Madelung constant which is characteristic of the structure of the solid, and z^* is an effective charge of the ion core. The fact that this quantity differs from z is associated with the difference between the pseudo wave function and the "real" wave function.

The third term in (61) also depends on the structure of the crystal. It is supposed to account for the difference between the actual band structure and that of the free electron gas. This difference is calculated with second order perturbation theory. The plane wave state $\eta(\bar{k},\bar{r})$ gets in this approximation the following energy

$$\varepsilon(\bar{k}) = \frac{1}{2}k^2 + \langle\bar{k}|V|\bar{k}\rangle + $$

$$+ 2\sum_{\bar{K}}{}' \frac{|\langle\bar{k}+\bar{K}|V|\bar{k}\rangle|^2}{k^2 - |\bar{k}+\bar{K}|^2}. \tag{67}$$

To use second order perturbation theory for this purpose is a rather questionable procedure. A great deal of attention has therefore been given to this particular difficulty. What emerges out of such discussions is that, while (67) does not give a good description of the band structure itself, it seems to yield a reasonable result for the total energy.

The potential V in (67) is the total pseudopotential in the crystal. The pseudopotential of the bare ions is obtained by super-imposing the ionic contributions which apart from a constant gives

$$\sum_{\bar{K}}{}' S(\bar{K}) v^{ion}(K) e^{i\bar{K}\cdot\bar{r}} , \tag{68}$$

in terms of the Fourier transform

$$v^{ion}(K) = \Omega^{-1}\int v^{ion}(r) e^{-i\bar{K}\cdot\bar{r}} dv , \tag{69}$$

of the free ion pseudopotential. Ω is the atomic volume and $v^{ion}(r)$ is in general more complicated than (65). The quantity S in (68) is the structure factor per atom

$$S(\bar{K}) = \frac{1}{n}\sum_j e^{-i\bar{K}\cdot\bar{R}_j} , \tag{70}$$

where the summation is carried out over the n atoms in the unit cell.

The bare ion pseudopotential is screened by the electron gas. This is taken into account by means of Lindhard's dielectric function (see e.g. ref.2), i.e. each Fourier component is divided by a screening factor ϵ , which can be calculated explicitly:

$$v(K) = v^{ion}(K)/\epsilon(K). \tag{71}$$

The "band structure correction" (67) must be summed over the occupied states

$$U_{bs} = \frac{2}{N} \sum_{k<k_F} \sum_{\bar{K}}{}' \frac{|\langle \bar{k} + \bar{K}|V|\bar{k}\rangle|^2}{k^2 - |\bar{k} + \bar{K}|^2} \tag{72}$$

The numerator in (72) is independent of $\bar{k}$ and the denominator can be explicitly summed:

$$\frac{2}{N} \sum_{k<k_F} \frac{1}{k^2 - |\bar{k} + \bar{K}|^2} = \chi(K). \tag{73}$$

The explicit expression for $\chi(K)$ is given e.g. in ref.2.

Finally we must take into account the fact that in a self-consistent field theory the total energy is not the sum of the orbital energies, but also contains a correction term to account for the fact that in the sum of orbital energies each interaction is counted twice. The final expression for the "band structure" contribution to the cohesive energy is therefore (as usual the prime indicates that $K \neq 0$)

$$U_{bs} = \sum_{\bar{K}}{}' |S(\bar{K})|^2 [v(K)]^2 \chi(K) \epsilon(K). \tag{74}$$

Here the last two factors depend only on the density of electron gas. The pseudopotential factor depends on the element under consideration, and the first factor depends on the structure of the unit cell. The structure of the crystal also enters indirectly, since the summation in (74) is carried out over the reciprocal lattice vectors.

The pseudopotential model of cohesion has been applied to the problem of why a certain crystal structure is preferred to another. In many cases results consistent with experimental data have been obtained, but less successful calculations are also known. Perhaps more important is that thanks to the relative simplicity of pseudopotential calculations a number of other "properties" can be studied, which so far have been unthinkable at the ab initio level. These include phonon dispersion relations, stacking fault energies, energies of formation of vacancies and elastic constants. In a certain sense pseudopotential theory can be expressed in terms of pair potentials,and one might therefore be tempted to believe that the Cauchy relations for the elastic constants would be satisfied within the pseudopotential formalism. This is not the case, however, since the term U_o in (61) contains volume dependent forces, which cannot be expressed as pairwise interactions.

A very important aspect of calculations of cohesive properties of ionic crystals is the short range repulsion between the ions, which prevents the crystal from collapsing under the influence of the electrostatic forces. This so-called Born-Mayer repulsion is obtained as a "by-product" of (40) and the parameters calculated quantum-mechanically agree quite well with those obtained semi-empirically in the Born-Mayer model. In principle the Born-Mayer repulsion between the ion cores should also be taken into account for the solids treated in this section. For those for which the pseudopotential model is applicable, in particular for the simple metals, one can, however, expect this repulsion to be negligible, since the corresponding cores are very small in comparison with the distances between the ions. In high pressure problems one can, however, expect such terms to play a more important part.

Summing up it is fair to say that the pseudopotential model has been quite successful in giving us more insight into the nature of cohesion in solids. On the other hand the most successful applications seem to have been carried out for perfect solids, where the results are already known from experiment. It would indeed be rewarding if the model could also be used in the vast area of defect problems, where experimental data are much more scarce. Unfortunately the situation here looks much less hopeful, since the use of perturbation theory, which is inherent in the pseudopotential model, seems to be limited to regular arrangements of atoms.

$X\alpha$ Calculations of Cohesive Properties

Most band calculations of ab initio type are carried out with so-called average exchange potentials, and it is only recently that some calculations of Hartree-Fock type, i.e. with the non-local exchange operator, have been published (cf. ref.8,12). These average exchange potentials are in general proportional to the third root of the density, but they may differ in the coefficients of that quantity. The "universal" coefficients suggested originally by Slater and later modified by Gáspár, Kohn and Sham, have been replaced by a coefficient α, which varies from case to case. Generally it is chosen such that the total $X\alpha$ energy (cf.(75)) for an atom agrees with the total Hartree-Fock energy (Schwarz), or such that the $X\alpha$ orbitals (see below) satisfy the virial theorem, when the kinetic and the potential energies are calculated as ordinary quantum mechanical expectation values (Berrondo-Goscinski); the two procedures give practically the same values for α (cf. ref.3).

A reason for discussing $X\alpha$ calculations of cohesive properties in a separate section, is that they represent an approach rather different from what has been discussed in previous sections. The density functional formalism is based on a theorem by Hohenberg and Kohn (cf. ref.3 and 8) according to which the total energy of the ground state of a system is a unique functional of the density. In the $X\alpha$ method the total potential energy is calculated from a density functional, but the kinetic energy is calculated as a quantum-mechanical expectation value. The total energy is given by the expression (h_1 , h_{12} given in (7))

$$E_{X\alpha} = \sum_i n_i \int u_i^*(1) h_1 u_i(1) dv_1 + \\ + \frac{1}{2} \int \rho(1)\rho(2) h_{12} dv_1 dv_2 + \\ + \frac{1}{2} \int \left[\rho_+(1) U_{X+}(1) + \rho_-(1) U_{X-}(1) \right] dv_1 . \tag{75}$$

Here the u_i's are orbitals (one electron functions) and n_i is the occupation number of u_i ; the possible values of n_i normally being 0, 1 or 2. The first term in (75) is the total kinetic energy and that part of the potential energy which is due to the nuclear attraction. Apparently the non-diagonal elements of the first order density matrix are needed to calculate the kinetic energy. The second term of (75) represents the classical electrostatic interaction of the charge distribution

$$\rho(1) = \sum_i n_i u_i(1) u_i^*(1) , \tag{76}$$

with itself, and the third term is the exchange (sometimes called exchange-correlation) contribution. The plus and minus signs used as indices in this third term refer to the two types of spin, and the quantity U is given by

$$U_{X\pm}(1) = -9\alpha \left[\frac{3}{4\pi} \rho_\pm(1) \right]^{1/3} , \tag{77}$$

with $\rho_\pm$ being those parts of (76) which are associated with up and down spin, respectively.

Variation of the total energy expression (75) with respect to the orbitals yields the effective one electron equation of the $X\alpha$ method

$$\left[h_1 + \int \frac{\rho(2)}{r_{12}} dv_2 + V_{X\pm}(1) \right] u_{i\pm}(1) = \varepsilon_{i\pm} u_{i\pm}(1) . \tag{78}$$

Here the average exchange potential is given by

$$V_{X\pm}(1) = -6\alpha\left[\frac{3}{4\pi}\rho_{\pm}(1)\right]^{\frac{1}{3}} \qquad (79)$$

(In this section we use Rydberg units for the energy to conform to the usage in ref.3)
The $X\alpha$ orbitals u_i are obtained as self-consistent solutions of (78).

The $X\alpha$ -method, i.e. primarily the effective one electron equation (78) can be applied to atoms, molecules or solids. For a solid solution of (78) means a band structure calculation and a number of procedures are available for obtaining these solutions. Once the $X\alpha$ orbitals u_i have been obtained, they can be used to calculate the total energy, by means of (75), and therefore also to get cohesive energies and other cohesive properties. The effective one electron equation (78) must be solved for each nuclear configuration, for which the cohesive energy is needed.

A few calculations of cohesive properties (cf. ref.3) using this method have been carried out. The results are sufficiently encouraging to make us look forward to other applications, but it is also desirable to have detailed studies of the limitations of the method. The advantage with this method is that it (hopefully) makes it possible to treat large groups of systems, for which the Hartree-Fock level is by no means sufficient, with a reasonable amount of computing effort. A drawback is the lack of systematic (at least in principle) procedures to go beyond the $X\alpha$ approximation.

Concluding Remarks

The three approaches to the study of cohesive properties of solids that have been discussed in this paper, were selected as examples of quantitative calculations of a type that is now relatively actively pursued. They are, however, by no means the only methods for handling such problems. One of the earliest methods for calculating band structures, the cellular method, has given rise to its own conceptual framework for describing cohesion[13)]. Relatively recently Samathiakanit[14)] has carried out a detailed study of the cohesive energy of the sodium metal. Linderberg[15)] has developed a method along the lines sketched in (14)-(27) and used it to calculate the cohesive energy of the neon crystal[16)]. Harris, Monkhorst and collaborators[12)] have carried out Hartree-Fock type calculations for alkali-metals as have also Stoll and Preuss[17)]. The A(lternant) M(olecular) O(rbital) method (see ref.1) has been explored in applications to alkali metals[18]. The AMO method is particularly suited to "one-dimensional solids", i.e. one-dimensional lattices of three-dimensional atoms[19].

It seems plausible that quite a few new calculations of cohesive properties of solids will appear within the next few years. The whole menagery of known procedures will probably be invoked and new methods will be developed. It is to be hoped that this will not result in a more and more diverging set of concepts to describe the same physical phenomena. An essential part of the research in this area lies in the study of the relations between the various procedures and in attempts to reduce the number of concepts and "explanations" as much as possible to such which are independent of the particular method of calculation.

References

1. P-O. Löwdin, Adv.Phys. 5, 1 (1956)
2. V. Heine and D. Weaire, Solid State Physics 24, 249(1970)
3. J.C. Slater, "Quantum Theory of Molecules and Solids" vol.IV, McGraw Hill Book Company, New York 1974.
4. P. Lindner and O. Goscinski, Int.J.Quantum Chem.,Symp.4, 251 (1971)
5. E.A. Hylleraas, Z.Phys. 63, 771 (1930)
6. R. Landshoff, Z.Phys. 102, 201 (1936)
   Phys.Rev. 52, 246 (1937)
7. P-O. Löwdin, "A Theoretical Investigation into Some Properties of Ionic Crystals",(Almqvist&Wiksell, Uppsala 1948)
8. J-L. Calais, Int.J.Quantum Chem. 9S, 497 (1975)
9. C. Froese-Fisher, Comp.Phys.Commun. 1, 151 (1969)
10. E. Clementi, IBM J.Res.Dev. 9, 2 (1965) (Supplement)
11. For a recent summary see P-O.Löwdin, Adv. Quantum Chem. 5, 185 (Ed. by P-O.Löwdin, Acad.Press, New York)
12. F.E. Harris in "Electronic Structure of Polymers and Molecular Crystals",(p. 453),(Ed. J-M.André and J.Ladik; Plenum Press, New York and London 1975); see also Harris' contribution to the present volume.
13. E.P. Wigner, F. Seitz, Solid State Physics 1, 97 (1955)
14. V. Samathiyakanit, Thesis, Göteborg (1969)
15. J. Linderberg, Arkiv Fysik 26, 323 (1964)
16. J. Linderberg, Arkiv Fysik 26, 383 (1964)
17. H. Stoll, H. Preuss, Int.J.Quantum Chem. 9, 775 (1975)
18. J-L. Calais and G. Sperber, Int.J.Quantum Chem.,7, 501 (1973) and references therein.
19. J-L. Calais in "Electronic Structure of Polymers and Molecular Crystals", (p. 389),(Ed. J-M.André and J.Ladik; Plenum Press, New York and London 1975)

SCATTERED WAVE CALCULATIONS FOR ORGANIC MOLECULES AND OTHER OPEN STRUCTURES

FRANK HERMAN

IBM Research Laboratory

San Jose, California 95193, U. S. A.

ABSTRACT

We will discuss the progress that is currently being made in going beyond the non-overlapping atomic spheres + muffin-tin model, with particular emphasis on open finite structures such as organic molecules. Topics include : replacement of non-overlapping atomic spheres by overlapping spheres, truncated spheres, and other types of atomic cells ; introduction of extra spheres ; replacement of the bounding sphere by a more snug-fitting surface such as an ellipsoid ; scattering by non-spherical potentials and atomic cells ; calculation of non-muffin-tin and non-spherical atomic corrections by perturbation theory ; redefinition of the uniform intersphere potential as a charge-density- rather than a volume-averaged potential. In the final section we will discuss in a preliminary way a new method that we have developed for dealing with open finite systems. The essential ingredients are a charge-density-averaged muffin-tin potential and simplified perturbation calculations.

I. INTRODUCTION

The cluster scattered wave method (1-3) has established itself over the past few years as a useful and flexible technique for determining the electronic structure of isolated molecules and molecular clusters (3-5). This method -- more properly the self-consistent statistical-exchange-correlation multiple scattering method -- is an outgrowth of the crystal scattered wave method (6-9) (KKR or Green's function method). Recently, the scattered wave method has also been adapted to thin film (10, 11) and polymer (12) geometries.

The traditional starting point for the scattered wave method is the non-overlapping atomic sphere muffin-tin model, henceforth model I. According to this model, atoms are represented by non-overlapping spheres, and, for a cluster, all the atomic spheres are in turn surrounded by an outer sphere. Such a model is shown in Fig. 1 for the TCNQ molecule (13). Inside each of the atomic spheres and outside the bounding sphere, electronic charge distributions and potentials are spherically averaged. In the intersphere (muffin-tin) region, these quantities are volume averaged.

The popularity and wide-spread use (3-5) of the cluster scattered wave method based on model I can be attributed to its computational and conceptual simplicity. The use of this model greatly simplifies the relevant boundary value problem, whether we are dealing with periodic structures in one, two, or three dimensions (polymers, thin films, crystals), or finite systems (molecules and clusters). In spite of the obvious physical and chemical deficiencies of model I, it is possible to obtain useful results having acceptable accuracy for many types of clusters with this model.

There are substantial computational advantages inherent in the scattered wave method (14), particularly when this is used in conjunction with model I. These advantages include small basis sets and energy-dependent basis functions that are just about as easy to construct for heavy atoms as for light ones. Moreover, the partitioning of space into mutually exclusive regions leads to the easy implementation of local density approximations for exchange and correlation effects (15-17). Relativistic effects are also readily included in these formalisms (18-20).

Because of these features, the combination -- scattered waves, non-overlapping atomic spheres, muffin-tins -- has been applied to a wide range of problems involving localized electronic states. We will not attempt to list all the references here : most can be found in recent reviews (3-5). We will be content to list some of the problems that are currently being investigated by the cluster scattered wave method based on model I (and closely related models) :

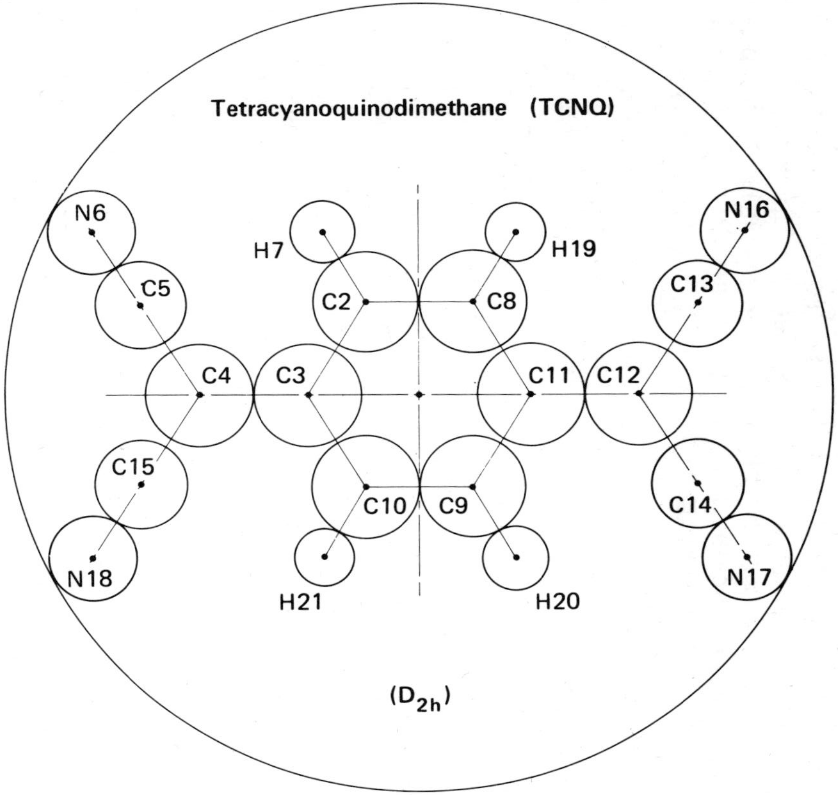

Fig. 1. Non-overlapping atomic spheres plus muffin-tin model for TCNQ molecule. D_{2h} symmetry. The outer sphere is assigned serial number 1 and the atomic spheres serial numbers 2 through 21. (13).

(a) inorganic chemistry : transition metal compounds ;
(b) organic chemistry : large molecules such as TTF, TCNQ, and their dimers ;
(c) bio-organic chemistry : copper and iron porphyrins ;
(d) defect solid state : vacancies, interstitials, and substitutional impurities in semiconductors and insulators ;
(e) solid state physics : complex clusters representing terrestrial and lunar materials as well as amorphous materials ;
(f) surface physics and catalysis : chemisorption of atoms and molecules on solid surfaces ; small metallic particles.

While the muffin-tin approximation has usually been satisfactory for closely packed, highly coordinated clusters such as Cu_{13} and Pt_{13} (21), this approximation has often been disappointing for open, loosely coordinated clusters such as large planar organic molecules, e. g. , TCNQ (13). This comparison is reminiscent of the situation in solid state physics, where KKR and APW (22-24, 9) energy band calculations based on the muffin-tin approximation have generally been more successful for close-packed than for loosely-packed structures. In most cases one could expect the muffin-tin approximation to be more successful the smaller the relative size of the intersphere region, and the more uniform the electronic charge distribution in this region.

In this paper we will discuss the progress that is currently being made in going beyond model I, with particular emphasis on open finite structures such as organic molecules. Among the attempts to go beyond model I we will consider are the following :

(a) replacement of non-overlapping atomic spheres by overlapping spheres, truncated spheres, and other types of atomic cells ;
(b) introduction of extra spheres to reduce the volume of the intersphere region ;
(c) replacement of the bounding sphere by a more snug-fitting outer boundary, such as an ellipsoid ;
(d) consideration of scattering by non-spherical atomic potentials including those associated with non-spherical atomic cells ;
(e) calculation of non-muffin-tin (non-constancy) and non-spherical atomic corrections by perturbation theory ;
(f) redefinition of the uniform intersphere potential as a charge-density- rather than a volume-averaged potential.

In the final section we will discuss in a preliminary way a new method that we have recently developed for dealing with open finite systems.

II. NON-OVERLAPPING ATOMIC SPHERES

This model has already been briefly described in the introduction and illustrated by a model for the TCNQ molecule (cf. Fig. 1). In order to minimize the size of the intersphere region, the atomic spheres are made as large as possible, so long as they do not overlap and so long as their radius ratios are chemically reasonable ; and the outer sphere, suitably centered, is drawn tangent to the outermost atomic spheres. Many different criteria are employed in practice for choosing the relative magnitudes of the atomic sphere radii when these are not uniquely defined by the cluster geometry (1- 5 ,13).

The principal difficulty with this model is that the electronic charge density is distinctly non-uniform in the muffin-tin region, especially for large open structures. Serious errors can arise when the electronic charge density and the molecular potential are volume averaged in the intersphere region to obtain their muffin-tin values. In the case of the TCNQ molecule, where the intersphere volume is about 30 times as large as the sum of the atomic sphere volumes, the muffin-tin approximation leads to excitation and ionization energies that differ from experiment by a factor of 2 (13,25-27).

These discrepancies are a consequence of two important features of the (self-consistent) model I solution : (a) there are significant differences between the potentials produced in the atomic spheres by uniform and non-uniform electronic charge distributions in the intersphere region ; (b) non-overlapping atomic spheres are too small to accommodate sufficient electronic charge to offset or mask the errors arising from item (a). In brief, model I places so much electronic charge in the unphysical muffin-tin region that the solution is overwhelmed by the resulting muffin-tin errors.

III. OVERLAPPING ATOMIC SPHERES

The simplest way to improve the treatment of open structures is to replace the non-overlapping atomic spheres by overlapping ones, as illustrated in Fig. 2 for the TCNQ molecule (13). By increasing the volume of the atomic spheres, more

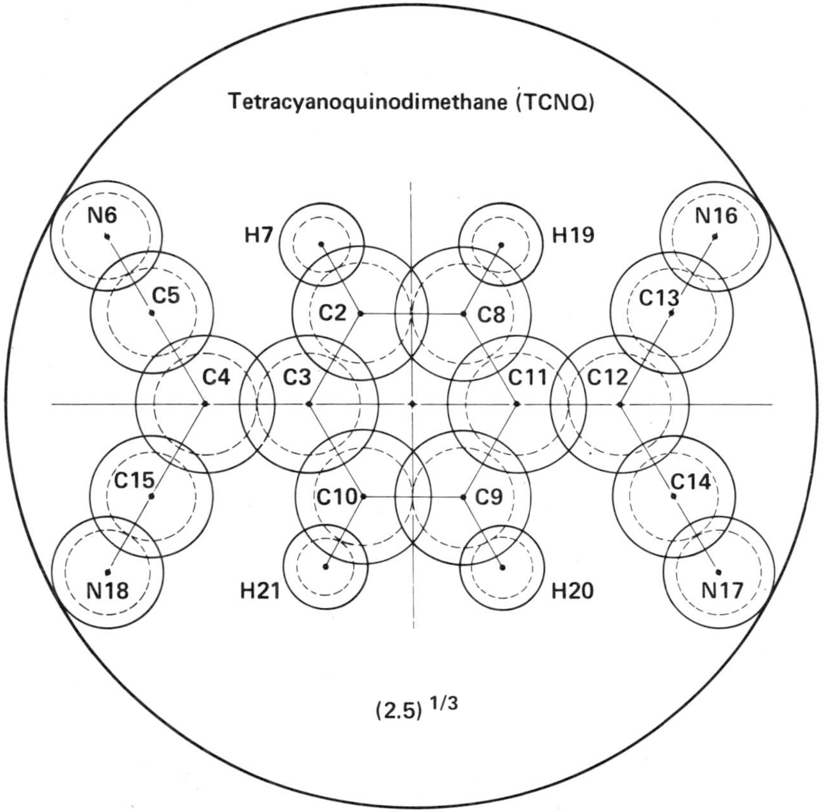

Fig. 2. Overlapping atomic spheres plus muffin-tin model for TCNQ molecule (solid circles). The original set of non-overlapping atomic spheres is shown by the dashed circles. Each of the original atomic sphere radii was multiplied by the scale factor $(2.5)^{1/3}$ to obtain the overlapping atomic spheres. (13)

charge is placed in physically realistic regions (spherical potential wells), at the expense of the unrealistic muffin-tin region. In this way the muffin-tin errors are reduced, and the quantitative results improved substantially (13, 25-27). The calculated energy level spectrum for the TCNQ molecule, based on an overlapping atomic sphere model (13), is compared with the experimental photoemission spectrum of TCNQ vapor in Fig. 3.

Overlapping atomic sphere models had been discussed earlier within the context of scattered wave band structure calculations (28, 29). As to cluster scattered wave calculations, it was recognized independently by many investigators (30-33, 13) that the same computer programs (34, 35) that had been used with non-overlapping atomic spheres could also be used, exactly as they stood, with overlapping spheres. The theoretical justification for this statement has been discussed (25).

There are many criteria for choosing the radii of the overlapping spheres, including arguments based on the virial theorem (25, 31) and electron counting (36). For some purposes it is sufficient to adjust the radii semi-empirically. One may require, for example, that the calculated first ionization energy agree with the experimental value, as was done in arriving at Fig. 3. Considering that only the first ionization energy was adjusted to experiment, the overall agreement between theory and experiment in Fig. 3 is satisfactory.

The principal advantage of the overlapping sphere model is that it leads to improved quantitative results with little or no increase of computer time and little or no computer program modification. Although the same computer programs can be used for overlapping as well as non-overlapping spheres, that is not to say that the solutions for the former are as rigorous as they are for the latter. If non-overlapping atomic spheres are replaced by overlapping ones, and if model I computer programs (34, 35) are employed without modification, then various types of overlap errors will be introduced. Some of these errors can be eliminated or minimized by modifying the programs in a minor way, but others cannot be avoided without substantial reprogramming.

Perturbation calculations for a number of diatomic molecules (37-39) indicate that the muffin-tin errors associated with model I arise primarily from the regions immediately outside the atomic spheres. By going to larger atomic spheres, these

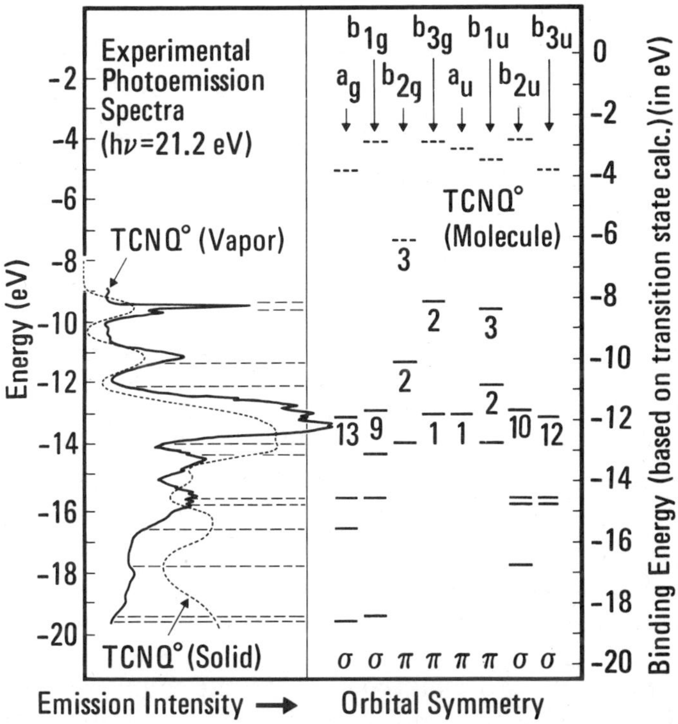

Fig. 3. Comparison of calculated energy level structure for the TCNQ molecule and experimental photoemission spectra for TCNQ vapor (solid line) and solid TCNQ (dashed line). See (13) for details.

muffin-tin errors are necessarily reduced. In practice, overlapping sphere models are more successful than non-overlapping sphere models because the muffin-tin error is reduced substantially, and this reduction far outweighs the overlap errors that crop up. Of course, this is true only so long as the atomic spheres overlap only slightly. If the overlap is too great, the overlap errors dominate the solution, and the model breaks down.

Three shortcomings of the overlapping sphere model are : (a) the intersphere region is still represented by a muffin-tin approximation ; (b) the choice of atomic sphere radii is somewhat arbitrary ; and (c) overlap errors are introduced. Fortunately, the calculated electronic structure is often not critically sensitive to the detailed choice of atomic radii, and the overlap errors can be kept manageably small by avoiding excessive overlap.

IV. EFFECTIVE SIZE OF MOLECULES

Before proceeding further, it is instructive to introduce the concept of the Van der Waals envelope of a molecule (40). This envelope is defined as the outer surface of a molecular model each of whose atoms is represented by a sphere having the appropriate Van der Waals radius. As is explained in texts on molecular crystallography (40), the packing of molecules in molecular crystals is determined to a large extent by the packing of their Van der Waals envelopes. Thus, the effective size of a molecule is to a large extent determined by its Van der Waals envelope.

In Fig. 4 we show the construction of the Van der Waals envelope for the TCNQ molecule, and in Figs. 5 and 6 this envelope is superimposed on two overlapping atomic sphere models, one corresponding to small overlap (Fig. 5) and the other to large overlap (Fig. 6). It will be seen that even for strong overlap the overlapping sphere envelope is still considerably smaller than the Van der Waals envelope.

It is also instructive to note that for the TCNQ molecule, 1/3 of the valence charge resides in the muffin-tin region for a non-overlapping sphere model (cf. Fig. 1) (13). This fraction is reduced by about a factor of 2 every time the atomic spheres are doubled in volume. For the models in Figs. 2 and 5, the muffin-tin fraction is about 1/6, and for the model in Fig. 6, about 1/12. Thus, even with strong overlap, there is

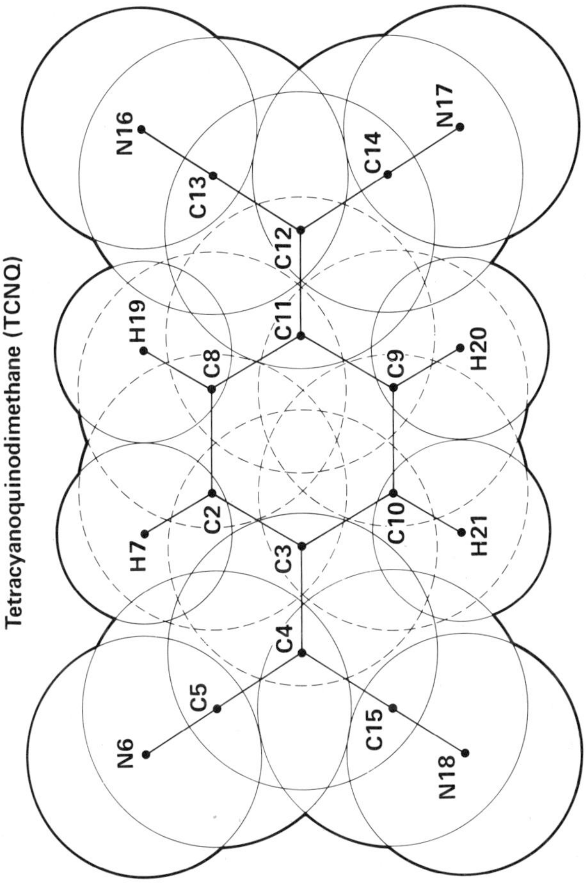

Fig. 4. Construction of Van der Waals envelope for the TCNQ molecule, using Van der Waals sphere radii as given by (40).

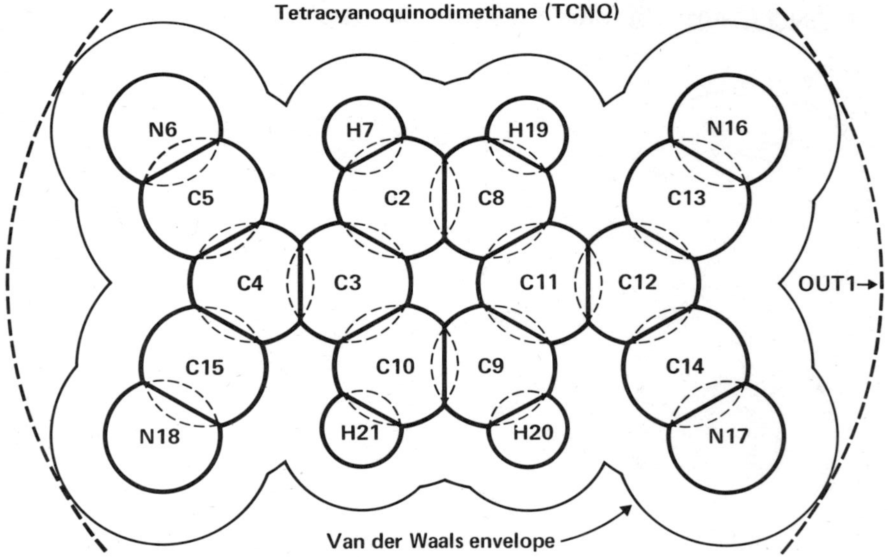

Fig. 5. Truncated atomic sphere model for the TCNQ molecule, weakly overlapping atomic spheres.

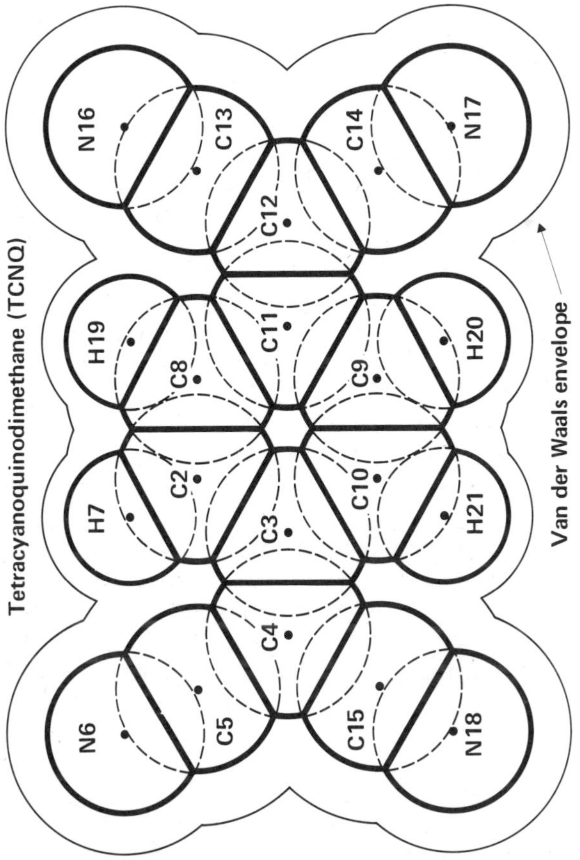

Fig. 6. Truncated atomic sphere model for the TCNQ molecule, strongly overlapping atomic spheres.

still a modest amount of valence electronic charge outside the overlapping sphere envelope. These considerations have an important bearing on the errors one should expect from the muffin-tin approximation for open structures.

V. TRUNCATED ATOMIC SPHERES

Conceptually, overlapping atomic spheres can be replaced by truncated atomic spheres, as suggested by Figs. 5 (small overlap) and 6 (large overlap). Each sphere is truncated by the planes that pass through the circles of intersection formed by the sphere and each of its immediate neighbors (33, 41, 42).

Let us continue to assume that we are content to take spherical averages of electronic charge distributions and potentials inside each atomic sphere. It is a simple matter to take these spherical averages (41, 42), provided no two atomic spheres overlap within the domain of a third, as in Figs. 5 and 6, but not Fig. 4. Under these conditions, the contributions of different neighbors are independent (additive). For a detailed mathematical treatment of the spherical harmonic expansion of a truncated atomic sphere, see the appendix to (41).

Confining ourselves to the case where no two atomic spheres overlap within the domain of the third, we can introduce a modulation function $\Theta(r)$ associated with each truncated sphere having the following property : $\Theta(r)$ is the fraction of the total solid angle that lies inside the truncated sphere at radius r. Clearly, $\Theta(r)$ will be unity between the origin of each sphere and the closest truncating plane, and will fall off in a simple algebraic manner at larger values of r (41, 42).

In computer programs for model I (34, 35) the various molecular orbitals are normalized on the assumption that the atomic spheres do not overlap. When these programs are used in conjunction with overlapping spheres, there is a double counting of charge in the overlap regions, and hence a (small) overlap error. A simple way to offset this double counting is to multiply each atomic charge distribution by the appropriate modulation function $\Theta(r)$, and then to renormalize the molecular orbital as a whole. Similarly, the calculated atomic potentials should be multiplied by the appropriate $\Theta(r)$'s to allow for the fact that one is now dealing with the spherical averages of <u>truncated</u> atomic spheres. In practice, the effects of introducing the modulation functions $\Theta(r)$ are rather minor for the relatively small atomic overlaps normally used (cf. Figs. 2 and 5).

A detailed mathematical treatment of truncated atomic sphere models and associated overlap corrections has recently been reported (43). In order to put overlapping and truncated atomic sphere models on a firmer theoretical basis, it will be necessary to modify the model I computer programs (34, 35) and incorporate the overlap corrections rigorously. Work in this direction is currently in progress (44).

VI. INTRODUCTION OF INTERSTITIAL (EXTRA) SPHERES

The volume of the intersphere region can be reduced by introducing extra atomic spheres (45-48, 27, 33) which are then treated like ordinary atomic spheres with zero nuclear charge. This is an attractive procedure for periodic structures, where voids can be readily filled with interstitial spheres. In the case of the silicon crystal, the unit cell can be represented by two atomic (silicon) spheres and two extra (interstitial) spheres (46, 48).

The essential advantage of introducing extra spheres is that portions of the intersphere region can be treated individually, the overall volume average being replaced by local spherical averages (27). In the case of molecules, the indiscriminate use of extra spheres can be hazardous. Unless the entire molecule is suitably coated, some molecular orbitals will be more strongly influenced by the extra spheres than others, and misleading results can be obtained. The safest procedure is to completely surround the molecule with extra spheres, and to fill all interior voids with interstitial spheres. Unfortunately, this greatly increases the computational effort. Nevertheless, in some applications it is useful to introduce extra spheres in order to test the sensitivity of the calculated electronic structure to the exact manner in which the molecular space is partitioned (47).

In Fig. 7 we have plotted the constant charge density contours for the $3b_{1u}(\pi)$ molecular orbital of TCNQ in a plane parallel to the molecular plane but lying 1.63 Bohr units away (49). This is 1/4 the distance between successive TCNQ molecules in a TCNQ stack in the TTF-TCNQ crystal (13, 25-27). The circles represent the intersections of the overlapping atomic spheres and the outer sphere with the molecular (not the plotting) plane.

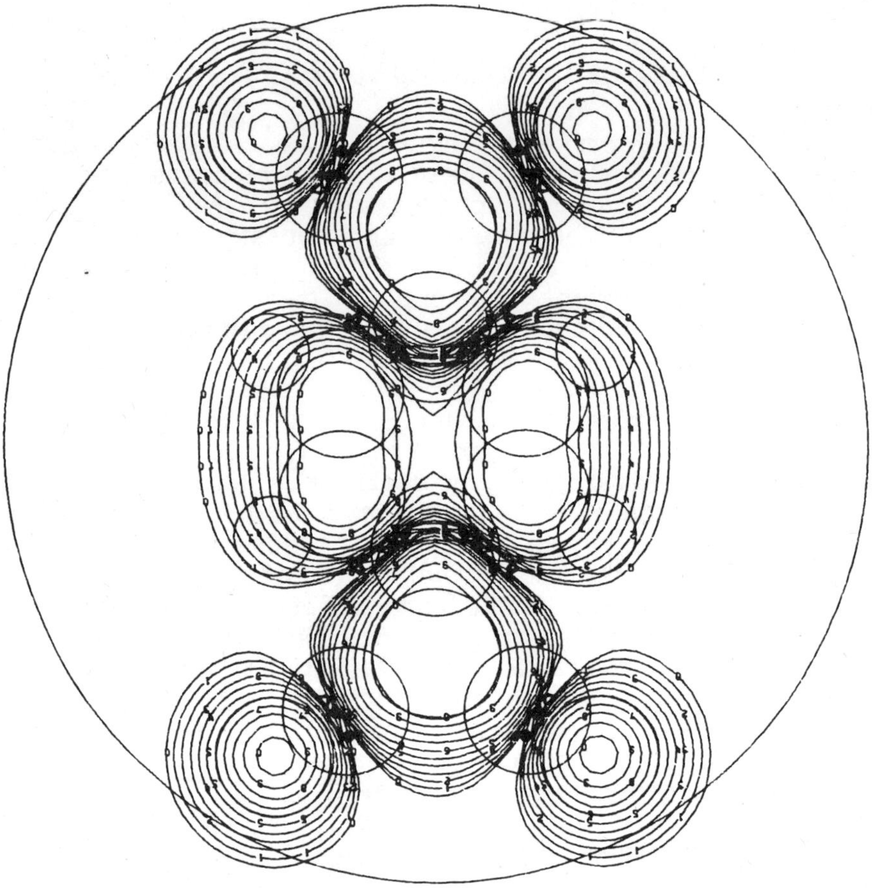

Fig. 7. Constant charge density contours for $3b_{1u}$ (π) TCNQ molecular orbital; no extra spheres (27, 49).

As part of the work leading to (27), we surrounded the atomic spheres in the molecular plane by a set of extra spheres, and carried out additional self-consistent scattered wave calculations. For this solution, the constant charge density contours for the $3b_{1u}$ (π) molecular orbital of TCNQ are shown in Fig. 8. The extra circles in this figure represent the intersections of the extra spheres with the molecular plane.

Roughly speaking, in the molecular plane each atomic sphere plus its neighboring extra spheres simulates a Van der Waals sphere. More complete simulation would require extra spheres above and below the molecular plane, in addition to those already present in the molecular plane. Considering how differently the molecular space is partitioned in Figs. 7 and 8, the similarity of the contour plots in these two figures is striking.

The reason the two plots are so similar is that the extra spheres are centered in the molecular plane, where the $3b_{1u}$ (π) molecular orbital vanishes by symmetry. Only a small fraction of the electronic charge of this orbital is enclosed by the extra spheres. Accordingly, the plot in the plane shown (where the π orbital reaches its peak amplitude, more or less) is not changed appreciably by the presence of the extra spheres, even though the potentials in the extra spheres are considerably deeper than the residual muffin-tin (average intersphere) potential (27).

The situation is quite different for the cyano group π' orbitals and the N lone-pair orbitals, for which a considerable fraction of the electronic charge lies in the extra spheres. For these and some other σ orbitals, the presence of extra spheres and their deeper potential wells (relative to the muffin-tin) leads to a marked redistribution of electronic charge(27), pointing to the desirability of using either larger atomic cells or a fully surrounding coat of extra spheres.

VII. NON-SPHERICAL OUTER BOUNDARIES

In the models considered thus far, the atomic spheres are surrounded by an outer sphere so that the molecular orbitals can be easily normalized. For rod-like and disk-like molecules, the use of a spherical boundary leads to an excessively large intersphere region, and, because of the volume averaging normally employed, to excessively large muffin-tin errors.

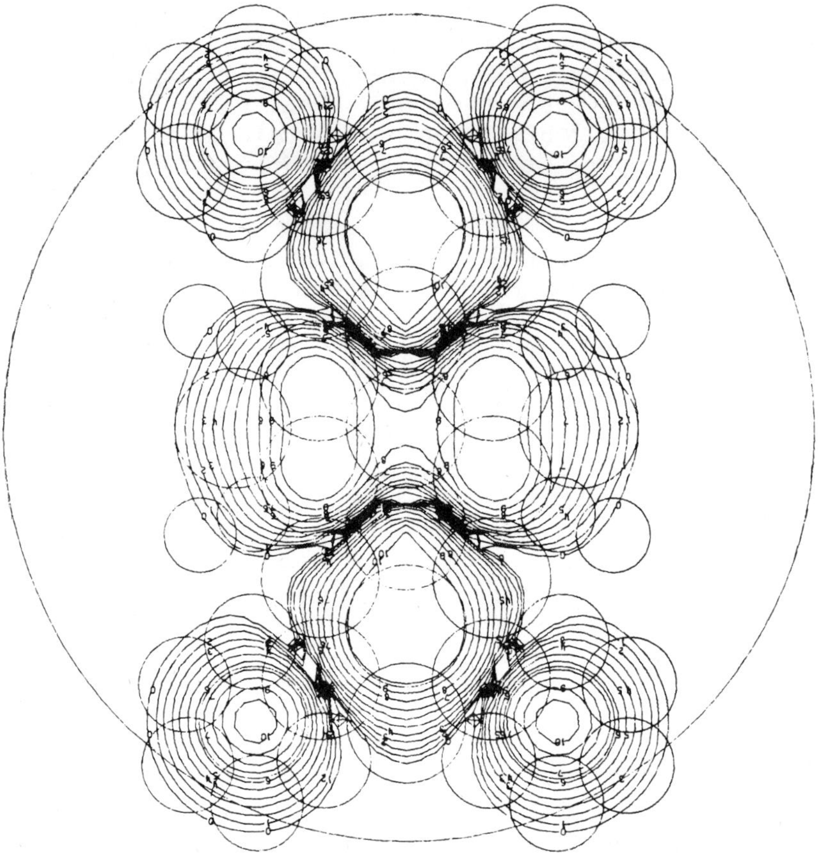

Fig. 8. Constant charge density contours for $3b_{1u}$ (π) TCNQ molecular orbital; extra spheres surround atomic spheres in the plane of the molecule (27, 49).

One way to reduce the intersphere volume is to replace the outer sphere by an ellipsoid or some other snug-fitting but mathematically tractable shape (50). This idea has received rather complete mathematical treatment, but it has not yet been implemented. In principle, this idea should lead to improved quantitative results.

An interesting question is whether the outer surface should be drawn externally tangent to the atomic sphere envelope or to the Van der Waals envelope. It has been common practice to circumscribe the atomic sphere envelope on the ground that this minimizes the muffin-tin volume and hence the muffin-tin error. Unfortunately, there are some orbitals, such as the lone-pair nitrogen orbitals on the TCNQ molecule, that are concentrated in directions away from the center of the molecule, so that a substantial fraction of their charge is to be found outside the outer sphere (13, 25-27, 49).

Unless very large basis sets are used in the exterior region, it is difficult to represent outward-pointing orbitals faithfully in electronic structure calculations. In considering snug-fitting external boundaries, one is between Scylla and Charybdis. If this boundary is drawn tangent to the atomic sphere envelope, one incurs exterior region errors which can be reduced only by the use of very large exterior basis sets. If this boundary is drawn tangent to the Van der Waals envelope, the exterior region errors are eliminated, but the muffin-tin errors are increased appreciably.

VIII. MOLECULAR ORBITAL PLOTS

In some problems there is greater interest in the spatial form of the molecular orbitals than in the detailed numerical results for the energy levels. For such problems one can use a small exterior basis set. If the molecular orbitals are plotted in the exterior region in terms of this basis set (spherical harmonic expansion for spherical outer boundary), the results are often quite unsatisfactory.

However, it is possible to get around this difficulty by plotting the molecular orbitals in the exterior region using the scattered wave expansions appropriate to the intersphere region (49). This extrapolation works well enough within the Van der Waals envelope (roughly speaking) but breaks down at sufficiently large distances from the outer boundary.

This technique has been used with considerable success in

plotting the electronic charge distributions and/or the molecular orbitals for all the occupied and some of the lower-lying unoccupied levels for the TCNQ and TTF molecules (49). Two illustrative charge density plots are shown in Figs. 7 and 8, and two illustrative molecular orbital plots in Figs. 9 and 10.

IX. NON-SPHERICAL POTENTIALS, CELLULAR MODELS, COUPLED WAVES

If the potential for a molecule is expanded in spherical harmonics with respect to each atomic center, non-spherical potential terms will occur inside each atomic sphere. If the atoms are represented by truncated atomic spheres (or other types of non-spherical atomic cells), the deviation from sphericity will affect the spherical harmonic expansion, introducing additional non-spherical terms and modifying the spherical average (41). In all the models that we have considered so far, the atoms have been represented only by their spherically averaged potentials. In general, however, it is necessary to consider the scattering by non-spherical atomic cells (29,46,14,41).

When the non-spherical terms are sufficiently small, they can be treated by perturbation methods (51). The non-spherical first-order perturbation corrections have recently been determined for a number of diatomic molecules represented by model I (37-39). It was found that the non-spherical corrections were considerably smaller than the muffin-tin corrections. This is likely to be a characteristic feature of model I calculations.

When molecules are represented by overlapping (truncated) atomic spheres, the non-spherical corrections will become larger and the muffin-tin corrections smaller. From the point of view of perturbation calculations, an advantage of using overlapping spheres is that the non-spherical corrections are considerably easier to determine than the muffin-tin corrections. However, if the atomic cells deviate too strongly from a spherical shape, first-order perturbation theory may no longer be adequate.

A more powerful approach to scattering by non-spherical atomic cells (and non-spherical potentials) is the method of coupled waves (52). Scattered wave calculations for molecules (and crystals) based on this method have been discussed by a number of authors (29,46,14,41,53). Such calculations have

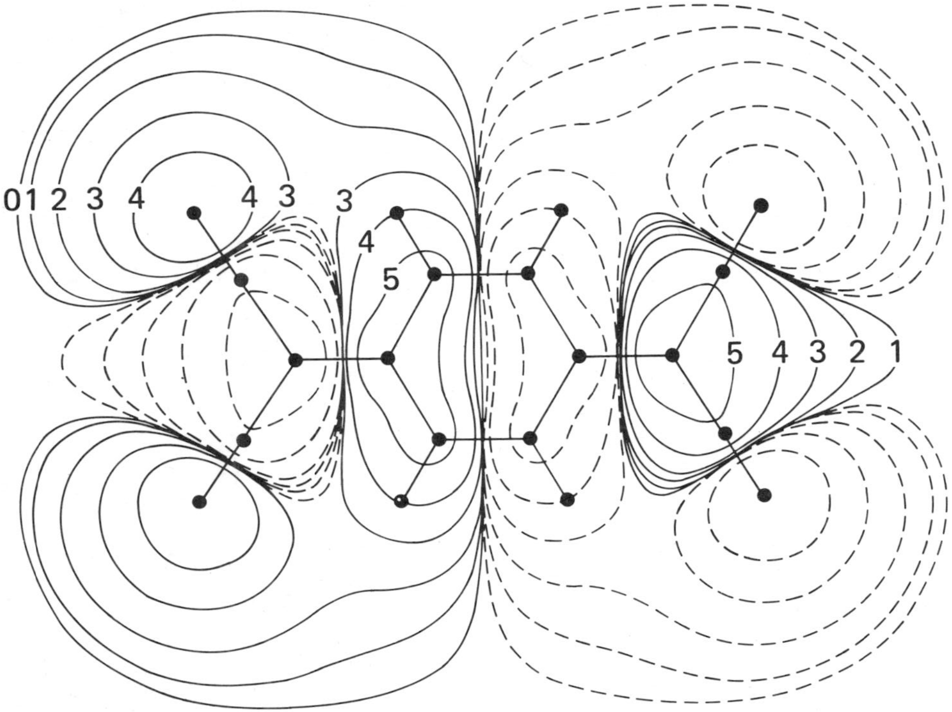

Fig. 9. Contours of constant wave function for lowest unoccupied level in TCNQ° (affinity level - $3b_{2g}{}^{*}$). Positive and negative values are distinguished by solid and dashed lines (49).

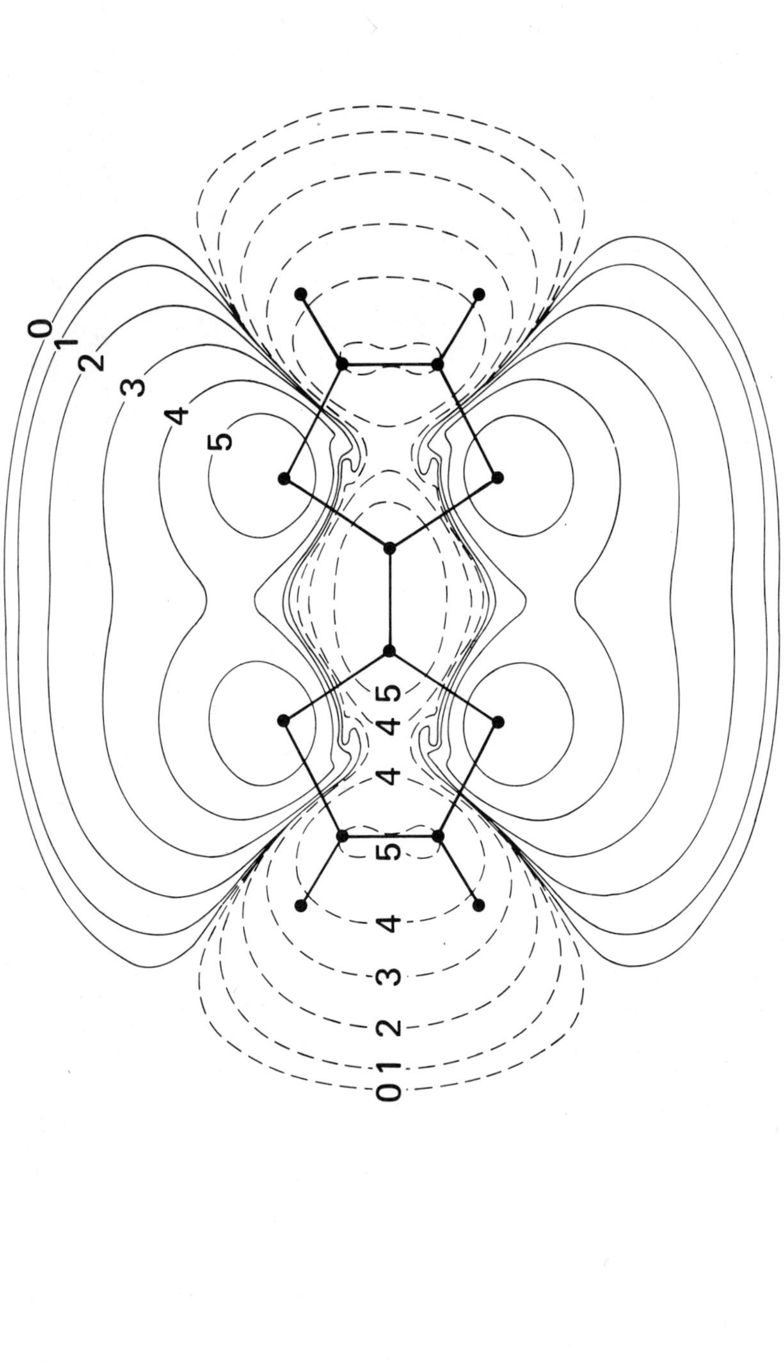

Fig. 10. Contours of constant wave function for highest occupied orbital in TTF°, $4b_{1u}$. Positive and negative values are distinguished by solid and dashed lines (49).

been carried out for crystals (46), but none have been reported so far for molecules.

Scattered wave computer programs for molecules based on the method of coupled waves are currently being developed in a number of laboratories. Work along these lines is in progress at the IBM Thomas J. Watson Research Center by A. R. Williams, at the University of Florida by D. M. Whitmore, and at the Ames Research Center by C. Y. Yang. It will be interesting to see how important the non-spherical contribution is to energy levels and total energies, and how the rigorous coupled wave results compare with approximate results based on various forms of perturbation theory (37-39, 51).

X. FULL-SCALE PERTURBATION CALCULATIONS

As already mentioned, first-order perturbation theory has recently been used to determine the non-spherical atomic corrections and the muffin-tin corrections for a number of diatomic molecules represented by model I. These calculations (37-39) are very important for the development of the subject because they demonstrate that improved potential energy surfaces can be obtained if the perturbation calculations are carried out carefully enough. However, the methods employed are not readily adaptable to different geometries and require considerable computer time. It would appear that the perturbation approach is a fruitful one, but that simplified treatments are necessary if one hopes to deal with large molecules.

XI. REDEFINITION OF THE MUFFIN-TIN POTENTIAL

Having considered several ways to improve the quantitative accuracy of scattered wave calculations based on model I, we are forced to the conclusion that the muffin-tin approximation itself must be re-examined if we hope to retain the advantages of such calculations.

Since the scattered wave approach to clusters is an outgrowth of the scattered wave approach to crystals, it is natural that the same type of muffin-tin approximation is made in both situations. In effect, it is assumed that the electronic charge density in the intersphere region can be represented by its volume average. If one carries out a variational calculation on this assumption (54), one finds that the proper choice of muffin-tin potential is the volume average over the intersphere region.

However, in open structures, the electronic charge density in the intersphere region is distinctly non-uniform, so one should abandon the earlier assumption about starting with its volume average and examine other possibilities.

A heuristic argument leading to a different choice of muffin-tin potential runs as follows : Suppose we want to carry out a self-consistent scattered wave calculation using a muffin-tin model, but we want to supplement the self-consistent solution by a perturbative treatment of non-spherical atomic and muffin-tin corrections. How do we choose the muffin-tin potential so that we incur the smallest possible overall error in the calculation of the electronic energy levels and/or the total energy ? This error can be defined as that given by first-order perturbation theory. For example, one might simply require that the sum of the energy level corrections for the occupied levels, suitably weighed for degeneracy, be zero.

If we assume that the principal error comes from the muffin-tin region, we can write

$$\sum_i \omega_i \Delta E_i = \sum_i \omega_i \int_{IS} (V(\mathbf{r}) - V_0)\, \psi_i^2(\mathbf{r})\, d\mathbf{r} = 0, \tag{1}$$

where the integration is carried over the intersphere region IS, the occupancy numbers of the molecular orbitals ψ_i are denoted by ω_i , and the sum is over all occupied levels i. The actual intersphere potential is denoted by $V(\mathbf{r})$ and the (as-yet unspecified) muffin-tin potential by V_0. It is obvious that the appropriate value to choose for V_0 is the charge-density-averaged value of $V(\mathbf{r})$ taken over the intersphere region :

$$V_0 = \frac{\int_{IS} V(\mathbf{r})\rho(\mathbf{r})\, d\mathbf{r}}{\int_{IS} \rho(\mathbf{r})\, d\mathbf{r}} \tag{2}$$

where

$$\rho(\mathbf{r}) = \sum_i \omega_i\, \psi_i^2(\mathbf{r}). \tag{3}$$

A somewhat better criterion is that the least-squares energy level error be a minimum. This leads to the following expression for V_0 :

$$V_0 = \frac{\sum_i \omega_i \int_{IS} \psi_i^2(\mathbf{r})\, d\mathbf{r} \int_{IS} \psi_i^2(\mathbf{r}) V(\mathbf{r})\, d\mathbf{r}}{\sum_i \omega_i \int_{IS} \psi_i^2(\mathbf{r})\, d\mathbf{r} \int_{IS} \psi_i^2(\mathbf{r})\, d\mathbf{r}} \tag{4}$$

In both cases it is clear that V_0 reduces to the volume average of $V(\mathbf{r})$ over the intersphere region if all the orbital charge densities are constant in this region. A similar analysis can be carried out for a least-squares total energy error. We have limited ourselves to the intersphere region for purposes of exposition, but the atomic cell and exterior regions can also be taken into account if desired.

In order to carry out the necessary calculations for V_0, it is necessary to know the molecular orbitals and the actual potential in the intersphere region. This information can be determined on a point-to-point basis on a suitably chosen mesh in the intersphere region, as in full-scale perturbation-type calculations (35-37) and in the discrete variational method (55). However, this would lead to considerable computational effort, which we would like to avoid.

In fact, our objective is to obtain improved quantitative results for open structures without significantly increasing the computer time required to get these results.

Our approach is to carry out self-consistent scattered wave calculations using an improved expression for V_0 (e.g., equation (4)) at all stages of the iteration (except the initial one). At the end of each stage of the calculation, the molecular orbitals are determined in the form of spherical harmonic expansions inside each of the atomic spheres and outside the outer sphere. In the intersphere region the molecular orbitals are represented by scattered wave expansions.

For the purposes of discussion, let us assume that the atomic spheres are weakly overlapping, as in Fig. 2 or 5. These overlapping (actually truncated) atomic spheres serve to define the boundary value problem. Starting with this truncated atomic sphere + muffin-tin model, we generate a set of so-called molecular muffin-tin orbitals (41). These are the "unperturbed" wave functions, and V_0 is part of the "unperturbed" potential. We may now surround each of the atomic spheres by a larger concentric sphere, thereby defining another set of truncated atomic spheres. We will call these two sets the inner and the outer atomic spheres, respectively. The outer spheres might be as small as those shown in Fig. 6, or as large as the Van der Waal spheres shown in Fig. 4. In any event, a substantial portion of the intersphere region, that containing the major fraction of the intersphere charge, will lie between the outer and inner sets of atomic spheres. This intermediate region can be decomposed into subcells, one corresponding to each inner atomic sphere. These cells are readily visualized if the inner spheres are like those in Fig. 5 and the outer spheres like those in Fig. 6.

In order to make further progress, we will approximate the orbital charge densities in the intermediate region by using a simple analytic expression in each sub-cell, for example

$$\rho_{ij\ell}(r) = a_{ij\ell}\, r^{\mu_{ij\ell}}\, e^{-\lambda_{ij\ell} r} \qquad (5)$$

where i denotes the molecular orbital, j the sub-cell (atomic center), and ℓ the partial wave. The three sets of parameters (a, μ, λ) can be determined by requiring $\rho_{ij\ell}$ and its radial derivative to be continuous at the radius of each inner sphere j and to agree with its value over a mesh on the outer sphere envelope in a least-squares sense. In this approach it is necessary to use a mesh over one surface, rather than over a volume. The analytic expression for the partial charge density $\rho_{ij\ell}$ is quite flexible and therefore sufficient for the purposes at hand.

The electronic charge lying outside the outer sphere envelope will be rather small, particularly if this envelope approaches the Van der Waals envelope in size. The simplest way to treat this small charge is to set it equal to zero and then renormalize the electronic charge in the intermediate region, keeping $\rho_{ij\ell}$ and its radial derivative continuous on the inner sphere envelope. This renormalization will have a negligible effect on the more important, in-close portion of the intermediate region, and only a minor effect on the less important, far-out portion.

Having established an electronic charge distribution over the entire molecule, the Coulomb and exchange-correlation potentials are readily determined, both their spherical averages and their non-spherical components in all inner and outer atomic spheres. It then becomes possible to carry out the necessary perturbation calculations in each of the outer (truncated) atomic spheres, into which the molecule has ultimately been partitioned.

In short, the inner (truncated) atomic spheres and the perturbation calculations are used to define a "reference" muffin-tin model. Then the outer (truncated) atomic spheres are used to obtain approximate electronic charge distributions and the approximate molecular potential in the intermediate region, which is the most important portion of the intersphere region. The perturbation calculations are easy to carry out. These are used not only to define an optimum muffin-tin potential, but also to evaluate the first-order corrections to the energy levels and the total energy.

The approach just outlined is currently being implemented by the author. Although not as rigorous as the full-scale perturbation calculations carried out for diatomic molecules (35-37), the present approach requires only slightly more computer time than model I calculations, and can be used for very large as well as very small isolated molecules and molecular clusters. It offers the promise of providing considerably improved quantitative results for large open clusters such as large organic molecules.

The present approach is similar in spirit to that adapted in recent calculations of the electronic properties of molecules in the scattered wave representation, i. e., the electronic charge distribution in the intersphere region is approximated in some computationally convenient manner (56-58). In contrast to other recently proposed large overlapping atomic sphere models based on cellular and LCAO methods (42, 59, 60), our proposal is based on the scattered wave method.

XII. ACKNOWLEDGMENTS

The author wishes to acknowledge fruitful discussions and correspondence with Dr. J. B. Danese, Prof. K. H. Johnson, Dr. R. P. Messmer, Dr. W. E. Rudge, Dr. D. M. Whitmore, Dr. A. R. Williams, and Dr. C. Y. Yang. He is also very grateful to Prof. P. Phariseau for the opportunity to present this work at the 1976 NATO Advanced Study Institute in Ghent.

REFERENCES

1. K. H. Johnson, J. Chem. Phys. 45, 3085 (1966).
2. K. H. Johnson, Adv. Quantum Chem. 7, 143 (1973).
3. K. H. Johnson, J. G. Norman, Jr., and J. W. D. Connolly, in Computational Methods for Large Molecules and Localized States in Solids, edited by F. Herman, A. D. McLean and R. K. Nesbet, (Plenum, New York, 1973), p. 161.
4. K. H. Johnson, Annual Revs. Phys. Chem. 26, 39 (1975)
5. N. Rösch, this volume, p.
6. J. Korringa, Physica 13, 392 (1947).
7. W. Kohn and N. Rostoker, Phys. Rev. 94, 1111 (1954).
8. B. Segall and F. Ham, Methods Computational Phys. 8, 251 (1968).
9. J. M. Ziman, Solid State Phys. 26, 1 (1971).
10. W. Kohn, Phys. Rev. B11, 3756 (1975).
11. N. Kar and P. Soven, Phys. Rev. B11, 3761 (1975).
12. N. Rösch and J. Ladik, Chem. Phys. 13, 285 (1976).
13. F. Herman and I. P. Batra, Nuovo Cimento 23B, 282 (1974); Phys. Rev. Letters 33, 94 (1974).
14. A. R. Williams, Intern. J. Quantum Chem. 8S, 89 (1974)
15. J. C. Slater, Adv. Quantum Chem. 6, 1 (1972); Quantum Theory of Molecules and Solids (Mc Graw-Hill Book Co., New York, 1974), vol. 4.
16. W. Kohn and L. J. Sham, Phys. Rev. 137, A1697 (1965); L. J. Sham and W. Kohn, Phys. Rev. 145, 561 (1966); N. D. Lang, Solid State Phys. 28, 225 (1973).
17. L. Hedin and S. Lundqvist, Solid State Phys. 23, 1 (1969) L. Hedin and B. I. Lundqvist, J. Phys. C. : Solid State Phys. 4, 2064 (1971); O. Gunnarsson and B. I. Lundqvist, Phys. Rev. B13, 4274 (1976).
18. C. Y. Yang and S. Rabii, Phys. Rev. A12, 362 (1975).
19. B. Cartling and D. M. Whitmore, Chem. Phys. Letters 35, 51 (1975).
20. W. V. M. Machado and L. G. Ferreira, Chem. Phys. Letters 37, 51 (1976).
21. R. P. Messmer, S. K. Knudson, K. H. Johnson, J. B. Diamond, and C. Y. Yang, Phys. Rev. B13, 1396 (1976).
22. J. C. Slater, Phys. Rev. 51, 846 (1937).
23. L. F. Mattheiss, J. H. Wood, and A. C. Switendick, Methods Computational Phys. 8, 64 (1968).
24. J. O. Dimmock, Solid State Phys. 26, 103 (1971).
25. F. Herman, A. R. Williams, and K. H. Johnson, J. Chem. Phys. 61, 3508 (1974).
26. I. P. Batra, B. I. Bennett, and F. Herman, Phys. Rev. 11, 4927 (1975); B. I. Bennett and F. Herman, Chem. Phys.

Letters 32, 334 (1975).
27. K. H. Johnson, F. Herman, and R. Kjellander, in Electronic Structure of Polymers and Molecular Crystals, edited by J. André, J. Ladik, and J. Delhalle (Plenum Publishing Corp., New York, 1975), p. 601.
28. A. R. Williams, Phys. Rev. B1, 3417 (1970); A. R. Williams, S. M. Hu, and D. W. Jepsen, in Computational Methods in Band Theory, edited by P. M. Marcus J. F. Janak, and A. R. Williams (Plenum Press, New York, 1971), p. 157.
29. R. Evans and J. Keller, J. Phys. C. : Solid State Phys. 4, 3155 (1971).
30. D. Liberman, unpublished.
31. N. Rösch, W. G. Klemperer, and K. H. Johnson, Chem. Phys. Letters 23, 149 (1973).
32. N. Rösch and K. H. Johnson, Chem. Phys. Letters 24, 175 (1974).
33. J. Keller, Intern. J. Quantum Chem. 9 583 (1975).
34. K. H. Johnson and F. C. Smith, Jr., M. I. T. scattered wave program (unpublished).
35. J. W. D. Connolly, University of Florida scattered wave program (unpublished) ; a modified version of Ref. 34.
36. J. G. Norman, Jr., J. Chem. Phys. 61, 4630 (1974) ; Molecular Phys., in press.
37. J. B. Danese and J. W. D. Connolly, J. Chem. Phys. 61, 3063 (1974).
38. J. B. Danese, J. Chem. Phys. 61, 3071 (1974).
39. J. B. Danese, Chem. Phys. Letters, in press.
40. A. I. Kitaigorodsky, Molecular Crystals and Molecules (Academic Press, New York, 1973).
41. O. K. Andersen and R. G. Woolley, Molecular Phys. 26, 905 (1973).
42. J. C. Slater, Intern. J. Quantum Chem. 9S, 7 (1975).
43. C. Y. Yang and K. H. Johnson, Intern. J. Quantum Chem. 10S, (1976), in press.
44. C. Y. Yang, Private Communication.
45. J. Keller, J. Phys. C. : Solid State Phys. 4, L85 (1971); A. R. Williams and J. van W. Morgan, ibid 5, L293 (1972).
46. A. R. Williams and J. Van W. Morgan, J. Phys. C. : Solid State Phys. 7, 37 (1974).
47. D. R. Salahub, R. P. Messmer, and F. Herman, Phys. Rev. B13, 4252 (1976).
48. J. R. Leite, B. I. Bennett, and F. Herman, Phys. Rev. B12, 1466 (1975).
49. F. Herman, W. E. Rudge, I. P. Batra, and B. I. Bennett,

Intern. J. Quantum Chem. 10S, (1976), in press.
50. L. Scheire and P. Phariseau, Intern. J. Quantum Chem. 8S, 109 (1974) ; Physica 74, 546 (1974) ; Chem. Phys. Letters 26, 149 (1974) ; Intern. J. Quantum Chem. 9S, 105 (1975).
51. D. G. Pettifor, J. Chem. Phys. 59, 4320 (1973).
52. A. D. Boardman, A. D. Hill, and S. Sampanthar, Phys. Rev. 160, 472 (1967). For a review of coupled wave methods, see : W. A. Lester, Methods Computational Phys. 10, 211 (1971).
53. D. M. Whitmore, to be published.
54. D. Liberman, Phys. Rev. 153, 704 (1967) ; see also D. Liberman and I. P. Batra, J. Chem. Phys. 59, 3723 (1973).
55. D. E. Ellis and G. S. Painter, Phys. Rev. B2, 2887 (1970).
56. D. Case and M. Karplus, Chem. Phys. Letters 39, 33 (1976).
D. Case, Ph. D. Thesis, Harvard University (in progress); D. Case, and M. Karplus, to be published.
57. R. P. Messmer, private communication.
58. L. Noodleman, Ph. D. Thesis, M. I. T., 1975 ; J. Chem. Phys. 64, 2343 (1976).
59. H. S. Fricker and P. W. Anderson, J. Chem. Phys. 55, 5028 (1971).
60. S. Antoci and L. Mihich, J. Chem. Phys. 64, 1442 (1976); S. Antoci and L. Barino, J. Chem. Phys. 65, 257 (1976) ; S. Antoci, J. Chem. Phys. 65, 253 (1976).

UNITARY GROUP APPROACH TO THE MANY-ELECTRON CORRELATION PROBLEM

J. PALDUS

Department of Applied Mathematics
University of Waterloo
Waterloo, Ontario, Canada

These lecture notes are intended to provide a rudimentary account of the unitary group approach to the many-electron correlation problem. They represent neither a review nor an original article, and the references are handled accordingly. In the limited space-time at our disposal, we cannot but briefly outline the basic concepts and procedures, completely avoiding any proofs or derivations.

I. INTRODUCTION

One of the mathematically and computationally very appealing approaches to the shell model (configuration interaction) calculations of the correlation effects in many-fermion systems, first proposed by Moshinsky [1], is based on the unitary group representation theory. Interestingly enough, some fundamental ideas of this approach were already pointed out by Jordan [2] in 1935, but it was not until rather recently that the necessary mathematical machinery has been sufficiently developed by Gelfand, Moshinsky, Biedenharn, Louck and others, extending thus the classical works by Weyl, Cartan, Casimir, Killing and others (cf., for example, Ref. [3,4]).

Recently, there has been a renewed interest in this and related approaches [5-14] in connection with large scale atomic and molecular CI (configuration interaction) calculations. It was found [8,9,11], for N-electron systems, that the relevant formalism may be considerably simplified and that efficient algorithms may be formulated. It is hoped that this brief exposé will stimulate further development and wider exploitation of this very general, simple and elegant approach.

II. RUDIMENTS OF THE UNITARY GROUP REPRESENTATION THEORY

1. All n×n unitary matrices U over the field of complex numbers $\mathbb{C}$ form a (matrix) Lie group : the unitary group U(n).

2. Any $U \in U(n)$ can be written in the form

$$U = \exp(A), \tag{1}$$

where A is a skew-Hermitian matrix. All n×n skew-Hermitian matrices over $\mathbb{C}$ form a real Lie algebra u(n) associated with the unitary group U(n), and any such matrix can be written in the form A=iS, where S is Hermitian.

3. The unitary group Lie algebra u(n) is a subalgebra of the general linear Lie algebra gl(n,$\mathbb{C}$) over $\mathbb{C}$ of all n×n matrices [in fact gl(n,$\mathbb{C}$) may be regarded as the complexification of u(n)].

A standard basis for gl(n,$\mathbb{C}$) consists of the matrix units e^{ij} whose (k,l)-entry is

$$(e^{ij})_{kl} = \delta_{ik}\,\delta_{jl}. \tag{2}$$

These matrix units satisfy the commutation relations

$$\left[e^{ij}, e^{kl}\right]_- = \delta_{jk} e^{il} - \delta_{il} e^{kj}\,, \tag{3}$$

which define the structure constants of the pertinent Lie algebra.

4. Every rep (representation) of a Lie group G induces a rep of the corresponding Lie algebra g, and vice versa, if G is connected (otherwise only a rep of the component connected to the identity is determined). Thus, the possible irreps (irreducible representations) of G [say, GL(n,$\mathbb{C}$)] may be determined by finding an irreducible set of linear operators (matrices) E_{ij} (i.e., endomorphisms of a finite dimensional inner product space) satisfying the same commutation relations as the matrix units. Moreover, for the unitary reps we require that

$$E_{ij}^{+} = E_{ji}\,. \tag{4}$$

The E_{ij}'s are referred to as the generators (infinitesimal operators) of the pertinent representation. (Note that in physics textbooks also the basis elements of the algebra, e.g., matrix units, are called generators since they possess the same structure constants as the infinitesimal operators.)

5. A unitary group is compact. In fact, it plays the "universal" role in the class of compact Lie groups (just as the symmetric group S(n) is "universal" for the class of finite groups in

view of Cayley's theorem) : Every compact Lie group is isomorphic to a subgroup of some U(n).

6. The representation theory of compact Lie groups is very similar to that for finite groups. Indeed, if G is a compact Lie group,

(i) G has a faithful linear rep,
(ii) all irreps of G have finite degree,
(iii) all irreps of G can be realized as tensor reps (over a linear space in which G has a faithful rep),
(iv) all finite-dimensional reps of G are equivalent to unitary reps and are fully reducible (decomposable), and
(v) the number of irreps of G (up to equivalence) is denumerable,

in addition to other remarkable properties (Peter-Weyl theorem, etc.) which will not be needed here.

7. Any irrep $\Gamma(m_n)$ of U(n) is uniquely labeled by a so-called highest weight vector m_n,

$$m_n = (m_{1n}, m_{2n}, \ldots, m_{nn}) \tag{5}$$

having non-increasing integer components

$$m_{1n} \geq m_{2n} \geq \cdots \geq m_{nn} \ . \tag{6}$$

When $m_{nn} \geq 0$ the integers m_{in} may be regarded as a partition of the integer N,

$$N = \sum_{i=1}^{n} m_{in} \ ,$$

and one can uniquely associate with each $\Gamma(m_n)$ a pertinent Young pattern, having m_{in} boxes (nodes) in the i-th row.

8. A representation $\tilde{\Gamma}$ is the conjugate of Γ if the Young patterns of Γ and $\tilde{\Gamma}$ are mutually conjugate (i.e., obtainable one from another by reversing the roles of rows and columns).

9. The canonical (orthonormal) basis for the carrier space of any irrep $\Gamma(m_n)$ of U(n) is represented by the Gelfand-Tsetlin basis, whose vectors are uniquely labeled by the triangular shaped tableaus $[m]$, called Gelfand tableaus (patterns),

$$[m] = \begin{bmatrix} m_{1n} \; m_{2n} & \cdots & m_{nn} \\ m_{1,n-1} & \cdots & m_{n-1,n-1} \\ & \cdots\cdots\cdots\cdots & \\ & m_{12} \cdot\cdot\, m_{22} & \\ & m_{11} & \end{bmatrix} . \quad (7)$$

The top row entries are the components of the highest weight vector m_n, Eq. (5), and the integer entries in subsequent rows (numbered from below) satisfy the so-called "betweenness conditions"

$$m_{i,j+1} \geq m_{ij} \geq m_{i+1,j+1} \quad (i \leq j,\ j=1,2,\ldots,n-1) . \quad (8)$$

A pattern (7) satisfying (8) is called a <u>lexical</u> pattern. The basis vectors are assumed to be lexically ordered ($[m]$ preceeds $[m']$ if the first nonvanishing component in the linearized difference array $(m_{1,n-1}-m'_{1,n-1}, \ldots, m_{n-1,n-1}-m'_{n-1,n-1}, m_{1,n-2}-m'_{1,n-2} \ldots m_{11}-m'_{11})$ is positive).

10. The number of possible lexical basis vectors equals the dimension of $\Gamma(m_n)$ and is given by Weyl's or Robinson's formulas.

11. The generators E_{ij} are classified into the raising, lowering or weight generators according to whether $i<j$, $i>j$ or $i=j$, respectively. Their matrix representatives in the canonical basis are strictly upper triangular, strictly lower triangular or diagonal, respectively. In view of (4) the real representation matrices of lowering and raising generators E_{ij} and E_{ji} $(i>j)$, respectively, are given by mutually transposed matrices. Finally, the raising (lowering) generator matrix representatives are determining by the representatives of corresponding elementary (primitive) generators $E_{i-1,i}$ $(E_{i,i-1})$, $i=2,3, \ldots, n$ since

$$E_{i,j\pm1} = \left[E_{ij}, E_{j,j\pm1}\right]_{-}, \quad (9)$$

as follows immediately from (3).

12. The matrix elements of weight generators in the Gelfand-Tsetlin basis are

$$<[m']|E_{ii}|[m]> = \delta([m'],[m])(k_i-k_{i-1}), \quad (10)$$

where

$$k_i = \sum_{j=1}^{i} m_{ji}, \quad (i=1, \ldots, n) ; k_o \equiv 0, \tag{11}$$

and

$$\delta([m'],[m]) = \prod_{i\leq j} \delta(m'_{ij}, m_{ij}). \tag{12}$$

(For typographical reasons we designate the Kronecker delta as $\delta(i,j)$, $\delta(i,j) \equiv \delta_{ij}$). The matrix elements of other than weight generators were given by Gelfand and Tsetlin and others [3]. These explicit expressions are rather complex but, for the cases pertinent in the N-electron problem, may be drastically simplified [8,9] as will be shown in Section IV.

III. RELATIONSHIP TO THE MANY-ELECTRON PROBLEM

1. Consider an N-electron system described by the Hamiltonian involving at most two-body forces

$$\hat{H} = \sum_{A,B} \langle A|\hat{z}|B\rangle \hat{X}_A^+ \hat{X}_B + \frac{1}{2} \sum_{A,B,C,D} \langle AB|\hat{v}|CD\rangle \hat{X}_A^+ \hat{X}_B^+ \hat{X}_D \hat{X}_C, \tag{13}$$

where $\hat{X}_A^+$ ($\hat{X}_A$) is the creation (annihilation) operator defined on some complete orthonormal set of single particle states $|A\rangle$, and

$$\langle A|\hat{z}|B\rangle = \int \Phi_A^*(x)\hat{z}\Phi_B(x)dx$$

$$\langle AB|\hat{v}|CD\rangle = \iint \Phi_A^*(x)\Phi_B^*(x')\hat{v}(x,x')\Phi_C(x)\Phi_D(x')dxdx' \tag{14}$$

are the matrix elements of the pertinent one particle (kinetic and external potential energy) and two-particle (Coulomb potential) operators, respectively. (Note that $\Phi_A(x) = \langle x|A\rangle$.)

2. In all practical applications, we restrict ourselves to the corresponding <u>model</u> problem defined by the Hamiltonian (13), where the summations are now restricted to a finite set of single particle spinorbital states $|A_i\rangle$, $i=1,2,\ldots, n'$, spanning some finite-dimensional subspace $V_{n'}$ of the pertinent single particle space, $\dim V_{n'} = n'$.

3. We then observe [2] that the operators

$$\hat{E}_{AB} = \hat{X}_A^+ \hat{X}_B \tag{15}$$

satisfy the same commutation relations as the matrix units [cf., Eq. (3)] and may thus be considered as generators of $GL(n,\mathbb{C})$ or of $U(n)$. Moreover, the model Hamiltonian (cf., III.2) may then be expressed in terms of these generators as

$$\hat{H} = \sum_{A,B} < A|\hat{z}|B > \hat{E}_{AB} + \frac{1}{2} \sum_{A,B,C,D} < AB|\hat{v}|CD > (\hat{E}_{AC}\hat{E}_{BD} - \delta_{BC}\hat{E}_{AD}) . \tag{16}$$

Thus, for each rep of U(n') we can get the corresponding matrix representative of $\hat{H}$ in the same representation space and the same basis.

4. This is particularly useful when $\hat{H}$ is spin independent [1], so that the pertinent spinorbitals $|A>$ may be represented by products of the orbital $|a>$ and spin $|\sigma>$ functions,

$$|A> = |a>|\sigma> \quad , \quad (\sigma = 1,2), \tag{17}$$

where σ=1 and 2 labels the spin-up and the spin-down states, respectively. The Hamiltonian then takes the form

$$\hat{H} = \sum_{a,b} <a|\hat{z}|b> \sum_{\sigma} \hat{X}^{+}_{a\sigma} \hat{X}_{b\sigma} + \frac{1}{2} \sum_{a,b,c,d} <ab|\hat{v}|cd> \sum_{\sigma,\tau} \hat{X}^{+}_{a\sigma} \hat{X}^{+}_{b\tau} \hat{X}_{d\tau} \hat{X}_{c\sigma} . \tag{18}$$

Assuming that the $|a>$ are orthonormal and span a finite-dimensional spinless one-particle space V, dim V=n, we can again consider the operators

$$\hat{e}_{a\sigma,b\tau} = \hat{X}^{+}_{a\sigma} \hat{X}_{b\tau} \tag{19}$$

as generators for U(2n). Moreover, the following partial traces

$$\hat{E}_{ab} = \sum_{\sigma} \hat{e}_{a\sigma,b\sigma} = \sum_{\sigma} \hat{X}^{+}_{a\sigma} \hat{X}_{b\sigma} \tag{20a}$$

$$\hat{\mathcal{E}}_{\sigma\tau} = \sum_{a} \hat{e}_{a\sigma,a\tau} = \sum_{a} \hat{X}^{+}_{a\sigma} \hat{X}_{a\tau} \tag{20b}$$

$$\left[\hat{E}_{ab}, \hat{\mathcal{E}}_{\sigma\tau}\right]_{-} = 0 \tag{20c}$$

also satisfy the same commutation relations as the matrix units, Eq. (3), and may thus be regarded as generators of U(n) and U(2), respectively. Since now $\hat{H}$, Eq. (18), is spin independent, it is expressible solely through the generators of U(n)

$$\hat{H} = \sum_{a,b} <a|\hat{z}|b> \hat{E}_{ab} + \frac{1}{2} \sum_{a,b,c,d,} <ab|\hat{v}|cd> (\hat{E}_{ac}\hat{E}_{bd} - \delta_{bc}\hat{E}_{ad}). \tag{21}$$

Consequently, instead of U(2n), we can consider its proper subgroup U(n) $\times$ U(2). The pertinent totally antisymmetric irreps of U(2n) [Pauli principle], considered as reps of this subgroup, are reducible, and decompose into the direct sums of irreps of U(n) × U(2), which in turn are given as (outer) direct products of mutually conjugate irreps of U(n) and U(2) components. In view of this conjugation property, the latter irreps may be uniquely labeled by the pertinent total spin quantum number. Thus, the irreps of U(n) determine completely the pertinent irreps of U(n) × U(2), and we can choose their irrep carrier spaces as derised spin-adapted subspaces for CI.

5. The pertinent irreps of U(n) must satisfy the conditions

$$0 \leq m_{in} \leq 2 \qquad \text{(Pauli principle)} \tag{22}$$

and

$$\sum_{i=1}^{n} m_{in} = N \qquad \text{(Total particle number conservation)} \tag{23}$$

and are uniquely determined by n, N and the desired spin multiplicity (2S+1) since

$$\sum_{i=1}^{n} \delta(m_{in}, 1) = 2S \tag{24}$$

Thus, the pertinent irreps are labeled by two column Young frames with N nodes, the first column of which has 2S more nodes than the second column. These irreps are conveniently labeled by the corresponding partition $\{2^a 1^b\}$, where a and b are given by N and S

$$a = N/2 - S \qquad b = 2S \tag{25}$$

6. We can, thus, summarize our main conclusions : The unitary group U(n) to be employed is determined by the dimension n of the single particle space V on which the model spin-independent Hamiltonian (18) is defined, i.e., by the number of orthonormal orbitals used. Then, the number of electrons N of the system considered and the desired multiplicity (2S+1) [or, equivalently, the total spin quantum number S] uniquely specify the irrep $\Gamma\{2^a 1^b\}$ of U(n) [cf., Eqs. (25)], and the canonical (Gelfand-Tsetlin) basis of its carrier space may be used as a spin-adapted N-electron basis for the pertinent CI problem.

We next outline an efficient algorithm for the generation and storage of such bases, the calculation of generator matrix representatives in these bases and, thus, the calculation of the desired CI matrices.

IV. N-ELECTRON FORMALISM

1. Restricting ourselves to an N-electron case, we immediately find from (8) and (22) that the entries m_{ij} of the Gelfand tableaux (7), which are pertinent to many-electron problems, may only equal 0,1 or 2 : we call them electronic Gelfand tableaus. Any electronic Gelfand tableau may thus be more economically represented by an $n \times 3$ matrix $[\chi_{ij}]$ (i=1, ..., n ; j=1,2,3) (called an ABC tableau*) , whose entries give the number of 2's, 1's and 0's in the corresponding row of the electronic Gelfand tableau, namely

$$\chi_{ij} = \sum_{k=1}^{i} \delta(j-1, m_{ki}). \tag{26}$$

The top (i.e., n-th) row defines the irrep considered $\Gamma(\chi_{nj}) \equiv \Gamma\{2^a 1^b\}$ where

$$\begin{aligned} \chi_{n3} &\equiv a = N/2-S \\ \chi_{n2} &\equiv b = 2S \\ \chi_{n1} &\equiv c = n-N/2-S. \end{aligned} \tag{27}$$

Moreover, since

$$\sum_{j=1}^{3} \chi_{ij} = i \tag{28}$$

any two columns determine the tableau uniquely. It is convenient to choose columns 3 and 1. The pertinent $n \times 2$ matrices are simply called tableaus (or AC tableaus).

2. The lexicallity conditions (8) require that the tableau columns χ_{ij} (j=1 or 3) form nondecreasing finite integer sequences, whose subsequent members differ by at most unity, i.e.

$$1 \geq \chi_{i+1,j} - \chi_{i,j} \geq 0 . \tag{29}$$

3. This suggests in turn that we define the difference tableaus Δ^k, (k=1,2), with entries

---

*) We call these matrices tableaus in order to remind us the unconventional numbering of rows and columns used : the rows are numbered from the bottom row upwards and the columns from right to left in order to achieve a simple relationship with the corresponding electronic Gelfand tableaus.

$$\Delta^k \chi_{ij} = \Delta^{k-1} \chi_{ij} - \Delta^{k-1} \chi_{i-1,j} \; ; \qquad (j=1,3 \; ; \; i=k,k+1, \; \ldots, \; n), \tag{30}$$

where we set

$$\Delta^o \chi_{ij} \equiv \chi_{ij} \quad \text{and} \quad \chi_{oj} = 0 \quad (j=1,3) \; . \tag{31}$$

Since, obviously

$$\chi_{ij} = \sum_{k=1}^{i} \Delta^1 \chi_{kj} \quad \text{and} \quad \Delta^1 \chi_{kj} = 0 \text{ or } 1 \; , \tag{32}$$

the first difference tableau Δ^1 again fully determines the corresponding Gelfand tableau, the number of 2's and 0's in each row being given by digital sums of appropriate tails [cf., Eq. (32)] of binary strings, forming the columns of Δ^1. Thus, we find that the spin-adapted N-electron basis states may be uniquely labeled by two binary strings forming the columns of the first difference tableau. Moreover, all binary strings $\Delta^1\chi_{ij}$, j=1 or 3, have the same digital sum given by the χ_{nj} (j=1 or 3, respectively). As we shall soon show, this representation also enables an efficient calculation of the necessary matrix elements.

4. The dimension of the irrep $\Gamma(\chi_{nj})$ of U(n) is

$$\text{Dim } \Gamma(\chi_{nj}) = \frac{n+1 - \chi_{n1} - \chi_{n3}}{n + 1} \binom{n + 1}{\chi_{n1}} \binom{n + 1}{\chi_{n3}} , \tag{33}$$

where $\binom{m}{n} = m!/[(m-n)!n!]$ is the usual binominal coefficient.

5. The necessary and sufficient conditions for a first difference tableau to be lexical (i.e., to represent a lexical electronic Gelfand tableau) are

$$\sum_{k=i}^{n} \Delta^1 \chi_{kj} \leq \chi_{nj} \qquad (j=1 \text{ and } 3) \tag{34a}$$

and

$$\sum_{k=i}^{n} (\Delta^1 \chi_{k1} + \Delta^1 \chi_{k3}) > \chi_{n1} + \chi_{n3} - i \; ; \qquad (i=1, \; \ldots, \; n). \tag{34b}$$

This particular form of the lexicallity conditions enables one to generate directly the lexically ordered canonical basis in terms of these tableaus. Indeed, for a given irrep $\Gamma(\chi_{nj})$ of U(n) we take subsequently for $(\Delta^1\chi_{i3}\,\Delta^1\chi_{i1})$, i=n, n-1, ..., 1 the values (01), (00), (11) and (10), in this order, and discard at each step the nonlexical patterns violating the conditions (34).

6. The matrix representatives of weight generators (occupation numbers) are given by

$$<[\Delta^1\chi'_{kj}]|\, E_{ii}\, |[\Delta^1\chi_{kj}]> = \delta(\Delta',\Delta)\ [1+\Delta^1\chi_{i3}-\Delta^1\chi_{i1}], \qquad (35)$$

where

$$\delta(\Delta',\Delta) = \prod_{j=1,3}\ \prod_{i=1}^{n}\ \delta(\Delta^1\chi'_{ij},\ \Delta^1\chi_{ij}). \qquad (36)$$

(Thus, the matrix representatives of weight generators are diagonal and their diagonal entries give the pertinent occupation numbers).

7. The matrix elements of elementary generators are given by the simple formula

$$<[\Delta^1\chi^{(i)}_{kj}]|E_{i-1,i}|[\Delta^1\chi_{kj}]> =$$

$$= \delta(\Delta^2\chi_{ij},j-2)\ \Big(h_i/(h_i-\Delta^2\chi_{ij})\Big)^{\Delta^2\chi_{i3}\Delta^2\chi_{i1}/2} \quad ; \qquad (37)$$

$$(j=1,3\ ;\ i=1,\ \ldots,\ n),$$

where

$$h_i = i - \sum_{k=1}^{i-1} (\Delta^1\chi_{k3} + \Delta^1\chi_{k1}) > \Delta^2\chi_{ij}\ . \qquad (38)$$

The entries $\Delta^1\chi^{(i)}_{kj}$ of the row-determining tableau $[\Delta^1\chi^{(i)}_{kj}]$ are the same as the entries of the column-determining tableau $[\Delta^1\chi_{kj}]$ except for k=i-1 and k=i, when

$$\Delta^1\chi^{(i)}_{kj} = \delta(\Delta^1\chi_{kj},0), \qquad (k=i-1,i). \qquad (39)$$

In order to better realize the simplicity of this algorithm, let us note that :

(i) The exponent $(1/2)\Delta^2\chi_{i3}\Delta^2\chi_{i1}$ in (37) is either 0 or ±1/2. Thus, the pertinent matrix elements equal either 1 or the square-root of a ratio of two integers differing by 1.

(ii) Any row (column) of an elementary generator matrix representative has <u>at most two</u> nonvanishing entries [corresponding to j=3 or j=1 in (37)].

(iii) Formula (37) enables one to determine a given column (row) for all elementary raising (lowering) generators simultaneously. The nonvanishing entries of the elementary generators $E_{i-1,i}$ in the column labeled by a given tableau $[\Delta^1\chi_{ij}]$ (j=3,1) are given by the row indices for which the entries of the second difference tableau equal $\Delta^2\chi_{i3} = 1$ and/or $\Delta^2\chi_{i1} = -1$.

(iv) When the exponent in (37) [cf., (i) above] does not vanish, the pertinent matrix element is simply determined by the parameter h_i [cf., Eq. (37)], which in turn is simply obtained from a digital sum of the appropriate tails $(\Delta^1\chi_{kj})$, $k=i-1, i-2, \ldots, 1$; $j=3,1$ of binary strings forming the columns of the first difference tableau $[\Delta^1\chi_{kj}]$.

Incidentally, the parameters h_i determine the intermediate spin coupling quantum numbers labeling the basis vectors of the Yamanouchi-Kotani genealogical basis, which is up to a phase equivalent to the Gelfand-Tsetlin basis.

(v) Finally, the corresponding row determining tableau $[\Delta^1\chi_{kj}^{(i)}]$ is obtained from the tableau $[\Delta^1\chi_{kj}]$ by a binary complementation of the entries k=i-1 and k=i for a pertinent j.

8. Example : Consider the quartet states of a 5 electron system in a minimum basis set approximation (i.e., n=N=5) and determine the canonical Gelfand-Tsetlin basis and the matrix representatives of elementary generators in this basis.

Thus, we have n=N=5 and S=3/2, so that we have to consider the irrep $\Gamma(1,3,1) \equiv \Gamma\{2\ 1^3\}$ of U(5) [or of SU(5) for that matter]. Following IV.5. we determine the canonical basis as shown in Table I. Clearly, we get Dim $\Gamma(1,3,1) = 24$ basis vectors [cf., Eq. (33)]. Using (35) we easily find the pertinent occupation numbers, shown in Table II, which determine the diagonal matrices representing the weight generators. Finally, designating the basis vectors (N-electron states) by their sequence member in the lexically ordered canonical basis (cf., bottom row in Table I), and using the following shorthand notation for the pertinent matrix elements,

$$\langle k|E_{i,i+1}|1\rangle = \langle k\{i\}1\rangle, \tag{40}$$

we easily determine all the nonvanishing matrix elements in the matrix representatives of elementary raising generators by using (37)-(39) [cf. Eqs. (41)].

Table I. Generation of the lexically ordered canonical basis for $\Gamma(1,3,1) \equiv \Gamma\{2\ 1^3\}$ of U(5). The basis vectors are labeled by the first difference tableaus, whose subsequent rows are generated as explained in IV.5. The last row numbers sequentially the basis vectors.

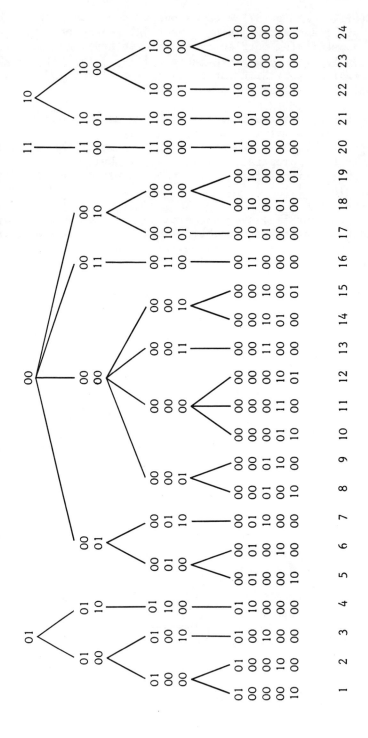

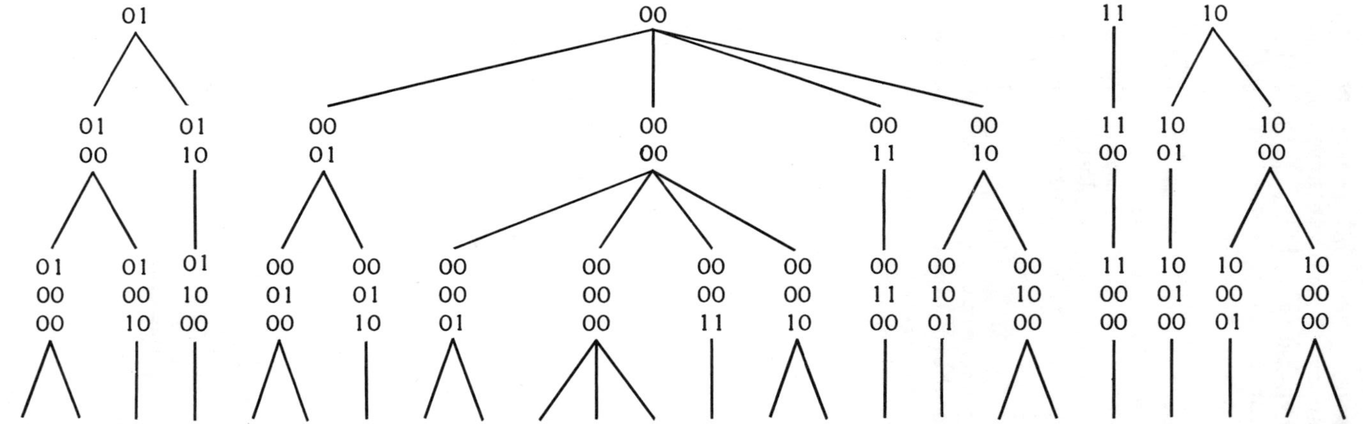

01	01	01	01	00	00	00	00	00	00	00	00	00	00	00	00	00	00	00	11	10	10	10	10
00	00	00	10	01	01	01	00	00	00	00	00	00	00	00	11	10	10	10	00	01	00	00	00
00	00	10	00	00	00	10	01	01	00	00	00	11	10	10	00	01	00	00	00	00	01	00	00
00	10	00	00	00	10	00	00	10	01	11	10	00	01	00	00	00	01	00	00	00	00	01	00
10	00	00	00	10	00	00	10	00	10	00	01	00	00	01	00	00	00	01	00	00	00	00	01
1	2	3	4	5	6	7	8	9	10	11	12	13	14	15	16	17	18	19	20	21	22	23	24

Table II. Occupation numbers (weight generator matrix elements) of canonical states listed in Table I

State	Occupation numbers	State	Occupation numbers	State	Occupation numbers
1	2 1 1 1 0	9	1 2 0 1 1	17	1 1 0 2 1
2	1 2 1 1 0	10	2 0 1 1 1	18	1 0 1 2 1
3	1 1 2 1 0	11	1 1 1 1 1	19	0 1 1 2 1
4	1 1 1 2 0	12	0 2 1 1 1	20	1 1 1 1 1
5	2 1 1 0 1	13	1 1 1 1 1	21	1 1 1 0 2
6	1 2 1 0 1	14	1 0 2 1 1	22	1 1 0 1 2
7	1 1 2 0 1	15	0 1 2 1 1	23	1 0 1 1 2
8	2 1 0 1 1	16	1 1 1 1 1	24	0 1 1 1 2

$$
\begin{aligned}
& \langle 1\{1\} 2 \rangle = \langle 2\{2\} 3 \rangle = \langle 3\{3\} 4 \rangle = \langle 1\{4\} 5 \rangle = \\
&= \langle 5\{1\} 6 \rangle = \langle 2\{4\} 6 \rangle = \langle 6\{2\} 7 \rangle = \langle 3\{4\} 7 \rangle = \\
&= \langle 5\{3\} 8 \rangle = \langle 8\{1\} 9 \rangle = \langle 6\{3\} 9 \rangle = \langle 8\{2\}10 \rangle = \\
&= \langle 12\{2\}15 \rangle = \langle 14\{1\}15 \rangle = \langle 14\{3\}18 \rangle = \langle 17\{2\}18 \rangle = \\
&= \langle 15\{3\}19 \rangle = \langle 18\{1\}19 \rangle = \langle 17\{4\}22 \rangle = \langle 21\{3\}22 \rangle = \\
&= \langle 18\{4\}23 \rangle = \langle 22\{2\}23 \rangle = \langle 19\{4\}24 \rangle = \langle 23\{1\}24 \rangle = 1 \; ; \\
& \langle 10\{1\}11 \rangle = \langle 11\{1\}12 \rangle = \sqrt{2} \; ; \; \langle 9\{2\}13 \rangle = \langle 13\{2\}14 \rangle = \sqrt{\tfrac{3}{2}} \; ; \\
& \langle 9\{2\}11 \rangle = \langle 11\{2\}14 \rangle = 1/\sqrt{2} \; ; \; \langle 7\{3\}13 \rangle = \langle 13\{3\}17 \rangle = \sqrt{\tfrac{2}{3}} \; ; \\
& \langle 7\{3\}16 \rangle = \langle 16\{3\}17 \rangle = 2/\sqrt{3} \; ; \; \langle 4\{4\}16 \rangle = \langle 16\{4\}21 \rangle = \tfrac{\sqrt{3}}{2} \; ; \\
& \langle 4\{4\}20 \rangle = \langle 20\{4\}21 \rangle = \tfrac{\sqrt{5}}{2} .
\end{aligned}
\qquad (41)
$$

The matrix representatives of the corresponding lowering generators are given by the transposed matrices (cf., II.11), i.e.

$$\langle k|E_{i,j}|l\rangle = \langle l|E_{j,i}|k\rangle . \tag{42}$$

Finally, the nonelementary generator matrix elements may be obtained by successively applying the recurrence relationship (9), which in the present context takes the form

$$\langle k|E_{i,j+1}|l\rangle = \sum_m (\langle k|E_{ij}|m\rangle\langle m|E_{j,j+1}|l\rangle - \langle k|E_{j,j+1}|m\rangle\langle m|E_{ij}|l\rangle). \tag{43}$$

It should be noted, in view of property (ii) of IV.7., that there are at most two nonvanishing terms in the summation on the right hand side of Eq. (43). In the same way, we can evaluate the matrix elements of the products of generators appearing in the two-electron part of our Hamiltonian [cf., Eq. (21)].

9. Let us mention that we have also developed [9] an algorithm enabling a direct calculation of nonelementary generator matrix representatives. This procedure is undoubtedly less efficient than the one based on the recurrence relationship (9). However, it might be useful whenever we have to restrict ourselves to some proper subspace of an N-electron space (limited CI).

V. COMMENTS ON APPLICATIONS

It should now be clear how we can use the above sketched formalism to calculate the full CI matrices in the Gelfand-Tsetlin canonical basis (which is equivalent, up to a phase, to the Yamanouchi-Kotani basis). Clearly, either an MO or AO one-electron basis can be used as long as it is orthonormal. Thus, in most ab initio approaches the MO basis will be preferable, even though this requires a rather expensive transformation of the two-electron integrals from the AO to the MO basis (cf., for example [15]). Moreover, when a molecule possesses a spatial symmetry, given by some point group, it is much easier to implement it using MO's which are symmetry adapted. This is a particularly simple task when the pertinent point group is Abelian, as is often the case.

We should note, however, that one can also construct efficient schemes based directly on the nonorthogonal AO basis [5,12]. Moreover, in the semiempirical approaches, the underlying (implicit) single-particle basis is usually assumed to be orthonormal anyway, so that a direct VB type approach based on AO's is straightforward. For model Hamiltonians based on the ZDO (zero differential overlap) approximation, the unitary group approach becomes particularly simple, since the two-particle part of the Hamiltonian contains only weight generators. For example, the widely exploited PPP (Pariser-Parr-Pople) Hamiltonian for the so called alternant systems takes the form

$$\hat{H}_{PPP} = \sum_{i,j}{}' \beta_{ij} \hat{E}_{ij} + \frac{1}{2} \sum_{i,j} \gamma_{ij}(\hat{E}_{ii}-1)(\hat{E}_{jj}-1), \tag{44}$$

where β_{ij} and γ_{ij} are the pertinent resonance and Coulomb integrals, and the first summation on the right hand side extends over nearest neighbors only (tight binding approximation for the short-range forces). However, even more sophisticated schemes based on, say, the NDDO (neglect of diatomic differential overlap) approximation enable a very simple VB-type formalism based on the unitary group approach to be developed [9].

Unfortunately, the dimensionality of the full CI matrices becomes prohibitive, even for rather modest problems. For example, using a double-zeta basis for a 10-electron problem possessing no spatial or other symmetry, the dimension of the full CI problem is 52,581,816 for singlets and 99,419,400 for triplets. It is, thus, essential that one restricts oneself to some subspace of an N-electron space of a much smaller dimensionality (limited CI). In doing so, however, we have to modify the straightforward algorithm outlined above. In particular, we cannot use directly the recurrence relationship (9) to calculate the nonelementary generator matrix elements or, similarly, the matrix elements of the products of generators appearing in the two-electron part of the Hamiltonian, since we cannot say a priori which states will appear as the intermediate states over which the summation must be carried out [cf., Eq. (43)].

Obviously, in order to calculate limited CI matrices, we need rectangular matrices for both lowering and raising generators (one only stores nonvanishing matrix elements), namely, the set of matrix elements $\langle k|E_{ij}|1\rangle$, where the state $|1\rangle$ belongs to the chosen subset of N-electron states, but where $|k\rangle$ is arbitrary. One possibile way to obtain these matrix elements is, of course, to implement the direct algorithm mentioned in IV.9.

Another tempting possibility is to select the appropriate N-electron states for the limited CI, so that the same algorithm, as in the full CI case described above, applies. An example of such a scheme is as follows :

(i) We start with the lexically maximal state $|\max\rangle$ of a given subproblem (using the MO basis, this state will usually represent the lowest energy state of a given multiplicity). We will refer to it as to the zero-level state, $|m^{(o)}\rangle \equiv |\max\rangle$.

(ii) We apply all possible elementary lowering generators to $|m^{(o)}>$; i.e., we calculate all the nonvanishing matrix elements of all the elementary generators $\hat{E}_{i,i-1}$

$$< m^{(1)} | E_{i,i-1} | m^{(o)} > \qquad (45)$$

for the column defined by $| m^{(o)}>$. This step is easily and efficiently carried out using our algorithm described in IV.7. The resulting states $| m^{(1)}>$ define the set of first level states.

(iii) Repeating the above procedure, we determine, in a general step, the (k+1)-level states from the k-level states and the pertinent matrix elements $< m^{(k+1)} | E_{i,i-1} | m^{(k)}>$, following IV.7.

(iv) A set of N-electron states, which contains all the states of the zeroth, first, etc., up to, say, the K-th level, will be called K-complete. We can then show that a K-complete set contains all the necessary intermediate states for the calculation of the matrix elements of all lowering generators using the recurrence relationship (9). In fact, the nonvanishing matrix elements for the generators $E_{i,i-1}$ can only occur between the states belonging to levels (k+1) and k, respectively.

(v) Moreover, using the K-complete set (or one of its subsets) as the basis of a limited CI approach also seems to be very reasonable from the physical viewpoint. Indeed, assuming that the single-particle orbital energies are equidistant, it easily follows that all the k-th level states have the same independent particle model energy, namely $k\Delta$, where Δ is the separation of the single-particle orbital energies. Thus, even when the assumption of equidistant spacing does not apply, this scheme will nevertheless pick up the states of approximately the same energy in the independent particle model, regardless of their excitation order.

(vi) Furthermore, if the k-th level state is an m-excited state, it easily follows that the corresponding (k+1)-level states are at least m-excited as well. Thus, an additional truncation, which depends on the excitation level, can be implemented in the scheme.

(vii) The necessary information, which has to be stored, is best represented in the form of a subgraph of a two-rooted graph, the roots being the lexically maximal and minimal states. Representing the i-th level states by the i-level vertices, we can represent all nonvanishing matrix elements by pertinent edges. The elementary generator matrix elements connect only vertices of neighboring levels. Applying successively the recurrence relationship (9) we then calculate the nonvanishing matrix elements of nonelementary generators, represented by the edges connecting more distant levels, until we reach the generator E_{1n}. Such a graph then contains all the necessary information (the values of the matrix elements label the pertinent edges) needed in a corresponding limited CI calculation. We refer to these graphs as harmonic excitation diagrams. Note also, that once this diagram is obtained with only first level edges (representing the elementary generator matrix elements), its vertices may be arbitrarily labeled by consecutive integers and the Gelfand tableaus (or, equivalent AC tableaus or first difference tableaus) associated with each state are no longer needed in the subsequent calculations.

Let us also mention that the procedure just outlined might prove to be very useful in the so called direct approaches [16-18] to the CI (or shell model) calculations. Here the essential step is to find for an arbitrary state $|\Phi_i\rangle$, given as a linear combination of chosen N-electron states, the state $|\Phi_i'\rangle = \hat{H}|\Phi_i\rangle$, without having available the matrix representative of $\hat{H}$ in the N-electron basis used, but only the pertinent one- and two-electron integrals. Indeed, with the appropriate harmonic excitation diagram we can carry out the above mentioned operation for an arbitrary selection of spin adapted N-particle states, contained in some K-complete set considered.

To conclude these notes, we would like to stress that in connection with various specific applications of the unitary group approach, there still remain many open problems, which will require a good deal of work and ingenuity to be solved, and that there is much room for improvement and development of new and efficient computational algorithms. We believe, however, that the universality and simplicity of this approach leaves little doubt about its usefulness and potentiality in the molecular electronic structure calculations.

Acknowledgements

The author is pleased to acknowledge a continued support of the National Research Council of Canada, which enabled him to pursue the research in this field.

Many thanks are due to Prof. A. Veillard, Directeur de Recherche au CNRS, and all members of his Laboratory, for their kind hospitality during the author's tenure of the visiting professorship at the University of Louis Pasteur in Strasbourg, where these notes were prepared. My sincere thanks are also due to Prof. S.G. Davison and to B.G. Adams, M. Math., for kindly reading these notes. Finally, I am very much obliged to Miss Monique Elsner for very patient and immaculate typing of the manuscript.

References

1. M. Moshinsky, Group Theory and the Many-Body Problems (Gordon and Breach, New York, 1968) ; first appeared in "Many-Body Problems and Other Selected Topics in Theoretical Physics" (Lectures of the Latin American School of Physics, University of Mexico, July-August 1965), M. Moshinsky, T.A. Brody and G. Jacob, Eds. (Gordon and Breach, New-York, 1966).

2. P. Jordan, Z. Physik 94, 531 (1935).

3. I.M. Gelfand and M.L. Tsetlin, Dokl. Akad. Nauk SSSR, 71, 825, 1017 (1950) ; I.M. Gelfand and M.I. Graev, Izv. Akad. Nauk SSSR, Ser. Mat. 29, 1329 (1965) [Amer. Math. Soc. Transl. 64, 116 (1967)] ; M. Moshinsky, J. Math. Phys. 4, 1128 (1963) ; G.E. Baird and L.C. Biedenharn, J. Math. Phys. 4, 1449 (1963); 5, 1723, 1730 (1964) ; 6, 1847 (1965) ; W.J. Holman, III and L.C. Biedenharn, in "Group Theory and Its Applications", Vol. 2, E.M. Loebl, Ed. (Academic Press, New-York, 1971), p.1; J.D. Louck, Amer. J. Phys. 38, 3 (1970).

4. H. Weyl, The Classical Groups, Their Invariants and Representations (Princeton University Press, Princeton, New Jersey, 1939) ; The Theory of Groups in Quantum Mechanics (Dover New York, 1931) ; cf. also a modern account by W. Miller, Jr., Symmetry Groups and Their Applications (Academic Press, New York, 1972) ; D.P. Želobenko, Compact Lie Groups and Their Representations (American Mathematical Society, Providence, Rhode Island, 1973).

5. M. Moshinsky and T.H. Seligman, Annals of Physics [N.Y.] 66, 311 (1971) ; T.H. Seligman, in "Second International Colloquium on Group Theoretical Methods in Physics" (University of Nijmegen, Holland, 1973).

6. J. Patera, J. Chem. Phys., 56, 1400 (1972).

7. P.E.S. Wormer and A. van der Avoird, J. Chem. Phys. 57, 2498 (1972) ; Int. J. Quantum Chem. 8, 715 (1974) ; P.E.S. Wormer, Ph.D. Thesis, University of Nijmegen, The Netherlands, 1975.

8. W.G. Harter, Phys. Rev. A8, 2819 (1973) ; W.G. Harter and C.W. Patterson, Phys. Rev. A13, 1067 (1976) ; C.W. Patterson and W.G. Harter, J. Math. Phys. (to be published) ; Unitary Calculus I and II (Springer Lecture Notes in Physics, to be published).

9. J. Paldus, J. Chem. Phys. 61, 5321 (1974) ; Intern. J. Quantum Chem. S9, 165 (1975) ; in "Theoretical Chemistry : Advances

and Perspectives", Vol.2, H. Eyring and D.J. Henderson, Eds. (Academic Press, New-York, 1976), p. 131.

10. F.A. Matsen, Intern. J. Quantum Chem. S8, 379 (1974) ; The Hückel-Hubbard Theory of Organic Chemistry (to be published); The Unitary Group Formulation of the Many-Body Problem, Intern. J. Quantum Chem. (to be published).

11. J.-F. Gouyet, R. Schranner and T.H. Seligman, J. Phys. A : Math. Gen. 8, 285 (1975) ; J.-F. Gouyet, Rev. Mex. Fis. (in print).

12. A.A. Cantu, D.J. Klein, F.A. Matsen, and T.H. Seligman, Theor. Chim. Acta 38, 341 (1975).

13. J. Drake, G.W.F. Drake, and M. Schlesinger, J. Phys. B8, 1009 (1975).

14. R.W. Wetmore and G.A. Segal, Chem. Phys. Letters 36, 478 (1975).

15. G.H.F. Diercksen, Theoret. Chim. Acta 33, 1 (1974).

16. T. Sebe and J. Nachamkin, Annals of Physics [N.Y.] 51, 100 (1969) ; R.R. Whitehead, Nucl. Phys. A 182, 290 (1972) ; R.R. Whitehead and A. Watt, Phys. Letters 41B, 7 (1972).

17. B. Roos, Chem. Phys. Letters 15, 153 (1972) ; in "Computational Techniques in Quantum Chemistry and Molecular Physics", Proceedings of the NATO Advanced Study Institute at Ramsau, Germany ; G.H.F. Diercksen, B.T. Sutcliffe and A. Veillard, Eds. (D. Reidel Publ. Co., Dordrecht-Holland, Boston-USA, 1975), p. 251, B. Roos and P. Siegbahn, in "Modern Theoretical Chemistry, Vol. II. Electronic Structure : Ab Initio Methods", H.F. Schaefer, Ed. (Plenum Publishing Corp.), in print ; A.H. Pakiari and N.C. Handy, Theoret. Chim. Acta 40, 17 (1975).

18. R.F. Hausman, S.D. Bloom and C.F. Bender, Chem. Phys. Letters 32, 483 (1975).

SUBJECT INDEX

LECTURERS

The lectures by Prof. Lidiard are not included in this volume. They can be found in "Defects and Their Structure in Nonmetallic Solids," edited by B. Henderson and A. E. Hughes (NATO Advanced Study Institutes Series (B--Physics), Volume 19), pp. 1-96, Plenum Press, New York, 1976, the proceedings of an Institute held at the University of Exeter in 1975.

J.-L. Calais
Quantum Chemistry Group, University of Upsala, Box 518,
S-751 20 Upsala, Sweden

J. Callaway
Department of Physics and Astronomy, Louisiana State University,
Baton Rouge, Louisiana 70803, U.S.A.

B.L. Gyorffy
H.H. Wills Physics Laboratory, University of Bristol,
Royal Fort, Tyndall Avenue, Bristol BS8 1TL, U.K.

F.E. Harris
Department of Physics, University of Utah,
Salt Lake City, Utah 84112, U.S.A.

F. Herman
IBM Research Laboratory,
San Jose, California 95193, U.S.A.

A.B. Lidiard
Theoretical Physics Division, A.E.R.E.
Harwell, Oxon, OX11 ORA, U.K.

N.H. March
Department of Physics, The Blackett Laboratory, Imperial College,
London S.W.7 2BZ, U.K.

J. Paldus
Department of Applied Mathematics, University of Waterloo,
Waterloo, Ontario, Canada

N. Rösch
Lehrstuhl für Theoretische Chemie, Technische Universität
München, Fed. Rep. Germany